海洋学专业英语

Academic English for Oceanography

周桂地　闫运伟　彭卫刚
王琳琼　姜仕军　宋翔洲　等 编著

海洋出版社
2024年·北京

图书在版编目(CIP)数据

海洋学专业英语/周桂地等编著．--北京:海洋出版社，2024. 12. -- ISBN 978-7-5210-1432-7

Ⅰ. P7

中国国家版本馆 CIP 数据核字第 2024HK9725 号

审图号:GS 京(2024)2588 号

责任编辑：程净净
责任印制：安　森
海洋出版社　出版发行
http://www. oceanpress. com. cn
北京市海淀区大慧寺路 8 号　邮编:100081
涿州市般润文化传播有限公司印刷　　新华书店经销
2024 年 12 月第 1 版　2024 年 12 月北京第 1 次印刷
开本:787mm×1092mm　1/16　印张:28
字数:780 千字　定价:138. 00 元
发行部:010-62100090　总编室:010-62100034

《海洋学专业英语》
编委会

主编：

总筹、物理海洋学（Physical Oceanography）部分和

通用知识技能（General Knowledge and Skills）部分：周桂地

海洋生物学（Marine Biology）部分：王琳琼

海洋地质与地球物理（Marine Geology and Geophysics）部分：彭卫刚、姜仕军

海洋技术（Ocean Technology）部分：闫运伟

统稿、审校：宋翔洲

编委（按姓氏笔画排序）：

于鹏飞　王毛毛　王琳琼　叶现韬　代富强　闫运伟

孙义程　孙高远　李　晗　吴　峰　吴　琼　何天辰

宋翔洲　张　帅　张鹏辉　张慧超　陈　欢　周桂地

姜仕军　谈明轩　曹海锦　彭卫刚

序

海洋是地球母亲的一个重要组成部分。人类的生存和发展离不开海洋。最近几十年，由于人类活动，全球的气候和生态环境正在发生重大变化。海洋科学技术正在成为人类认识、保护和适应地球家园的重要前沿。

英语是科学界的通用国际语言，因此，青年学者必须能够熟练运用英语与其他国家的学者进行交流。河海大学周桂地及其同仁经过多年的努力收集了海洋科学技术中最常用的词汇和表达方式，涵盖多个学科分支，并总结了科技英语的特点和难点，编著了这本《海洋学专业英语》。这本书以课文的形式引进海洋学科中的常用词汇，注重阅读理解，并从语言和科学角度进行了详实而深入的注释，是一种新颖的科技英语教学方式。

期望本书的出版能为青年学者提供一个学习英语的好工具，为他们在攀登科学高峰时助一臂之力。

黄瑞新

2024 年 12 月于伍兹霍尔

前　言

与科学的其他领域一样，海洋学的发展离不开国际交流与合作。人类只有一个地球，一个世界。海洋作为人类命运共同体的重要组成部分，以它为对象、载体、媒介的科学研究日益丰富、深化，且常常需要全球涉海国家携手合作。鉴于英语已经成为包括海洋学在内的科学界的通用国际语言，我们在坚持深耕基础研究、实现自主创新的同时，也应具有国际视野，学好、用好英语以服务于我国海洋事业的发展。海洋学专业的本科生、研究生、科研工作者们应该具有阅读英文科技论文和听懂英语科技报告的能力，也应具有撰写英文科技论文和用英语进行口头交流的能力。因此，海洋学专业英语知识和能力的培养应该从本科生阶段抓起，并在后续的学习和深造过程中不断强化。有鉴于此，海洋学专业英语课程目前已在国内涉海高校的本科生培养方案中成为十分重要的一环。

作为我国较早开设海洋科学专业的高校之一，河海大学海洋学科的历史可追溯到1957年设置的海洋水文专业。2015年10月，在学校“河向海延伸”的战略支持下，齐义泉教授回归母校创建海洋学院，实现了独立办学。迄今已形成集物理海洋学、海洋生物学、海洋地质学、海洋技术、大气科学和资源与环境等本科生和研究生教学单位于一体的独特办学体系，汇聚了80多名专任教师，其中拥有海外学位或留学访问经历的占95%以上，普遍具有较高的专业英语水平和丰富的国际交流经验。在此种背景下，学院十分重视本科生的专业英语教育工作，形成了多学科融合的海洋学专业英语教学团队，矢志为国家培养具有坚定的报国志向、雄厚的专业能力、高超的交流水平、深远的国际视野的新一代海洋人才。

根据我们的教学实践经验，我们深感一本优秀教材对教学的重要性。据非正式调查，现阶段的海洋学专业英语教学多采用一本特定英文专著作为教材，或由教师或学生自行选择的数篇科技期刊论文作为阅读材料。因这些材料并非为专业英语教学而作，常难避免结构零散、重点不明确、涵盖面不够完整等问题，且与专业理论课程、普通英语课程、科技论文写作等课程区分度不高，使教学过程出现流于宽泛、缺少抓手的困难。因此，一本内容丰富却又重点明确的教材十分难得，特别是能融合海洋学的物理、生物、地质、技术等不同方向的教材更是少之又少。肩负时代重任，我辈勉力为之。河海大学海洋学院的海洋学专业英语教学团队在前期教学积累的基础上尝试承担起编写这样一本教材的重任，以期助力我国海洋学领域教材市场的繁荣。

我们秉承的理念是，由于专业英语的书面化、正式化特征，英语的听说读写四种能力中应该以阅读能力为基础，写作能力建立在阅读能力基础之上，听说能力为前两者的自然延伸,最终目的是实现四种能力的全面提升。然而,四种能力中任何一种的提升绝不是在短短十几个课时的周期中能明显见效的,需要持续不断地学习积累。况且课堂教学受到课时量的限制往往难以兼顾全部四个方面,而仅能择其要者进行传授。因此,专业英语课程的作用更多是带领学生入门、掌握规律和方法,而教材则不仅在课上发挥作用,更可供学生课后反复详细研读使用。因此,本书以丰富的阅读材料来培养语感、扩充词汇量,辅以关于语言和科学内容的注解来帮助理解,兼顾学术英语的一般特征、口语发音难点、书面科技语言的规范等必要知识。我们建议教师在使用本书时首先投入一部分时间讲授本书第五部分(General Knowledge and Skills),然后再使用前四部分相关方向的阅读材料进行教学。可由教师根据课时量带领学生精读、精析几篇,然后由学生自主训练并由教师进行指导,课下学生可继续自行使用本书进行强化和拓展,相信认真学习必有裨益。

本书的雏形为河海大学海洋学院多个方向的专业英语课程历年所积累的素材、讲义和学生反馈意见等,编写组成员都是活跃在教学一线的中青年教师,受到了学校和学院的大力支持,也得到了国家自然科学基金杰青项目(42425601)、江苏省优势学科建设经费,以及河海大学大禹教学团队“面向国家业务需求的海气相互作用教学团队”等相关资助。我们有幸得到了许多专家学者的指导和改进意见,在此表示感谢。书中的阅读材料大都取自公认的优秀英文教材、专著、论文等,并包括一些线上资源和自编文章,原文的来源在各部分最后进行了汇总。由于编者水平有限,书中错漏在所难免,望使用本书的教师和学生不吝指正。

本书编委会

2024 年 9 月

本书的内容和用法

本书汇集了物理海洋学、海洋生物学、海洋地质与地球物理、海洋技术四个方向的近80篇专业英语范文作为阅读分析材料。每篇文章中都用下划线标出了重要词汇,并在文章后面附有单词表(Vocabulary)。词汇的选择以专业术语名词和词组为主,也包含特别重要的动词、形容词等,并适当收录了有助于提升专业英语阅读和写作能力的普通英语高级词汇。单词表中采用了以下符号进行标记。

n:名词或名词词组

n pl:以复数形式出现的名词

adj:形容词

adv:副词

vt:及物动词

vi:不及物动词

pl:复数

sing:单数

=:同义词或等效写法/拼法

syn:同义词或近义词

ant:反义词或对照词

~:引申词或拓展词

单词表中的单词如果包含了需要解释的词根或词缀,则用下划线标出,并在下方用方括号括起进行解释,如:

radiometer:

[-meter]:(仪器)……计、……仪

同一篇文章中的生词一般只标注在第一次出现的位置,且后面的文章中一般只标注前面文章中没有出现过的单词。

同时,文章中涉及重要科学概念的句子或段落、语法语义较难理解的地方,或者写作方式灵活以及用了高级词汇或表达方式的地方等都通过上标进行编号,并在文后的注释(Notes)部分逐一进行解释。

通过学习文章原文、单词表、注释,读者可以从词汇量、语法、语感、科学理解等方面得到提升。

本书的第五部分简要介绍了专业英语的一般知识,包括学术英语的特点、重点学术词汇列表等,并补充了英语口语发音和科技论文写作方面的知识和技能。根据实际教学需要,可先学习此部分内容,然后进入前四部分不同方向的训练。

另外,本书提供丰富的线上附加资源,包括章节练习题、AWL完整列表、术语词汇列表等,请读者使用。

Contents

PART I

PHYSICAL OCEANOGRAPHY

1. Ocean Currents

Text

Currents are cohesive streams of seawater that circulate through the oceans. Some are short-lived and small, some are long-lived and vast flows, and some can take centuries to complete a circuit of the globe[1]. The Gulf Stream Current, which shows up in the temperature-coded satellite image (Figure Ⅰ-1) as a broad dark orange swath[2], has followed this course through the North Atlantic for millennia. In contrast, the small eddy currents that spin off the Gulf Stream die out within a few months[3].

Currents are caused by winds, gravity, and variations in water density in different parts of the ocean. Most currents belong to the surface circulation, which exist in a relatively thin upper layer (500–1000 m) of the ocean; there are also relatively strong currents in the deep ocean[4].

The dominant pattern of surface circulation is the gyre—a well-organized, roughly circular flow. Five enormous gyres spin in subtropical waters, two in both the Atlantic and Pacific Oceans, and one in the Indian Ocean. Smaller subpolar gyres stir the northern Atlantic and Pacific. One surface current circles endlessly around Antarctica[5].

These gyres are made up of currents set in motion by winds and gravity, and steered by the geometry and topography of the basins and the rotation of the Earth. Wind is the most important cause of surface currents. When strong, sustained winds blow across the ocean, friction drags a

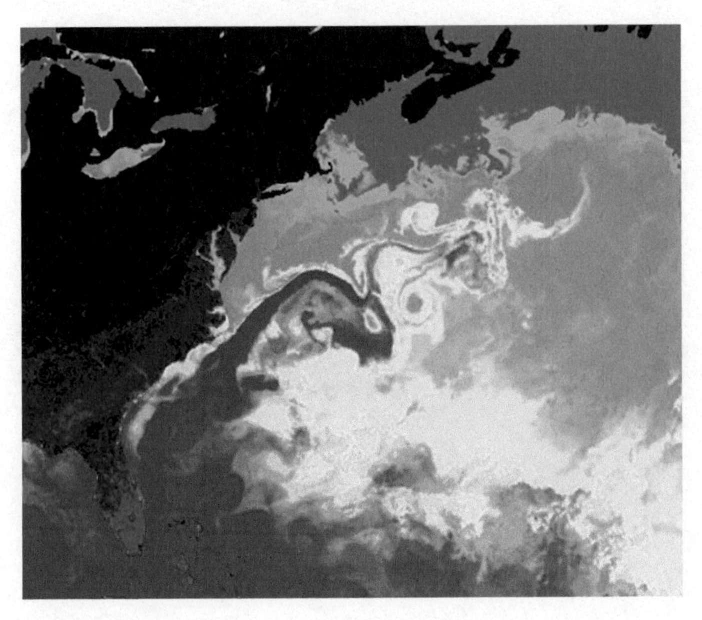

Figure Ⅰ-1: Satellite radiometer image of sea surface temperature, showing the Gulf Stream and its meanders and rings. [Source: NASA]

thin layer of water into motion[6].

The movement of the very topmost layer of the ocean pulls on the water just beneath, which then in turn starts the layer under it moving. Energy from the wind is mostly dissipated within the upper ocean and turbulent stress driving the currents diminished with depth. Consequently, wind-driven currents slow down with depth, and it is confined to a few hundred meters of the upper ocean[7].

Surface currents are also caused by gravity. The top of the ocean is not flat but has broad hills and valleys. Where currents converge or run into a continent, water piles up. The major ocean gyres circle around a low mound a meter or so high[8]. And in summer, intense sunlight can heat and expand seawater, raising the surface by "several" centimeters in the tropics[9].

Currents run down these gentle slopes under the pull of graviy. However, in large-scale dynamic oceanography, gravity's role is mostly balanced out through hydrostatic approximation, leaving out a horizontal pressure gradient related to the surface height difference. Winds and gravity start water moving, but the currents that form do not flow parallel to the wind or straight down the steepest surface. Instead, the currents move at an angle to the force that generates them—the Coriolis effect[10].

The Coriolis Effect occurs because the earth's surface rotates faster at the equator than at the poles. It influences the paths of moving objects that are only loosely in contact with the ground, from currents to winds to airplanes[11].

When objects move toward higher, slower moving latitudes, they outpace the rotation of the surface, and seem to veer toward the east. When objects move toward lower, faster moving latitudes, they lag behind the rotation of the surface. When objects move to the east, the increase of angular velocity around Earth's axis causes increased centrifugal force, hence their orbits normal to the axis expand, thus leaving a southward component in the local coorindate whose z-axis is pointing to the geocenter. The Coriolis deflection is to the right in the northern hemisphere, and to the left in the southern hemisphere.

The movements of currents are also constrained by the shape of the ocean basins. When a current runs into a continent, it must turn aside. The complex interplay between wind, gravity, Coriolis effect, and topography determines the location, size, shape, and direction of the surface current gyres. For example, the North Atlantic gyre, like all the subtropical gyres, is driven by 2 of Earth's prevailing winds—the trade winds and the westerlies[12].

The trade winds start a current that is turned by the Coriolis Effect into a westward flow along the equator. The Equatorial Current gets warmer and warmer as it travels across the tropics[13]. On the other side of the gyre, winds known as the westerlies, combined with the Coriolis deflection, push water to pile up toward the center latitude of the gyre, which then forces a eastward mid-latitude flow[14]. Called the North Atlantic Current, this flow loses heat to the atmosphere[15].

The eastern and western currents of the gyre begin where the equatorial and mid-latitude currents are blocked by land. On the Atlantic's western boundary, the Gulf Stream moves away from the equator and flows north. The Gulf Stream is the strongest, deepest, and fastest part of the gyre, and it transports an enormous amount of heat toward the pole[16]. Finally, the slow and very shallow Canary Current runs south along the eastern edge of the Atlantic, carrying cold water toward the equator to complete the gyre. A single trip around this circuit takes about 10 years.

Although the gyres dominate, a number of other currents also make important contributions to surface circulation. For example, the very warm Equatorial Countercurrent, which flows eastward, can help set up the unusual weather pattern called El Niño[17]. A much colder flow, called the Labrador Current, travels along the west side of Greenland. This current is notorious for flushing icebergs, including the one that sank the Titanic, into the heavily traveled North Atlantic shipping lanes.

Many surface currents—the ones with names—have been in constant motion for millennia. Other currents are temporary—longshore, rip, and upwelling currents only run in certain seasons or weather conditions[18]. Longshore currents flow along coastlines when waves run into the shore at an angle. They bulldoze great volumes of sand along the shore, causing beaches to disappear and harbors to fill in. Rip currents form where obstacles channel water away from the shoreline. Many an unwary swimmer and beachcomber has been swept out to sea after stumbling into a rip.

Upwelling occurs when winds push surface water away from the shore, and deeper water rises to fill the gap. These cold currents bring nutrients to the surface and stimulate high plant and animal productivity. Deep-water circulation has a scale, pace, and power very different from surface circulation. Deep currents twist together into a continuous stream that loops through all the oceans, called the global conveyer belt[19]. With a volume more than 16 times the combined flow of all the world's rivers, the conveyer belt slowly but steadily empties one ocean into another, and over the course of 1,000 years, turns the water in them upside down.

This vast, global circulation is driven by density variations in the ocean. Sometimes called thermohaline circulation because it depends on temperature and salinity, the conveyer begins on the surface of the ocean near the poles[20]. There, the water gets very cold, chilled by low air temperatures to freezing and below. Polar seawater also gets saltier, because when sea ice forms, the salt is left behind[21]. As seawater gets colder and saltier, its density increases, and it starts to sink toward the bottom. Surface water is pulled in to replace the sinking water, and in its turn, eventually becomes cold and salty enough to sink. Thus, a current begins.

The conveyer belt begins at the surface of the North Atlantic where great amounts of water cool and sink off the coast of Greenland. Hemmed in by the continents, this new deep water can only flow south, past the equator, all the way to the far ends of Africa and South America. As the current travels around the edge of Antarctica, fresh streams of cold water sink into and recharge the conveyer belt.

Two sections split off and turn northward, one into the Indian Ocean, the other into the Pacific. Both these currents warm up and become less and less dense as they travel, enough that they eventually rise back toward the surface. Drawn by the inexorable pull of the conveyer belt, these now warm waters loop back the way they came and eventually return to the North Atlantic to begin the long journey all over again.

Currents are an integral and dynamic part of the world's oceans—they help determine the characteristics and behavior of seawater, and the distribution and abundance of marine life[22]. But currents are surprisingly important to landlocked creatures like us as well because they partially regulate the global climate and govern the productivity of fishing grounds.

Upwelling, the rise of deeper water to the surface, occurs only on 10% of the ocean. But that small area makes up half of the world's fisheries. The cool, nutrient-filled water in upwelling currents support blooms of algae and seaweed, the base of the food chain for many clams, crustaceans, and fish. Herring, anchovy, and sardines, three of the most widely harvested fish, are especially concentrated in upwelling zones. Such sea life is an increasingly large component of man's food supply.

Currents play an important role in the Earth's climate system. Overall, ocean currents moderate the planet's temperature extremes. Warm flows, like the western boundary currents, carry heat from the tropics toward the poles[23]. Cold flows, such as the eastern boundary currents, bring cooler temperatures to low latitudes. On a regional scale, some areas are even more strongly affected. Because Western Europe is bathed in warm waters and winds coming east across the Atlantic[24], its climate is much warmer and milder than other areas at the same latitude, such as northern Canada and Alaska.

Although the ocean currents that affect climate are large and vigorous, scientists are beginning to suspect that they are surprisingly easy to disrupt. It is possible that global warming could severely alter current patterns, at least in the short term. If there is more rainfall in the North Atlantic, and significant melting of glacial and sea ice, a layer of warm fresh water could form at the sea surface. This layer could block the formation and sinking of cold salty water there, and turn off the global conveyer belt.

Once the conveyer belt, and its northward pull on warm surface currents, shuts down, average temperatures in much of Europe would plunge 5 - 10℃. Unlike many other causes of climate change, catastrophic cooling due to the loss of the global conveyer belt could be quite rapid, taking just a few years or decades.

Vocabulary

current:(n) 海流
stream:(n) 水流,海流
flow:(n) 流动,海流
circuit:(vt) 绕行;(n) 环路
Gulf Stream:(n) 湾流
　或称墨西哥湾流
satellite image:(n) 卫星图像
swath:(n) 长条,条带
　卫星遥感中特指卫星单一传感器扫描过的条带宽度,即刈(yì)幅
millennia:(n) 几千年
　sing: millennium 一千年
　adj: millennial 千年尺度的
eddy:(n) 涡,涡旋
　syn: vortex 涡旋,漩涡
spin:(vi) 旋转
radiometer:(n) 辐射计
　[radio-]:无线电,射线,辐射,放射性
　[-meter]:(仪器)……计,……仪
sea surface temperature:(n) 海面温度
　缩写为 SST
meander:(n/vi) 弯曲,蜿蜒
gravity:(n) 重力
variation:(n) 变化,变动
　vi/vt: vary(使)变化
　n: variability 变率,变化性
　adj: variable 可变的,多变的
density:(n) 密度
surface circulation:(n) 表层环流
upper layer:(n) 上层
　layer:(n) 层;(vt) 分层
dominant:(adj/n) 最主要的,占支配地位

的(事物)

vt: dominate 主导,支配

gyre:(n) 流环,流圈,流涡

subtropical:(adj) 亚/副热带的

n: subtropics 亚/副热带(地区)

[sub-]: 次,亚,副

subpolar:(adj) 亚/副极地的

polar:(adj) 极地的

motion:(n) 运动

geometry:(n) 几何学,几何特征

[geo-]: 与地球有关的

[-metry]: 测量(学)

topography:(n) 地形,地形学

[-ography]:(学科)……学

~bathymetry:(n) 水深,测深学

~relief:(n) 地形,陆高海深

~orography:(n)(山的)地形

friction:(n) 摩擦,摩擦力

adj: frictional

energy:(n) 能量

~energetics: 能量学,能量学特性

~momentum: 动量

dissipate:(vt/vi) 耗散

n: dissipation 耗散(过程)

adj: dissipative 耗散性的

turbulent:(adj) 湍流的,紊流的

n: turbulence 湍流,紊流

stress:(n) 应力

wind-driven current:(n) 风生流

drive:(vt) 驱动,推动,驱使

converge:(vi) 辐聚,辐合

n: convergence 辐聚度

ant: divergence 散度

adj: convergent 辐合的

ant: diverge 辐散

sunlight:(n) 光照

expand:(vt) 使膨胀

n: expansion

tropics:(n) 热带(地区)

dynamic:(adj) 动态的,变化的,动力学的

syn: dynamical

n: dynamics 动力学

hydrostatic approximation:(n) 静力近似

[hydro-]: 水,液体,流体

~hydrology:(n) 水文学

~hydrostatic balance: 静力平衡

parallel:(adj) 平行的;(n) 地球上的纬线或纬圈

~meridian: 经线,经圈

force:(n) 力

generate:(vt) 产生,生成

Coriolis effect:(n) 科氏效应

~Coriolis force:(n) 科氏力

equator:(n) 赤道

adj: equatorial 赤道的,赤道上的

pole:(n) 地极,极地,极地区域

latitude:(n) 纬度

adj: latitudinal 纬度的

~longitude:(n) 经度

angular velocity:(n) 角速度

orbit:(n) 轨道;(vt) 绕轨道旋转

adj: orbital 轨道的,旋转的

coordinate:(n) 坐标(系)

geocenter:(n) 地心

deflection:(n) 偏转,转向

vt: deflect

hemisphere:(n) 半球

~northern hemisphere 北半球

~southern hemisphere 南半球

ocean basin:(n) 海盆

prevailing：(adj) 盛行的，主导的
vi：prevail
trade wind：(n) 贸易风，信风
简称 trade
westerly：(n) 西风(从西方来的风)
[-erly]：描述位置或方向。描述风时表示风的来向，否则表示去向(向西)或位置(在西方)
Equatorial Current：(n) 赤道流
North Atlantic Current：(n) 北大西洋流
transport：(n/vt) 输运
Canary Current：(n) 加纳利流
Equatorial Countercurrent：赤道逆流
[counter-]：反，逆
weather pattern：(n) 天气形态
El Niño：(n) 厄尔尼诺
西班牙语词，英语环境中可用普通的 n
~La Niña：拉尼娜(也可用 n)
Labrador Current：(n) 拉布拉多流
iceberg：(n) 冰山
constant：(adj) 不断的，连续的；(n) 不变性，不变的事物，常数，常量，恒量
longshore current：(n) 沿岸流
rip current：(n) 裂流，离岸流
upwelling (current)：(n) 上升流
ant：downwelling 下降流
coastline：(n) 海岸线
wave：(n) 波，浪
shore：(n) 海岸
volume：(n) 体积，量，分量，份额
nutrient：(n) 营养物质
productivity：(n) 生产力
scale：(n) 尺度
global conveyer belt：(n) 全球输送带
thermohaline circulation：(n) 热盐环流
[thermo-]：热的，热学的
[-haline]：(n) 咸水；(adj) 咸的
syn：saline
temperature：(n) 温度
salinity：(n) 盐度
salty：(adj) 咸的，高盐的
sea ice：(n) 海冰
sink：(vi) 下沉；(n) 汇
fresh：(adj) 淡的，低盐的
rise：(vi) 上升
characteristic：(n) 特征，特性；(adj) 特有的，独特的
vt：characterize 描述……的特征
be characterized by 成为……的特征
distribution：(n) 分布
abundance：(n) 丰度
adj：abundant 丰富的
marine：(adj) 海洋的，海生的
syn：maritime
climate：(n) 气候
adj：climatic
adv：climatically
component：(n) 组成部分，成分，(矢量的)分量，(力的)分力
moderate：(vt) 中和，调和
extreme：(n) 极端；(adj) 极端的
western boundary current：(n) 西边界流
regional：(adj) 区域性的
global warming：(n) 全球变暖
short term：(n) 短期；(adj) 短期的
可用连词符连接
ant：long term (long-term)
rainfall：(n) 降雨
~precipitation：(n) 降水，包含雨、雹、雪等各种形式

glacial：（adj）冰川的
  n：glacier 冰川
℃：摄氏度，读作 degree(s) Celsius，
  或 degree(s) centigrade
  ~°F：华氏度（Fahrenheit）
  ~K：开尔文（Kelvin）

climate change：（n）气候变化
catastrophic：（adj）灾难性的
cooling：（n）冷却，降温
  ~heating：（n）加热
decade：（n）十年，年代
  adj：decadal

Notes

[1] 此句描述了海流的时空尺度特征。short-lived 指时间尺度短，small 指空间尺度小。vast 指广阔的空间尺度，因为此种海流环绕全球海洋。相应地，这种大尺度海流具有几百年的时间尺度。这反映了自然界的一个基本规律，即运动的时间尺度小则往往空间尺度也小，空间尺度大则往往时间尺度也大，可以总结为“it takes time to travel”。这里用到了“海流”的多种说法，即 current、stream、flow，其中 current 一般特指海流，stream 和 flow 可指包括江、河、溪流和管道水流等在内的各种流动。下文提到的 circulation 和 gyre 特指大尺度可闭合的环状流动。

[2] temperature-coded satellite image 意指用颜色代表温度的卫星图像，即将不同的颜色与温度的高低对应起来，习惯上用红色代表高温，蓝色代表低温。湾流作为暖流，其温度比同纬度周围水体高，因此在图上表现为一支深入黄、绿色区域的红色条带（swath）。此句用 which 引导定语从句，在全句中做插入语。

[3] eddy current 是指涡旋的流动。这里的涡旋是指从湾流主轴上脱落（spin off）下来的环形流，是所谓的中尺度（mesoscale）现象，空间尺度约为数十至数百千米，与几千千米尺度的湾流相比是比较小的。这些涡旋在几个月的时间尺度上存留，能够携带其内部的水体移动，然后消亡（die out）。

[4] 表层环流的垂直范围大约从海表到几百米深度，相对于深度为几千米的海洋来讲，是一个浅层（shallow layer）。

[5] gyre 通常指水平尺度几千千米甚至几万千米尺度的环状流，可译为“流环”“流圈”或“环流”。全球有 5 个副热带（南、北纬 40°—45° 以内）流环，分别位于南、北副热带太平洋（northern and southern subtropical Pacific），南、北副热带大西洋，南副热带印度洋。印度洋的北半球部分很小，没有发展成熟的大尺度流环。副极地（南、北纬 45°以外）流环只在北半球有，且受海盆形状的影响，范围比副热带流环小。南半球有贯通纬圈的南极绕极流（Antarctic Circumpolar Current，ACC），即文中所述“one surface current circles endlessly around Antarctica”。ACC 极地一侧有两个副极地流环，即威德尔环流（Weddell Gyre）和罗斯环流（Ross Gyre）。

[6] 海气界面上的摩擦力（friction）是海洋从大气获取动量的方式，通常用风应力（wind

stress）表示。风应力作用的深度通常用流速降低至表层流速的 4.3%（即 $e^{-\pi}$）之处来表示，此层叫作埃克曼层（Ekman layer），也称摩擦层（frictional layer），深约十几米至百米左右。

[7] 此段简单解释埃克曼层的成因，详细论述见“Ekman Transport and Pumping”（Text 9）。

[8] 海面的起伏与海水的辐聚（convergence）或辐散（divergence）有关。海水辐聚即堆积，则海面隆起，同时形成下降流，辐散则相反。大尺度环流水平方向距离很长或者范围很大，但垂直方向水位起伏仅有米的量级。由于重力的作用，垂向堆积的水在其维持力撤去后，一定会向水平方向散开，形成水平流。

[9] 海水温度变化造成的海面起伏叫作比容海平面变化（steric sea level change），量级通常较小。不同时间尺度上比容海平面变化的原因不同，此处讲述的是季节尺度，即由太阳辐射的变化导致的。全球变暖引起的海水增温也能导致比容海平面变化（上升）。

[10] 科氏效应（Coriolis effect）由地球自转导致的惯性力引起，称为科氏力（Coriolis force），科氏力也叫地转偏向力。地转（geostrophy）运动是海洋和大气——地球流体（geophysical fluid）——大尺度运动的重要属性。

[11] 只有当物体相对地球运动时，才会产生科氏力。科氏参数的量级为 $10^{-5}\ s^{-1}$，因此，科氏力通常是一个非常微弱的力，与地面接触的物体受的摩擦力通常远大于科氏力。但大气和海水由于具有流动性而与地面接触不严，因此，只要科氏力作用的时间足够长（即缓慢运动，也即大尺度运动）就能受到它的明显影响。

[12] trade wind，直译为贸易风，即信风，因常年稳定吹拂而得名。发生在赤道两侧，北侧为东北风，南侧为东南风。westerly 意为西风，即从西方吹来的风，可单独做名词，也可以用作形容词，如 westerly wind、westerly jet。注意 -erly 后缀表示风的来向，而另有 -ward 后缀表示去向。westerly wind 等于 eastward wind。-ward 后缀也可作 -wards，通常用作形容词时不加 s，用作副词时加 s，但二者没有严格区分。此处 trade wind 和 westerly 都用了复数，因为各大洋、南北半球合在一起有多个贸易风和西风。

[13] equatorial current 是大洋环流（general circulation）的组成部分。世界大洋主要表层环流包括：

- 寒流（cold current）：东格陵兰流（East Greenland Current），拉布拉多流（Labrador Current），加纳利流（Canary Current），本格拉流（Bengula Current），亲潮（Oyashio），加利福尼亚流（California Current），秘鲁流（Peru Current），西澳大利亚流（West Australia Current）；
- 暖流（warm current）：挪威流（Norwegian Current），湾流（Gulf Stream），巴西流（Brazil Current），阿拉斯加流（Alaska Current），黑潮（Kuroshio），东澳大利亚流

(East Australia Current), 鲁汶流 (Leeuwin Current); 厄加勒斯流 (Agulhas Current),莫桑比克流(Mozambique Current);

- 纬向流(meridional current):北大西洋/太平洋流(North Atlantic/Pacific Current),南大西洋/太平洋/印度洋流(South Atlantic/Pacific/Indian Current),三大洋都有的北赤道流(North Equatorial Current)、南赤道流(South Equatorial Current)、赤道逆流(Equatorial Countercurrent),南极绕极流(Antarctic Circumpolar Current),南极副极地流(Antarctic Subpolar Current);
- 北冰洋环流(Arctic Circulation):穿极流(Transpolar Current),波弗特环流(Beaufort Gyre)。

[14] 根据 Sverdrup、Stommel、Munk 等的风生大洋环流理论(wind-driven circulation theory),风应力旋度(wind stress curl)驱动副热带环流。mid-latitude 意为中纬度,实为名词,但此处用作形容词,等于 mid-latitudinal。mid 后面的短线可以省略,合成一个单词。

[15] lose heat to 意为将热量向……传递而导致自身热量流失,也可翻译为加热……,表示海气之间的热交换(air-sea heat exchange)。北大西洋流(North Atlantic Current)携带低纬度暖水而来,因此,比同纬度的大气要热,向大气传递热量。

[16] 湾流作为西边界流(western boundary current)是整个北大西洋副热带流环中最强的部分。强海流往往出现在海盆西侧的现象叫作西向强化(western/westward intensification)。transport 译为输运或输送,通常指通过水体的运动而携带的属性(如温度、密度)和物质(如盐分、叶绿素)。

[17] "countercurrent" 即"逆流"。weather pattern 即天气形态,pattern 指空间上特定的分布形式。厄尔尼诺(El Niño)和拉尼娜(La Niña)现象分别对应热带东太平洋增暖和变冷,二者都是 ENSO(El Niño-Southern Oscillation)现象的形态。实际上此处使用 weather pattern 的说法并不严格,因为天气的时间尺度是几天,而 ENSO 的时间尺度是几年(即年际 interannual)。ENSO 这一概念除涉及海温变化外,还包括大气气压的变化,即南方涛动(Southern Oscillation)。oscillation 一词常见于大尺度气候模态(climate mode)的名称,对于大气常译为涛动,对于海洋常译为振荡。除 ENSO 外,常见的气候模态还包括北太平洋年代际振荡(Pacific Decadal Oscillation)、北大西洋涛动(North Atlantic Oscillation)、大西洋多年代际振荡(Atlantic Multidecadal Oscillation)、北极涛动(Actic Oscillation)、印度洋偶极子(Indian Ocean Dipole)、南半球环状模(Southern Annular Mode)等。

[18] longshore current 意为沿岸流,近义词有边缘流(rim current)、近岸流(coastal current)等,含义并不完全相同。rim current 通常比 longshore current 更贴近边界,coastal current 则距离稍远。海流离岸的距离越大则越有可能不总是沿着海岸。而边界流(boundary current)通常指更大尺度的海盆边界流,如湾流,距离海岸更远。

[19] global conveyer belt 意为全球传送带或输送带,是热盐环流的比喻用法,即将连接全球海洋的环流系统比作传输货物的传送带(conveyer belt)。

[20] thermohaline circulation 即热盐环流,由 thermo(热)和 haline(盐)两个词根组成,反映了人们对热盐环流的传统认知:热(温度)和盐[盐度(salinity)]决定密度(density),密度重的水下沉,密度轻的水上升导致热盐环流。实际上热盐环流不仅与密度变化有关,机械能(mechanical energy)也有很重要的作用。详见"Thermohaline Circulation"(Text 13)。

[21] 这叫作海水的结冰析盐现象,即 brine rejection。brine 意为咸水、卤水,与 saline 基本同义,但后者可以作形容词,意为咸的、高盐的,与 salty 同义。

[22] marine 是形容词,意为海洋的、海里的、与海洋有关的、(生物)海生的、与航海有关的、海事的,等等。海洋科学可以叫作 oceanography,也可以叫作 marine science。

[23] "向极地方向"可以说 toward the poles,也可以说 poleward(s)。向赤道方向则为equatorward(s)。

[24] 此处所谓 warm waters 指湾流和北大西洋流带到欧洲方向的暖水,而 warm winds 指大气的西风急流,经过海洋时被加热,因此相对暖一些。coming east 意为向东。

2. Hydrologic Cycle

Text

The complex, constant movement of water on Earth—from the oceans to the air, across the landscape, and through plants and animals—is called the hydrologic cycle or water cycle[1] (Figure Ⅰ-2). It is powered by solar energy[2], and aided by gravity. The water cycle includes all the processes that shift water—both those that physically move water, and those that convert water between solid, liquid, and gaseous states[3].

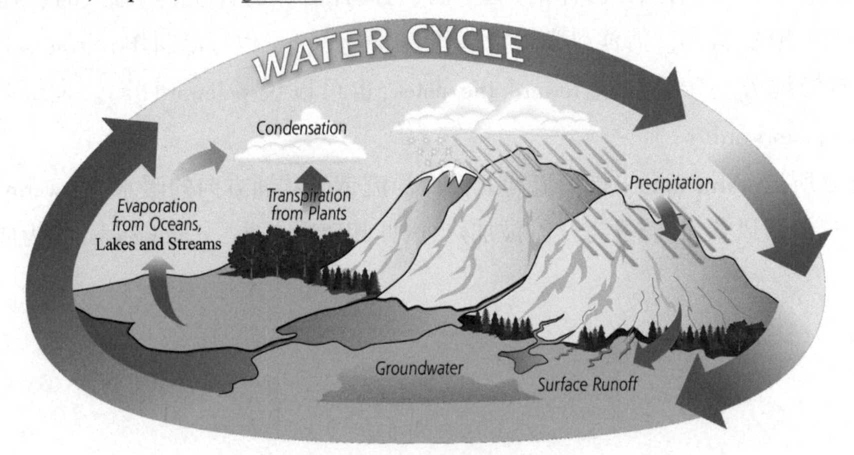

Figure Ⅰ-2: Schematic for global hydrologic cycle. [Source: NASA]

The cycle also contains many reservoirs[4], where water accumulates. Within these reservoirs, water is spread unevenly across the Earth's surface. The great majority—more than 97%—lies in the oceans. Just over 2% is frozen in glaciers and ice caps. The rest—a meager half percent or so—is divided between the atmosphere, lakes and streams, and the ground.

Because most water is in the oceans, most water is salty, and stays salty for a long time. Within the vastness of the ocean, water molecules may linger for millennia. Tides and waves move ocean water over short distances, while currents circulate the oceans around the globe. Carrying warm water away from the equator, and cold water away from the poles, oceanic circulation moderates the planet's temperature extremes.

Wherever the sun shines on the ocean, evaporation creates fresh water vapor out of salty seawater. Winds lift the moist air high into the atmosphere, and blow it about the globe. As air moves away from the warm ocean it starts to cool off. With sufficient cooling, water vapor changes into drops of liquid water—a process called condensation[5].

With continued condensation, tiny drops of water grow bigger and bigger and gather into clouds. The average airborne water molecule stays aloft for only 10 days or so, until it joins a water droplet heavy enough to fall from the sky. Most precipitation, be it rain, snow, sleet, or hail, drops right back into the ocean[6]. There, the water circulates until evaporation claims it once again.

Some snow and rain also lands on the continents. In cold regions, at high latitudes or altitudes, where snow builds up in the winter and doesn't melt in the summer, ice caps and snowfields grow and persist over thousands of years. The clean ice in glaciers and ice caps, most frozen well before the age of man, is both the purest and the largest storehouse of fresh water on Earth[7].

When ice and snow melt, or rain falls on land, water is pulled swiftly downhill by gravity. Some of it flows across the top of the ground, a process called runoff. Surface water gathers into rivers, pauses for a time in lakes, and rushes down to the ocean. As it flows across the ground, running water cuts into the earth, wearing down and reshaping the ground. Moving water is the most powerful geologic force sculpting the landscape.

Although the handiwork of runoff is visible everywhere on land, more precipitation actually infiltrates, or soaks into, the earth than runs off. More than 95% of the planet's liquid fresh water is groundwater—water held within the ground. Some shallow groundwater doesn't last long. Evaporation directly from the soil, and transpiration through plants, both transfer moisture back to the air. The process of transpiration is part of plant metabolism. Liquid water is absorbed by roots, and lifted to the leaves, where it is converted to vapor and lost to the sky.

Most groundwater is neither evaporated nor transpired. It slowly drains downward, slipping through tiny pores between soil grains, following cracks and caves in the bedrock, to seep into streams, lakes, and eventually, the ocean. Along the way, the water interacts with the ground. Some pollutants, such as bacteria, are filtered out of the water, while some minerals, like sodium and arsenic, are picked up by the water. Earth's oceans are salty because groundwater has been carrying dissolved mineral salts down to the ocean for billions of years.

As the water cycle spins, the earth's water moves from oceans to the atmosphere to the land, and back to the ocean, over and over, and over again. However, the water cycle is not a closed system. Tectonic activity inside the planet pulls water out of the system when seawater is dragged down inside the planet at subduction zones, and also adds water into the system when steam erupts from volcanoes. Some moisture also leaves the outer edge of the water cycle, when vapor high in the atmosphere "leaks" into space. And, water can also enter the system from above, when icy comets collide with earth[8].

The water cycle is the most fundamental system operating on the surface of the earth. The varied processes of the water cycle control the global climate, shape the landscape, and allow life to exist.

The movement and distribution of water across the planet determines broad climate patterns. Evaporation powers vast weather systems like hurricanes and cyclones, while uneven precipitation nourishes rainforests or parches deserts. Other weather events, such as floods, droughts, and blizzards, are all aspects of the water cycle.

Runoff is unequalled in its ability to erode rock and soil, and reshape the face of the land. Mountains are worn low and canyons are carved deep, grain-by-grain from the relentless force of moving water.

For mankind, perhaps the most immediate impact of the water cycle is its influence on the quantity and quality of fresh water. Because of the unequal distribution of water around the cycle, fresh water is relatively scarce—only 0.65% of Earth's total water supply is neither salty nor frozen. And the quality of that vital water is at risk, because as it moves through the water cycle, it is exposed to a host of natural and manmade pollutants. As clouds drift above cities, and rivers run past factories and fields, their waters pick up industrial and agricultural wastes. As groundwater infiltrates through the soil, it gathers up both mineral compounds and buried toxic chemicals.

The global demand for clean, fresh water is increasing rapidly. At the same time, the supply is threatened by pollution and climate change. Scientists hope that by studying the complex workings of the water cycle, we will learn how to predict and preserve sufficient water resources to sustain the ever-growing world population.

Vocabulary

landscape:(n) 风景,地形地貌
hydrologic cycle:(n) 水循环
  hydrologic:(adj) 水的,与水相关的,水文的
  hydrology:(n) 水文学
  cycle:(vt/n) 循环
  ~carbon cycle:(n) 碳循环
water cycle:(n) 水循环
solar:(adj) 太阳的
  ~lunar:(adj) 月亮的
process:(n) 过程
  ~physical process:物理过程

shift:(vt/n) 移动,转移
physically:(adv) 物理上的
  adj:physical 物理的,物理学的
  n:physics 物理学
gaseous:(adj) 气体的
schematic:(n) 示意图;(adj) 简略的,概要的,示意的
reservoir:(n) 水库,蓄水池,容器
accumulate:(vt) 积累,聚集
  n:accumulation
ice cap:(n) 冰冠
meager/meagre:(adj) 微小的,很少的

molecule：(n) 分子

adj：molecular 分子的

~atom：(n) 原子

~ion：(n) 离子

tide：(n) 潮汐

adj：tidal 潮汐的

evaporation：(n) 蒸发

adj：evaporative 蒸发的

water vapor：(n) 水蒸气

或省略 water

vapor 英式拼写为 vapour

seawater：(n) 海水

与下文的 freshwater 相对

moist：(adj) 潮湿的

n：moisture 水汽，潮气，湿气

vt/vi：moisturize 加湿

sufficient：(adj) 足够的，充足的

adv：sufficiently

condensation：(n) 凝结

adj：condensational 凝结的

vi：condense 凝结

airborne：(adj) 由空气携带的

aloft：(adv) 在高空

droplet：(n) 小水滴

[-let] 小的

precipitation：(n) 降水(包括雨、雪、雹等各种形式)

~rainfall：(n) 降雨

~snowfall：(n) 降雪

sleet：(n) 雨夹雪，冻雨

hail：(n) 冰雹

altitude：(n) 高度

snowfield：(n) 雪原

persist：(vi) 延续，持续，保持

adj：persistent

n：persistence

fresh water：(n) 淡水

可连成一个词，即 freshwater，与 seawater 相对

runoff：(n) 径流

geologic：(adj) 地质的，地质学的

= geological

n：geology 地质学，地质特征

infiltrate：(vt)(液体)(使)渗入

groundwater：(n) 地下水

transpiration：(n) 蒸腾作用

moisture：(n) 湿气，水汽，潮湿

adj：moist 湿的，潮湿的

metabolism：(n) 新陈代谢

absorb：(vt) 吸收

n：absorption

pore：(n) 孔隙

grain：(n) 颗粒

crack：(n) 裂缝，裂纹

bedrock：(n) 基岩

interact：(vi) 相互作用

n：interaction

pollutant：(n) 污染物

bacteria：(n, pl) 细菌

单数形式为 bacterium

filter：(vt) 过滤，滤波；(n) 过滤器，滤波器

mineral：(n/adj) 矿物质(的)

vt：mineralize 使矿物化

sodium：(n) 钠

arsenic：(n) 砷

dissolved：(adj) 溶解的

vt/vi：dissolve

n：dissolution

closed system：(n) 闭环系统

tectonic：(adj) 地质构造的
　n：tectonics 地质构造，地质构造学
subduction：(n)(板块的)俯冲
　在物理海洋学中指海水的潜沉
volcano：(n) 火山
fundamental：(adj) 基本的，根本的，核心的，本质的，基础的；(n) 基本法则，根本原则
weather system：(n) 天气系统
hurricane：(n) 飓风
　专指大西洋一侧，太平洋一侧的相同现象即台风 typhoon
cyclone：(n) 气旋，气旋式海洋涡旋
　大气里分为热带气旋(tropical cyclone)、温带气旋(extratropical cyclone)等
　adj：cyclonic 气旋式的
　~anticyclone：(n) 反气旋，反气旋式海洋涡旋
　　adj：anticyclonic 反气旋式的
flood：(n) 洪水；(vi) 发洪水
drought：(n) 干旱
blizzard：(n) 暴风雪
aspect：(n) 方面，特征，考虑事物的方式
erode：(vt) 侵蚀
canyon：(n) 峡谷
impact：(n) 效果，影响；(vt) 产生影响
quantity：(n) 数量，数值，参量
　vt：quantify 量化
quality：(n) 质量，品质，性质，水质
　vi/vt：qualify(使)有资格/胜任
scarce：(adj) 稀少的，缺乏的
vital：(adj) 至关重要的，必要的
expose：(vt) 使暴露，使显露，使遭受
a host of：许多
　= hosts of
natural：(adj) 自然的
manmade：(adj) 人为的，人工的
　syn：anthropogenic 人类活动引起的，人类学的
drift：(vi/n) 漂流
industrial：(adj) 工业的
　n：industry
agricultural：(adj) 农业的
　n：agriculture
compound：(n) 化合物，混合物
toxic：(adj) 有毒的
chemical：(n) 化学物质；(adj) 化学的
　n：chemistry 化学
predict：(vt) 预报，预测
　n：prediction 预报，预测
　adj：predictable 可预测的
　n：predictability 可预报性

Notes

[1] cycle 意为循环。各种物质在全球的循环再分配都可叫作 …cycle，如碳循环 carbon cycle。水循环可用 hydrologic cycle 或 water cycle，但前者更精确。

[2] solar energy 即太阳能量，也叫太阳辐射(solar radiation)。

[3] process 意为过程，这里包括物理上 (physically) 移动水的过程，以及将水在固、液、气三相 (state) 间转换的过程。实际上水的相变也是物理过程，此处所谓 physically 更倾向于表示在“形体上”发生移动，即采用了 physically 的表示身体的意思。

[4] reservoir 本义为蓄水的水池、水库等,可提供一定的缓冲能力。引申为水或其他物质、性质的储存处或缓冲处。例如,carbon reservoir 碳库、energy reservoir 能量池等。

[5] 蒸发是水由液体转变为气体的相变过程,水汽携带的热量要在高空遇冷凝结时才释放出来,因此叫作潜热 latent heat。蒸发使海洋失去热量引起海面冷却的过程叫作 evaporative cooling。

[6] precipitation 是降水的总称,包含液态和固态的各种形式。其中降雨(rainfall)在中低纬地区是最主要的形式,因此,有时不甚严格地将 rainfall 与 precipitation 混淆使用。

[7] 冰冻圈(cryosphere)的主要组成部分:冰川(glacier)、冰原(icefield)、冰冠(ice cap)、冰盖(ice sheet)、冰架(ice shelf)、永久冻土(permafrost),以及海洋中的海冰(sea ice)。其中,glacier 是陆地冰川的通用说法,而 icefield、ice cap、ice sheet 是不同大小的冰川(由大到小),ice shelf 是冰探出陆地、覆盖在海洋上的部分。

[8] 上述几个改变地表水量的过程相对于整个海洋来讲非常微小,因此,全球水循环虽不严格闭合,但在海洋学研究范畴内可认为基本闭合。

3.Ocean Waves

Text

Within a wave, individual water particles move in circles that get smaller with depth and eventually stop altogether[1]. Boats or bottles or other debris floating on the surface do not go anywhere either, but simply bob up and down. The only thing waves do transmit across the ocean is energy[2].

Wave anatomy is very simple. The highest surface part of a wave is called the crest, and the lowest part is the trough. The vertical distance between the crest and the trough is the wave height. The horizontal distance between two adjacent crests or troughs is known as the wavelength[3]. All waves have this same basic anatomy. But wave behavior is a complicated dance, choreographed by the forces that cause them and the ocean around them[4].

Many things, from moving ships to earthquakes, to a bird skipping across the water, cause waves, but most are generated by wind. When wind blows across the ocean surface, the friction between the air and water kicks up a series of small ripples. These bumps on the ocean give the wind something to push against, and soon the ripples grow into waves. The waves grow higher, longer, and faster, reaching their maximum size when they nearly match the speed of the wind.

The longer and further the wind blows, the bigger and faster the waves become[5]. Small waves characterize areas where islands interrupt the flow of wind and water. The largest waves on Earth form where strong winds blow steadily across kilometers of open sea. In the long empty stretch between Antarctica and the Indian Ocean, even the average wave is 7 meters high[6].

In deep water, a group of wind-driven waves, called a wave train, develops into a series of harmonious, rounded swells[7]. The train keeps moving even as it leaves behind the wind that formed it. In the open sea, wave trains soon encounter other sets of waves traveling in different directions and with different speeds, heights, and wavelengths. Interference between wave trains can produce a confused, highly irregular sea.

Sometimes, wave energy can be focused by the interaction of wave trains with currents or perhaps other waves to produce a freakishly large, or rogue wave. The phenomenon of rogue waves is poorly understood, but it appears these waves can reach a height of 30 meters and are responsible for many shipwrecks.

When waves move into shallow water, their behavior changes dramatically. The definition of shallow water depends on the size of the wave. At a depth of half the wavelength, the wave starts to "feel the bottom"—the deepest circling water particles come in contact with the seafloor. Friction between the seafloor and the water profoundly changes the speed, direction, and shape of waves. First, waves slow down as they drag across the bottom. The wavelength decreases, and the waves in the train start to bunch up.

If a wave is coming toward land at an angle, or the shoreline is uneven, some parts of a wave will feel bottom before other parts and slow down first. This causes the wave to bend, or refract, so that waves turn toward the shore or wrap around islands or headlands[8]. As the water shallows, the energy in the wave cannot move downward and instead is forced upward, increasing the wave height.

The deepest part of the wave slows down more than the top of the wave. The wave begins to lean forward as the crest rushes ahead of the base. Eventually the wave topples over and breaks against the shore. The wave collapses into foaming sheets of water, called swash, that roll up and then down the beach, carrying along sand and gravel.

Waves caused by the wind can erode the coast, bulldoze nearshore sediments, and generate strong currents[9]. These effects are the cumulative result of wave after wave, day after day, endlessly pounding against the shore.

Another type of wave is much more powerful, able to remodel the coast, not over the space of years, but in just a few hours. These waves are the largest and most energetic on Earth and are called tsunami[10].

Tsunami have the same basic anatomy as wind driven waves, but their scale is quite different. In deep water, wind waves travel between 8 and 105 kilometers per hour, may reach heights of 14 meters, and are no more than a few tens of meters apart.

Tsunami are barely noticeable in the open sea—their height is just one meter or less, and successive wave crests are kilometers apart. These waves travel very fast, racing across an ocean at the speed of a jet[11].

Tsunami are caused by geologic events that push away a mass of water. Underwater landslides, volcanic eruptions, even asteroids falling into the ocean from space, can spawn tsunami. But most are caused by earthquakes.

When a quake suddenly shoves a large piece of seafloor up or down, the entire overlying ocean moves too. The displaced water rushes away from the disturbance in the form of waves that spread in all directions.

When tsunami come onshore, they behave generally like breaking wind waves, but with vastly magnified results. The water motion in a tsunami involves the entire water column, not just a thin layer at the top of the ocean. When all of that enormous energy is compressed into shallow water, the wave grows to towering heights, as much as 30 meters or more.

As the wave rises, water is sucked into the crest from ahead of the wave, and the ocean actually pulls away from the shore. When the crest does arrive, the tsunami breaks with enormous force, much higher than the normal surf line, and the water from the collapsed wave can sweep far inland. A tsunami, like a wind wave, is usually not a single wave but part of a series. Because their length is so great, successive waves take several minutes to almost 2 hours to arrive.

Tsunami hammer the coast with great energy and can dramatically reshape islands and shorelines in minutes. Because they reach areas well away from the beach, they can engulf entire cities and then drag the debris and victims back into the ocean.

Thankfully, tsunami are rare. But ironically, their infrequency increases the destruction and death they cause. In between disasters, people forget. Reefs, marshes, and mangroves that dissipate wave energy are destroyed for development. Buildings are constructed in danger zones. When the next tsunami does arrive, many people do not flee to safety, but instead let their curiosity draw them to the shore and to their deaths.

Coastlines are among the most energetic environments on Earth. There, waves concentrate the power of the wind, of earthquakes, and of undersea volcanoes on the thin strip of rock and sand that guards the edge of the continent.

Some attempts have been made to harness the relentless power of waves and turn it into electricity. Generator designs vary, but all rely on the up and down motion of waves to spin turbines and produce electricity. The idea of wave power is appealing—waves are free and non-polluting. Although the technology is still under development, it holds great promise and scientists estimate wave power could supply two times the electricity the world currently consumes[12].

Even as energy companies struggle to use wave power, the waves themselves continue their eternal work. Everywhere the land meets the ocean, waves are working away to smooth and straighten and drive back the coastline. Waves slam ashore with great force, fracturing cliffs and widening caves. Gravel and sand rolled and flung against the shore abrade away rocks.

Wave refraction focuses the power of waves into protruding rocks and headlands, wearing them back quickly. Then the sediments made by wave action are rolled onto beaches and across bays. With time, waves turn jagged coastlines into smooth stretches of sand.

Although the cumulative effect of waves is dramatic, individual waves are not particularly powerful or remarkable. The exception to this is the tsunami. Tsunami are the most spectacular of waves. They also stand alone in their ability to spread death and destruction to every corner of an ocean. Spawned by events hidden from their victims by distance and the depths of the ocean, tsunami strike with little warning and tremendous force.

Over all of recorded history, perhaps 1 million people have died in tsunami. The most deadly struck in 2004, killing more than 250,000 people from Indonesia to Africa. Spurred by this and earlier disasters, many governments are working to install warning systems and rebuild natural coastal barriers to tsunami. Unfortunately, the incredible power of these waves guarantees that tsunami will continue to extract a high toll from those who live and play along the shore.

Vocabulary

particle:(n) 颗粒,微粒,质点
debris:(n) 碎片,残骸(集合名词,无复数)
float:(vi) 漂浮,漂流;(n) 漂浮物
transmit:(vt) 传送,传递,传导
  n: transmission
  ~transfer:(vt/vi) 转移,传递
  ~transport:(vt) 输运
anatomy:(n) 解剖学,结构
  ~structure:(n) 结构
crest:(n) 波峰,山峰
  syn: peak
trough:(n) 波谷,山谷
wave height:(n) 波高
  ~significant wave height 有效波高
  syn: amplitude 振幅
adjacent:(adj) 邻近的,相邻的
wavelength:(n) 波长
  ~wavenumber: 波数(波长的倒数)
series:(n) 系列,一连串的,序列
  pl: series
ripple:(n) 波纹
maximum:(adj) 最大的;(n) 最大值
  pl: maxima
  syn: maximal
  vt: maximize 使最大化
  ant: minimum
open sea:(n) 开阔海洋
  ~open ocean 开阔大洋
  ant: coastal sea 近岸海域
  ant: marginal sea 边缘海
  ~ marginal:(adj) 边缘的
  n: margin 边缘
wave train:(n) 波列
harmonious:(adj) 谐波的
  syn: harmonic
  n: harmonic 谐波
swell:(n) 涌浪
encounter:(vt/n) 遇到,面临,偶遇
interference:(n)(波的)干涉
irregular:(adj) 不规则的
  n: irregularity
interaction:(n) 相互作用
  vi: interact
rogue wave:(n) 畸形波,疯狗浪
phenomenon:(n) 现象
  pl: phenomena
shipwreck:(n) 船难
definition:(n) 定义
  vt: define
seafloor:(n) 海床,海底
shoreline:(n) 海岸线
refract:(vt) 折射
  n: refraction
headland:(n) 海岬
foaming:(adj) 形成泡沫的
  n: foam 泡沫
swash:(n/vi) 冲刷
sediment:(n) 沉积物
  adj: sedimentary

cumulative：（adj）累积的，渐增的

energetic：（adj）高能的，强烈的

tsunami：（n）海啸

pl：tsunami 或 tsunamis

jet：（n）喷气式飞机，射流，急流

mass：（n）物质，质量

landslide：（n）山崩，泥石流

spawn：（vt/vi）产卵，大量产生

物理海洋学中常指波动的产生

displace：（vt）取代，移动

n：displacement

disturbance：（n）扰动

onshore：（adj/adv）陆上的/地，向岸的/地

ant：offshore

generally：（adv）一般地，总体上地

adj：general

n：generality 一般性

water column：（n）水柱

compress：（vt）压缩

n：compression

adj：compressible 可压缩的

n：compressibility 可压缩性

ant：incompressible

surf line：（n）激浪线

即波浪最高能淹没的水位线

~surf zone：激浪带

即从波浪开始破碎的位置至激浪线之间的区域

inland：（adj/adv）内陆的；陆地方向地

reef：（n）礁

marsh：（n）沼泽，湿地

mangrove：（n）红树，红树林

construct：（vt）建筑，建造，建立，构成，形成，组成；（n）观念，概念，构想，构成物，组成物，建造物

adj：constructive 建设性的，有用的

n：construction 建造，构造，建设，建筑物，构造物，解释，说明

estimate：（vt）估计，估算，评估

n：estimation

ashore：（adv）向岸地

bay：（n）海湾

~表示海湾的名词还有 embayment、gulf 等，意思没有严格区分

remarkable：（adj）值得注意的，引人注目的，显著的，突出的

Notes

[1] 本文所说的波动（wave）主要是指表面重力波，实际上 wave 一词可以表示任何形式的波动，包括海洋内部的重力波——内波（internal wave），以及毛细波（capillary wave，由表面张力形成）、惯性重力波（inertial-gravity wave，考虑科氏力的重力波）、罗斯贝波（Rossby wave，由 β 效应引起）、开尔文波（Kelvin wave）、声波（sound wave 或 acoustic wave）等。线性假设下，深水表面重力波的水质点做圆周运动，半径随深度减小。按照传播方向与介质振动方向垂直还是平行，可分为横波（transverse wave）和纵波（longitudinal wave），后者又称疏密波（rarefaction wave）。

[2] 线性波动（linear wave）仅传播能量，不传播物质。非线性波动（nonlinear wave），如斯托克斯波（Stokes wave）的水质点轨迹在一个周期内不闭合，而是有净的位移，因

此有物质传递。

[3] 波峰与波谷之间的距离除了叫作波高,也可以叫作振幅(amplitude)。wave length 可以合并成一个词 wavelength。波长的倒数(或倒数的 2π 倍)叫作波数(wavenumber)。从时间的角度看,水质点先后两次到达波峰或波谷位置的时间间隔叫作周期(period),而周期的倒数叫作频率(frequency),频率的 2π 倍叫作角频率或圆频率(angular frequency)。

[4] 这是比喻的说法,将波动比喻成舞蹈,将产生它们的力和周围的海洋比喻成编舞者。

[5] 风吹拂的时长叫作风时(wind duration),风吹拂的距离叫作风区(fetch)。

[6] 南大洋(the Southern Ocean),尤其是南大洋的印度洋部分,是世界上海浪最大的地方,因为风时和风区都足够长。

[7] 涌浪的波面比较平滑,更接近正弦或余弦曲线,因此,此处描述为 harmonious。

[8] 此处所讲的弯曲(bend)、转向(turn toward)、环绕(wrap around)都是针对波峰线来说的。波峰线(crest line)是波峰的连线,与波向线(wave ray)垂直。波浪传播靠近海岸的过程中,波峰线倾向于与等深线平行,原因是浅水波波速与水深有关,水深越深波速越快,因此,水深处的波峰逐渐追上水浅处的波峰,最终形成平行于等深线的波峰线。等深线如果弯曲,则波峰线弯曲。等深线环绕岛屿,则波峰线也倾向于环绕岛屿。

[9] 海浪形成海流的典型例子是裂流(rip current),中文又称离岸流。

[10] tsunami 一词来源于日语"津波",字面原意"港口波浪"。单复同形,或加 s 都可以。

[11] 此处 jet 是指喷气式飞机。流体力学里常见意思是喷射流,海洋和大气中指强流或强风,常译为急流。如急流(jet stream)、西风急流(westerly jet)、黑潮延伸体急流(Kuroshio extension jet)。

[12] promise 此处并非指承诺,而是指前景(名词)。相应的形容词为 promising,意为有前景的、值得期待的,可用于形容某理论、技术或科学发现。

4. Tides

Text

Tides are the regular, alternating rise and fall of sea level caused by the gravitational pull of the moon and sun[1]. The changing of the tide is often rapid and dramatic. On a smaller scale, similar motions occur on large lakes, in the atmosphere, and even within the solid earth.

During a tidal cycle in the oceans, sea level rises, or floods, until it reaches its highest normal point, called high tide. Sea level drops, or ebbs, to its lowest point at low tide. The difference between high and low tide is the tidal range. These tides occur because forces generated by the gravitational pull of the sun and the moon, and by the rotation of the earth, tug seawater into two enormous bulges. One bulge is located on the edge of the planet closest to the moon. The other bulge forms on the opposite side of the planet[2].

The sun and the moon both contribute to the formation of these tidal bulges. The tidal force generated by each body is determined by its mass and its distance from Earth. The sun is 27 million times more massive than the moon, but it is also almost 400 times farther away. As a result, the moon exerts more than twice as much tidal force on the oceans, so that a tidal bulge forms beneath and follows the moon. Solar gravity acts only to reinforce or diminish the moon's pull[3].

The other tidal bulge forms on the far side of the planet. As the earth rotates through space, the water on it tends to keep moving in a straight line due to inertia[4]. This tendency is able to tug the ocean up into a bulge where the moon's gravitational force is weakest. These bulges are the basis of Earth's tides, but many other factors also influence the behavior of tides.

On a global scale, the dynamic interaction between the pull of the sun and moon determines the range and magnitude of the tides[5]. When the sun and moon align, their gravitational forces combine to produce the very highest and lowest tides, called spring tides. When the sun and moon are at right angles, they pull at the ocean from different directions and moderate tides, with a smaller range, form. These are called neap tides[6].

On a more local scale, the timing, size, and speed of tides varies with the shape of the coastline, seafloor topography, river discharge, and even wind and weather conditions. For example, many areas have two, equal high tide-low tide cycles each day. In some regions though, the two cycles have different heights, and in still others, there is only one high and low tide each day[7].

In all of these circumstances, the once or twice daily rise and fall of sea level poses enormous challenges and opportunities for the plants and animals that live in the dynamic space between high and low tides. This challenging habitat is known as the intertidal zone. It has four subdivisions—the spray zone, and the upper, middle, and lower intertidal zones.

The spray zone is wetted by the splash of breaking waves, but rarely submerged. This zone supports very limited life. The upper intertidal zone is underwater only during high tide. The middle intertidal zone is mostly submerged except for brief periods once or twice a day during low tide. The lower intertidal zone is only exposed to air during the lowest spring tides.

The organisms in the two highest zones are out of the water for long periods of time—up to several weeks at a stretch in the spray zone, and several hours once or twice a day in the upper intertidal. In order to survive here, they must be able to avoid drying out, endure temperature extremes, withstand intense sunlight, and survive a wide range of salinities. They also must escape predation by land animals during their long exposure.

In contrast, the plants and animals living in the deeper intertidal zones are less affected by the challenges posed by exposure, because they are usually protected by the cover of water. However, the lower, middle, and even upper intertidal zones are relentlessly pounded by every breaking wave. They must be able to bear up under these repeated blows, as well as maintain their position despite the powerful surge of tides, waves, and currents.

The rise and fall of the tides has powerful effects on both deep ocean and coastal environments. Twice a day, tides roll through the open sea and tidal currents bounce off of sea floor ridges and shelves. The resulting turbulence stirs and mixes ocean waters, redistributing heat and nutrients and helping to regulate climate[8]. Some recent studies suggest that approximately half of the energy needed to power the ocean's global circulation pattern comes from the tides.

However important the role of tides in the deep ocean, their impact on human activity is most apparent along the shore. Many of these effects are expensive, and some are dangerous. The constant shifting of local sea level poses a number of problems for coastal residents, and the designers of facilities such as marinas, ports, bridges, and offshore oil platforms. In areas where the tidal range is large, tides can worsen the damage caused by storms and tsunamis. If these destructive waves come onshore at high tide, flooding and erosion can occur much further inland than if they arrive at low tide.

Although tides can be destructive, they also have the potential to become an important source of reliable, renewable energy. Electricity can be generated by both the up and down motion of sea level, and by the sideways flow of tidal currents. There are technological barriers to harnessing tidal power, as well as environmental concerns. One of the most promising designs involves the use of offshore turbines, which function much like underwater windmills.

The economic impact of tides on coastal inhabitants is significant. But tides also play a powerful role in the emotional relationship between people and the sea. Tidepools, small pockets of water and life that appear at low tide, provide a uniquely accessible, intimate means of experiencing the mystery and wonder of the ocean realm[9]. These transient windows into the sea have made many of those who explore them more aware, more appreciative, and more protective of oceanic ecosystems.

Vocabulary

alternating: (adj) 交替的,间隔的
  vt/vi: alternate(使)交替、替换
  adj: alternative 另外的,可替换的,可供选择的
sea level: (n) 海平面,此处指潮位
gravitational: (adj) 引力的,重力的
  n: gravitation
  syn: gravity (n) 重力
  gravitation 指任何两个物体之间的引力,而 gravity 特指地球上的物体受到的重力(地心引力与惯性离心力的合力)
tidal cycle: (n) 潮周期,潮循环
flood: (n/vi) 涨潮
high tide: (n) 高潮
ebb: (n/vi) 退潮
  ~the tide is on the ebb: 正在退潮
low tide: (n) 低潮
tidal range: (n) 潮差
bulge: (n/vi) 隆起,凸起
massive: (adj) 大而重的,大量的,剧烈的,严重的
exert: (vt) 施加(影响、压力)
tidal force: (n) 引潮力
inertia: (n) 惯性
  adj: inertial
  ~thermal inertia 热惯性(即热容)
tendency: (n) 趋势,倾向
factor: (n) 因素,要素,因数,系数
magnitude: (n) 规模,尺寸,大小,数量,数值
align: (vi) 排成直线,对齐
spring tide: (n) 大潮
right angle: (n) 直角
neap tide: (n) 小潮
  ~new moon: 新月,朔
  ~full moon: 满月,望
  ~first/third quarter of the moon: 上弦月/下弦月
local: (n) 局地的
river discharge: 径流
  syn: runoff 或 run-off
daily: (adj) 逐日的,每天的
syn: diurnal 每天的,日周期的
intertidal zone: (n) 潮间带
spray zone: (n) 飞沫带
submerge: (vt/vi) 淹没,潜入水中
organism: (n) 生物,有机体
in contrast: 相反地
  ~contrast: (n) 差别
maintain: (vt) 保持,维持
  n: maintenance
surge: (n/vi) 汹涌翻腾,急剧上升
  ~storm surge: (n) 风暴潮
ridge: (n) 山脊,脊状突起带,高压脊
  ~Mid-Atlantic Ridge: 大西洋中脊
shelf: (n) 大陆架,陆棚,突出的岩石或

土地

pl: shelves

mix: (vt) 混合

n: mixing

approximately: (adv) 大约,大概,近似地

vt/vi : approximate 近似等于,估算

adj: approximate 近似的,接近的

n: approximation 近似

apparent: (adj) 明显的,清晰可见的

resident: (n/adj) 常住居民,定居者

n: residence 居住,停留,住所

vi: reside 居住,位于,存在于

marina: (n) 船坞

offshore: (adj/adv) 离岸的/地

storm: (n) 风暴,暴风雨

erosion: (n) 侵蚀

tidal current: (n) 潮流

significant: (adj) 显著的,重要的

n: significance 显著性,重要性

tidepool: (n) 潮池,满潮湖,蓄潮池,潮水潭

syn: rock pool

uniquely: (adv) 独一无二地,独特地

adj: unique

accessible: (adj) 可达到的,可进入的

n/vt: access 接近,进入,通路

intimate: (adj)(知识)详细的,深刻的,彻底的

realm: (n) 界,领域,范围

transient: (adj) 短暂的,瞬变的

explore: (vt) 考察,探究,勘探,探讨,检查,评估

n: exploration

adj: exploratory

ecosystem: (n) 生态系统

~ecology: (n) 生态学

~marine ecology: (n) 海洋生态学

Notes

[1] tide 可做可数名词,指不同类型的潮汐,也可做集合名词,指潮汐现象。中文的“潮”和“汐”含义略有不同,即“昼涨称潮,夜涨称汐”,英文无此区分。潮汐最直观的表现是海面 (sea level) 的涨落,即潮位 (tidal level) 的变化,而潮流 (tidal current) 指潮汐引起的海水水平流动。sea level 不止用在潮汐中,实际上可以指各种尺度的海面变化,时间尺度较大的海面变化通常叫作海平面变化,如海平面上升 (sea level rise) 常指海面随气候变化长期上升的趋势。本文所讲的由天体引潮力产生的潮汐叫作天文潮 (astronomical tide),中文语境中常与风暴潮 (storm surge) 区分。

[2] 这里描述的是引潮力和潮汐静力理论。引潮力是地球上的物体受到月球和太阳的万有引力 (gravitational force) 和由于地球绕地月或地日公共质心旋转产生的惯性离心力 (centrifugal force) 的合力,以月球引潮力为主。潮汐静力理论 (static tide theory) 即平衡潮理论 (equilibrium tide theory) 是牛顿在《自然哲学的数学原理》中提出的,假设地球是完美球体并全部被等深水覆盖,忽略海水惯性,即海水可以无延迟响应引潮力的变化。在此基础上海面被引潮力拉成潮汐椭球,在近月点和远月点凸

起,即文中说的 enormous bulges。由于地球的自转,地球上每个点在一天之内经历潮位起伏。此理论可以解释一些最基本的潮汐现象,但也存在局限。后来拉普拉斯提出了潮汐动力理论(dynamic tide theory)。

[3] 由 A 向 B 施加一个力或其他形式的强迫(如热量),可用动词 exert。solar gravity 是指太阳产生的引潮力,由于引潮力与引力有关,而引力与重力相似,因此这里直接用了 solar gravity 这个不严格的说法。

[4] 正是由于惯性的存在导致物体倾向于做匀速直线运动,因此有脱离旋转中心向外的趋势,由此产生的力即离心力(centrifugal force),也需要有一个力将物体拉回来,即向心力(centripetal force)。词根 centri-意为"center"。

[5] 全球尺度上的潮位变化,即忽略海陆分布时,月球和太阳引潮力的相对变化决定潮差。这里 range 和 magnitude 基本同义,都是指潮差的大小,即潮汐的强度。

[6] 此即潮汐的月不等现象(monthly inequality),即朔望大潮、两弦小潮,与月相(phase of the moon,moon phase 或 lunar phase)有关。月相:新月/朔(new moon)、上弦月(first quarter of the moon)、满月/望(full moon)、下弦月(third quarter of the moon)。如果仅有月球引潮力,则不管何种月相潮汐强度应该都相同,正是由于地、月、日相对位置的变化才有了潮汐的月不等现象。此句的 the very highest and lowest tides 并非语病,very 是用来强调严格意义上的最高和最低,不是"很""非常"的意思,不然后面的形容词不可能用比较级。此句最后的 with a smaller range 是插入语,为了理解句子结构可将它去掉,即可明白讲的是…and moderate tides form,即潮差较小的潮汐生成了。

[7] 潮周期可分为日潮(diurnal tide)、半日潮(semidiurnal tide)、混合潮(mixed tide)。正规和不正规潮汐分别用 regular 和 irregular 形容,如 irregular semidiurnal tide。

[8] 此处所讲的潮流撞击海底山脊和陆坡产生湍流的过程与内潮(internal tide)密切相关。由于引潮力作用在整个水体上,不仅海面有潮位起伏和潮流变化,海洋内部也有潮汐。潮汐与海底地形相互作用产生内波(internal wave),这种内波也称为内潮。内潮破碎产生湍流和混合,是大洋环流的重要组成部分。

[9] 注意此处的 means 是方法或手段的意思,而不是动词"意味着",或名词"平均"。

5. Hurricanes

Text

Hurricanes are severe storms defined by high velocity winds that rotate around a central, low-pressure core[1]. Similar tropical storms are called cyclones or typhoons in other parts of the world[2]. They form over tropical oceans between about 5 and 20 degrees of latitude, where seawater is hot enough to give them strength, and the rotation of the Earth makes them spin[3].

These storms are spawned by the interaction of the sea and the air. Their creation requires just a few simple conditions—a weather disturbance, such as a thunderstorm, that pulls in surface air from all directions and about 60 m of water 26.6℃, or warmer, at the top of the ocean.

Hurricanes start simply with the evaporation of warm seawater, which pumps water into the lower atmosphere. This humid air is then dragged aloft when converging winds collide and turn upwards. At higher altitudes, water vapor starts to condense into clouds and rain, releasing heat that warms the surrounding air, causing it to rise as well[4]. As the air far above the ocean rushes upward, even more warm moist air spirals in from along the surface to replace it[5].

As long as the base of this weather system remains over warm water, and its top is not sheared apart by high altitude winds[6], it will strengthen and grow. More and more heat and water will be pumped into the air. The pressure at its core will drop further and further, sucking in wind at ever increasing speeds. Over several hours to days, the storm will intensify, finally reaching hurricane status when the winds that swirl around it reach sustained speeds of 33 m/s or more.

A fully developed hurricane is a highly organized and complex system of wind, clouds, and rain. Although meteorologists can break a hurricane down into just a few major elements, they are far from understanding the formation and behavior of these features. At the center of a hurricane is the eye, a nearly circular area of eerily fair weather, 8 to 190 km across. Surrounding the eye is the eyewall, a towering ring of clouds that contains the hurricane's most powerful winds and heaviest rainfall. Spiraling into the eyewall are the rain bands, wide zones of intense thunderstorms that sweep outward for 80 to 480 km. In between these bands, rainfall and wind speed diminish.

Even as the familiar rotating spiral structure of a hurricane begins to form, the storm is pushed into motion by the Earth's prevailing winds. In the tropics, the trade winds blow to the west. Thus, storms that form in the prolific hurricane breeding grounds of the eastern Atlantic are driven slowly but inexorably towards the Caribbean, Gulf of Mexico, and Central and North America.

Eventually, hurricanes turn away from the tropics and into mid-latitudes. Once they move over cold water or over land, and lose touch with the hot water that powers them, these storms weaken and break apart. The lifespan of a hurricane may be as brief as day or as long as several weeks. Most of it is spent in the empty reaches of the open ocean where the wind and waves it generates menace the occasional sailor or delight distant surfers.

But when hurricanes come onshore, they change from weather curiosities to natural disasters. In a single day, the average hurricane releases 200 times the energy generated by all the world's electrical power plants combined. This energy can devastate both the human and natural landscape. Hurricane force winds often extend across an area 240 km wide. Gale force winds may occur over a 480–640 km wide stretch[7]. The force of these winds and the debris they hurl can demolish buildings, power lines, and trees.

But although hurricanes and high winds are synonymous, water causes the most death and destruction. The strong winds and low air pressure in a hurricane create a dome of water 4.5–6 m high and 80–160 km wide, called a storm surge. When the storm makes landfall, so does the surge. Topped by crashing waves and driven with enormous force, a storm surge can devastate low-lying coastal areas[8]. Dangerous flooding comes not just from the ocean, but also from the skies. Torrential rains often fall over a wide area for several days, leading to both flash and long-term flooding.

Hurricanes hold the dubious distinction of causing both the deadliest and the most damaging natural events in history. In 1900, the Galveston Hurricane killed more than 8,000 people. Just over 100 years later, Hurricane Katrina racked up damages of 100 billion dollars or more. In that same interval, cyclones and typhoons cost hundreds of thousands of lives and many billions of dollars worldwide.

The human misery and economic impact inflicted by hurricanes has spurred intense research into predicting storm development and formation. Although scientists are still unable to forecast the track of individual storms with precision, they have made strides in understanding hurricane dynamics and seasonal storm patterns[9].

Recent studies have shown a clear link between ocean surface temperatures and tropical storm intensity—warmer waters fuel more energetic storms. As ocean temperatures have warmed over the last several decades, the frequency of Category 4 and 5 hurricanes has increased.

Most studies agree that over the next one or two decades, hurricane frequency will increase substantially as a result of natural climate cycles[10]. Some research also suggests that storm intensity will increase as well, due to warmer sea surface temperatures induced by global warming. The implications of these predictions are unsettling for the population living in coastal and near coastal areas that are vulnerable to hurricane damage.

Vocabulary

velocity: (n) 风速、流速

low-pressure core: (n) 低压中心

core: (n) 中心,核心

~trough: (n) 低压槽

~high-pressure core: (n) 高压中心

~ridge: (n) 高压脊

tropical storm: (n) 热带风暴

~extratropical storm: 热带外风暴

typhoon: (n) 台风

thunderstorm: (n) 雷暴

humid: (adj) 潮湿的,湿润的

n: humidity 湿气,湿度

~relative humidity: 相对湿度

~absolute humidity: 绝对湿度

~specific humidity: 比湿

syn: moist

n: moisture 湿气,水汽

vt: moisturize 加湿

spiral: (n/vi) 螺旋

~Ekman spiral 埃克曼螺旋

~β-spiral β 螺旋

shear: (n/vi) 剪切,切变

~shear instability: 切变不稳定

intensify: (vi) 强化,加强

n: intensification 强化,加强

n: intensity 强度

swirl: (vi) 旋转

sustain: (vt) 维持,保持,支撑,支持

adj: sustained 持续的

meteorologist: (n) 气象学家

n: meteorology 气象学

adj: meteorological 气象学的

element: (n) 组成部分,元素

feature: (n) 特征

eyewall: (n)(台风的)眼墙

rain band: (n) 雨带

structure: (n) 结构

lifespan: (n) 生命周期

hurricane force wind: 飓风级大风,蒲福风力等级里的第 12 级

~Beaufort wind scale: 蒲福风力等级

extend: (vi) 延伸,占据

n: extent 范围,程度

gale force wind: 狂风级大风,蒲福风力等级里的 7~10 级

dome: (n) 穹顶,穹状物;(vi) 向上凸起

storm surge: (n) 风暴潮

~storm surge inundation: 风暴潮漫滩

landfall: (n) 登陆

long-term: (n/adj) 长期(的)

interval: (n) 间隔,间歇,间距

forecast: (n/vt) 预报,预测

~numerical forecast: 数值预报

~hindcast: (n/vt) 后报

~nowcast: (n/vt) 临近预报,即时预报

track: (n) 路径;(vt) 追踪,跟踪

syn: trajectory

precision: (n) 精度,准确性

adj: precise 精确的,准确的

~accuracy: (n) 精度,准确度

adj: accurate

dynamics: (n) 动力学

adj：dynamical

~thermodynamics：(n) 热力学

seasonal：(adj) 季节(尺度)的

subseasonal：季节内的，次季节的

pattern：(n) 图案，花样，样式，形态

category：(n) 类型

vt：categorize 分类

substantially：(adv) 大幅地，可观地

adj：substantial

natural climate cycle：(n) 自然气候循环

induce：(vt) 导致

implication：(n) 隐含意义

vt：imply 暗示、暗指、意指

Notes

[1] 风暴 (storm) 是一个统称，可以泛指具有大风暴雨的天气现象。飓风或台风就是风暴的一种，通常生成在热带海域。

[2] 飓风 (hurricane) 发生在北美和欧洲附近，即北大西洋和东北太平洋。台风 (typhoon) 发生在东亚，即西北太平洋。发生在澳大利亚两侧，即印度洋和南太平洋的称为 cyclone。这三者仅名称不同，实际上是同一种现象——热带气旋 (tropical cyclone)。中国气象局将热带气旋按强度从弱到强分为 6 个等级：热带低压 (tropical depression)、热带风暴 (tropical storm)、强热带风暴 (severe tropical storm)、台风 (typhoon)、强台风 (severe typhoon) 和超强台风 (super typhoon)。

[3] 台风或飓风生成的必要条件之一是海温足够高，因此，仅在热带海域生成。同时要有足够大的科氏力 (Coriolis force) 的作用，因此，不能太靠近赤道。

[4] 水蒸气在高空凝结放热的过程叫作凝结加热 (condensational heating)，或者潜热释放 (latent heat release)，是对流层大气非绝热加热 (diabatic heating) 最主要的组成部分。

[5] 此处描述的是流体的连续性 (continuity)，空气上升离开海面，此处的空气不可能空缺，一定有别处的空气过来补充。描述此过程的方程即连续性方程 (equation of continuity)。

[6] 风的剪切 (shear)，即水平风速的垂向梯度 (vertical gradient of horizontal velocity)。剪切太强则上层形成的湿空气很快被带离原来的位置，热带气旋不能发展。

[7] 蒲福风力等级 (Beaufort wind scale) 如下，其中 0 级(无风)为 Calm：

1 级 Light Air

2 级 Light Breeze

3 级 Gentle Breeze

4 级 Moderate Breeze

5 级 Fresh Breeze

6 级 Strong Breeze

7 级 High Wind / Moderate Gale / Near Gale

8 级 Gale / Fresh Gale

9 级 Strong Gale / Severe Gale

10 级 Storm / Whole Gale

11 级 Violent Storm

12 级 Hurricane

[8] 当风暴潮与天文大潮高潮同时出现时灾害最甚。

[9] predict 与 forecast 都表示预报,通常二者是通用、可互换的,但 forecast 在表示基于严格计算的预报时稍微更常用一些。forecast 本身有动词和名词两种用法,同时也可用 forecasting 作名词,二者意义相同。precise 与 accurate 都是“精确的”,但含义细分有所不同。precise 指每次测量之间的稳定程度,即对同一观测对象的多次测量给出的结果在一个小的误差范围内,不会出现大的变化,但不保证测量值与真实值接近。accurate 指观测值与真实值接近,测量给出的结果能较好地反映真实情况,但不保证多次测量结果稳定。最佳的情况是既 precise 又 accurate。动力学(dynamics)是研究在力的驱动下物体的运动(即动量或动能)变化的学问。热力学(thermodynamics)是研究在热源或冷源的驱动下物体的温度(即内能)变化的学问。所谓 seasonal storm pattern 是指风暴的季节性空间分布形态,即每个季节的风暴频率、个数、强度等特征的空间分布。pattern 一词常用于指代某属性或特征的空间分布形态,如西高东低、带状分布、偶极子等任何可能的情况。

[10] natural climate cycle 意为自然气候循环,是指与人类活动无关、由自然因素引起的气候循环变化,如冰期(glacial period)与间冰期(interglacial period)的变化。目前全球变暖(global warming)的一部分原因是处在上一个冰期,之后温度逐步回升的过程中。与此相对的是人为气候变化(anthropogenic climate change),即工业革命(industrial evolution)后化石燃料(fossil fuel)的燃烧释放二氧化碳(CO_2,carbon dioxide)等温室气体(greenhouse gas)导致温度上升。

6.Physical Properties of Seawater

Text

One of the most important physical characteristics of seawater is its temperature. Temperature was one of the first ocean parameters to be measured and remains the most widely observed[1]. In most of the ocean, temperature is the primary determinant of density; salinity is of primary importance mainly in high latitude regions of excess rainfall or sea ice processes[2]. In the mid-latitude upper ocean (between the surface and 500 m), temperature is the primary parameter determining sound speed. As a parcel of water is compressed or expanded, its temperature changes. The concept of "potential temperature" has been introduced, which is an intrinsic property of a water parcel, independent of the environment pressure.

Temperature is a thermodynamic property of a fluid, due to the activity or energy of molecules and atoms in the fluid. Temperature is higher for higher energy or heat content[3]. Heat and temperature are related through the specific heat capacity[4]. The range of temperature in the ocean is from the freezing point, which is around −1.7℃ (depending on salinity)[5], to a maximum of around 30℃ in the tropical oceans. This range is considerably smaller than the range of air temperatures.

The ease with which temperature can be measured has led to a wide variety of oceanic and satellite instrumentation to measure ocean temperatures[6]. Mercury thermometers were in common use from the late 1700s through the 1980s. Reversing (mercury) thermometers, invented by Negretti and Zamba in 1874, were used on water sample bottles through the mid-1980s. These thermometers have ingenious glasswork that cuts off the mercury column when the thermometers are flipped upside down by the shipboard observer, thus recording the temperature at depth[7]. Satellites detect thermal infrared electromagnetic radiation from the sea surface; this radiation is related to temperature[8]. Satellite sea surface temperature (SST) accuracy is about 0.5−0.8 K, plus an additional error due to the presence or absence of a very thin (10 mm) skin layer that can reduce the desired bulk (1−2 m) observation of SST by about 0.3 K[9].

Seawater is almost, but not quite, incompressible[10]. A pressure increase causes a water parcel to compress slightly. This increases the temperature in the water parcel if it occurs without exchange of heat with the surrounding water (adiabatic compression)[11]. Conversely if a water parcel is moved from a higher to a lower pressure, it expands and its temperature decreases. These changes in temperature are unrelated to surface or deep sources of heat. It is often desirable to compare the temperatures of two parcels of water that are found at different pressures. Potential temperature is defined as the temperature that a water parcel would have if moved adiabatically to another pressure. This effect has to be considered when water parcels change depth. As opposed to potential temperature, the in-situ temperature is termed the absolute temperature[12].

Seawater is a complicated solution containing the majority of the known elements. While the total concentration of dissolved matter varies from place to place, the ratios of the more abundant components remain almost constant. This "law" of constant proportions was first proposed by Dittmar (1884)[13], based on 77 samples of seawater collected from around the world during the Challenger Expedition, confirming a hypothesis from Forchhammer (1865).

The dominant source of the salts in the ocean is river runoff from weathering of the continents. Weathering occurs very slowly over millions of years, and so the dissolved elements become equally distributed in the ocean as a result of mixing. The total time for water to circulate through the oceans is, at most, thousands of years, which is much shorter than the geologic weathering time. However, there are significant differences in total concentration of the dissolved salts from place to place. These differences result from evaporation and from dilution by freshwater from rain and river runoff[14]. Evaporation and dilution processes occur only at the sea surface.

Salinity was originally defined as the mass in grams of solid material in a kilogram of seawater after evaporating the water away; this is the absolute salinity as described in Millero et al. (2008)[15]. For example, the average salinity of ocean water is about 35 grams of salts per kilogram of seawater (g/kg), written as "$S = 35‰$" or as "$S = 35$ ppt" and read as "thirty-five parts per thousand."[16] Because evaporation measurements are cumbersome, this definition was quickly superseded in practice. In the late 1800s, Forch, Knudsen, and Sorensen (1902) introduced a more chemically based definition: "Salinity is the total amount of solid materials in grams contained in one kilogram of seawater when all the carbonate has been converted to oxide, the bromine and iodine replaced by chlorine, and all organic matter completely oxidized." This chemical determination of salinity was also difficult to carry out routinely. The method used throughout most of the twentieth century was to determine the amount of chlorine ion (plus the chlorine equivalent of the bromine and iodine) referred to as chlorinity[17], by titration with silver nitrate, and then to calculate salinity by a relation based on the measured ratio of chlorinity to total dissolved substances.

These definitions of salinity based on chemical analyses were replaced by a definition based on seawater's electrical conductivity, which depends on salinity and temperature. This conductivity-based quantity is called practical salinity, sometimes using the symbol "psu" for "practical salinity units", although the preferred international convention has been to use no units for salinity. Salinity is now written as, say[18], $S= 35.00$ or $S=35.00$ psu. The algorithm that is widely used to calculate salinity from conductivity and temperature is called the practical salinity scale 1978 (PSS 78). Electrical conductivity methods were first introduced in the 1930s. The accuracy of salinity determined from conductivity is ±0.001 if temperature is very accurately measured and standard seawater is used for calibration[19]. This is a major improvement on the accuracy of the older titration method, which was about ±0.02.

Density, usually denoted ρ, is the amount of mass per unit volume and is expressed in kilograms per cubic meter (kg/m^3). A directly related quantity is the specific volume anomaly, usually denoted α, where $\alpha = 1/\rho$[20]. If there is range of pressures, the effects of adiabatic compression should be included when comparing water parcels. A more appropriate quantity is potential density, which is the same as in situ density but with temperature replaced by potential temperature and pressure replaced by a single reference pressure that is not necessarily 0 dbar[21].

Seawater density is important because it determines the depth to which a water parcel will settle in equilibrium—the least dense on top and the densest at the bottom. The distribution of density is also related to the large-scale circulation of the oceans through the geostrophic/thermal wind relationship[22]. Mixing is most efficient between waters of the same density because adiabatic stirring, which precedes mixing, conserves potential temperature and salinity and consequently, density. More energy is required to mix through stratification. Thus, property distributions in the ocean are effectively depicted by maps on density (isopycnal) surfaces, when properly constructed to be closest to isentropic[23].

The relationship between the density of seawater and temperature, salinity, and pressure is the equation of state for seawater. Today, the most common version is "EOS 80" (Millero & Poisson, 1980; Fofonoff, 1985). EOS 80 uses the practical salinity scale PSS 78. The formulae may be found in UNESCO (1983), which provides practical computer subroutines and are included in various textbooks such as Pond and Pickard (1983) and Gill (1982). EOS 80 is valid for $T = -2$ to 40℃, $S = 0$ to 40, and pressures from 0 to 10,000 dbar, and is accurate to 9×10^{-3} kg/m^3 or better. A new version of the equation of state has been introduced (IOC, SCOR, and IAPSO, 2010), based on a new definition of salinity and is termed TEOS-10.

Density values evaluated at the ocean's surface pressure are shown in Figure Ⅰ-3 (curved contours) for the whole range of salinities and temperatures found anywhere in the oceans. The shaded bar in the figure shows that most of the ocean lies within a relatively narrow salinity range. More extreme values occur only at or near the sea surface, with fresher waters outside this range (mainly in areas of runoff or ice melt) and the most saline waters in relatively confined areas of high evaporation (such as marginal seas). The ocean's temperature range produces more of the ocean's density variation than does its salinity range. In other words, temperature dominates oceanic density variations for the most part. As noted previously, an important exception is where surface waters are relatively fresh due to large precipitation or ice melt; that is, at high latitudes and also in the tropics beneath the rainy Intertropical Convergence Zone of the atmosphere.

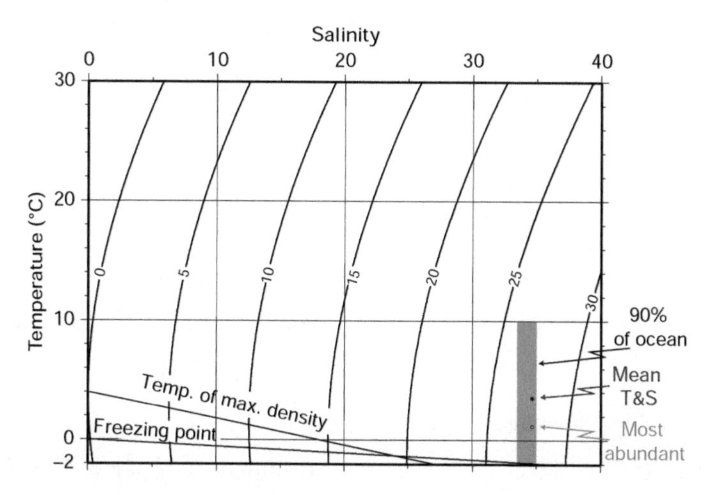

Figure Ⅰ-3: Values of density σ_t[24] (curved lines) and the loci of maximum density and freezing point (at atmospheric pressure) for seawater as functions of temperature and salinity. The full density ρ is $1000+\sigma_t$ with units of kg/m^3. [Source: Talley et al., 2011]

Vocabulary

parameter: (n) 参数

vt: parameterize 参数化

n: parameterization

measure: (vt) 测量,观测

n: measurement

observe: (vt) 观测,观察

n: observation

adj: observational

primary: (adj) 主要的

determinant: (n) 决定因素

vt: determine

excess: (adj/n) 过量(的),过多(的)

syn: extensive

upper ocean: (n) 上层海洋

parcel: (n) 水块,水团,微团

potential temperature: (n) 位温

potential: (adj) 潜在的,可能的

~potential energy: 势能

~potential vorticity: 位涡

~conservative temperature: 保守温度

intrinsic: (adj) 内在的,固有的,本质的

property: (n) 属性

fluid: (n/adj) 流体(的)

atom: (n) 原子

adj: atomic

heat content: (n) 热含量

specific heat capacity: (n) 比热容

这里 specific 指"单位质量的",译为"比"字

capacity: (n) 容量

freezing point: (n) 冰点(温度)

~boiling point: 沸点(温度)

~dew point：露点(温度)

dew：(n) 露水

considerably：(adv) 相当大地，相当多地，相当重要地

adj：considerable

instrumentation：(n) 仪表，仪器(集合名词)

mercury：(n) 水银，汞

thermometer：(n) 温度计

reversing thermometer：(n) 颠倒温度计

sample：(n) 样本

shipboard：(adj) 船载的，船基的

~airborne：(adj)(飞)机载的

~shore-based：(adj) 岸基的

infrared：(n/adj) 红外线，红外的

[infra-]：在……之下/后

~ultraviolet：(n/adj) 紫外线，紫外的

[ultra-]：超过，极其

~visible light：(n) 可见光

electromagnetic：(adj) 电磁的

radiation：(n) 辐射

adj：radiative

vt/vi：radiate

~infrared radiation：红外辐射

~long-wave radiation：长波辐射

~solar radiation：太阳辐射

~short-wave radiation：短波辐射

error：(n) 错误，差错，误差

skin layer：(n) 皮层

bulk：(n)(巨大的)体积、容量，大部分，主体

指具有一定体积的水体的某属性或对其进行的测量，相对于针对薄层进行的测量而言

~bulk formula：块体公式

slightly：(adv) 轻微地，少量地

adj：slight

adiabatic：(adj) 绝热的

ant：diabatic 非绝热的

~diabatic heating：非绝热加热

absolute temperature：绝对温度

solution：(n) 溶液

concentration：(n) 浓度

matter：(n) 物质

abundant：(adj) 丰富的，大量的

n：abundance

proportion：(n) 比例

expedition：(n) 科学考察，探险

hypothesis：(n) 假设，假说

vt：hypothesize / hypothesise

syn：(n/vt) conjecture

weathering：(n) 风化(过程/作用)

dilution：(n) 稀释

vt：dilute

absolute salinity：(n) 绝对盐度

et al.：……等(人)

拉丁语缩写词，相当于 and others。注意 al 后面的句点也是单词的一部分，不能省略。

~etc.：……等(物)

相当于 and others 或 and so on，后面同样要有一个句点。

carbonate：(n) 碳酸盐

[-ate]：(化学)盐

~nitrate：(n) 硝酸盐

~carbon：(n) 碳

oxide：(n) 氧化物

vt：oxidize 氧化

n：oxidation 氧化(过程)

~carbon dioxide：二氧化碳

[di-]：二重，二倍，二次

~oxygen：(n) 氧

bromine：(n) 溴

iodine：(n) 碘

chlorine：(n) 氯

organic：(adj) 有机的

n：organism 生物体，有机体

ant：inorganic 无机的

equivalent：(n) 等效性，等效物，当量；(adj) 等效的，等价的，相当的

chlorinity：(n) 氯度

titration：(n) 滴定

nitrate：(n) 硝酸盐

~nitrogen：(n) 氮

substance：(n) 物质，材料，重要性，本质

adj：substantial 重要的，可观的，大量的，有价值的

analysis：(n) 分析

pl：analyses

vt：analyze / analyse

adj：analytical 分析的，解析的

~analytical solution 解析解

ant：synthesis 综合

conductivity：(n) 电导率

practical salinity：(n) 实用盐度

practical：(adj) 实用的，实际的

ant：impractical

unit：(n)(计量)单位；(adj) 单位的，每单位量的

convention：(n) 准则，惯例，公约，大型会议

algorithm：(n) 算法

practical salinity scale：(n) 实用盐标

standard seawater：(n) 标准海水

calibration：(n) 标定，校准

denote：(vt) 表示，标示

specific volume：(n) 比容

anomaly：(n) 异常，距平

adj：anomalous

appropriate：(adj) 合适的，恰当的

potential density：(n) 位密

in situ：(adj/adv) 现场的，原位的

拉丁语借词。虽由两个词组成，但作为单个词使用，也可用连字符组合

dbar：(单位)分巴，即 deci-bar

1 分巴 = 1/10 巴 = 10^4帕

equilibrium：(n) 平衡，平衡态

pl：equilibria

adj：equilibrial

geostrophic：(adj) 地转的

ant：ageostrophic

~geostrophic balance 地转平衡

~geostrophic current：地转流

n：geostrophy 地转平衡，地转性

thermal wind relationship：(n) 热成风关系

conserve：(vt) 保持恒定，使守恒

n：conservation 守恒

~mass conservation：质量守恒

~angular momentum conservation：角动量守恒

~potential vorticity conservation：位涡守恒

consequently：(adv) 因此，结果，所以

adj：consequent 作为结果的，随之发生的

n：consequence 结果，后果，影响

stratification：(n) 层结

vt/vi：stratify

isopycnal：(adj) 等密度的，(沿)等密面的

ant：diapycnal 跨等密面的

~isopycnal surface 等密面

~isothermal 等温面
~isobar 等压面
~isobath 等深线
isentropic：(adj) 等熵的
~isentropic surface 等熵面
~entropy：(n) 熵
equation of state：(n) 状态方程
formula：(n) 公式，方程
pl：formulae
subroutine：(n) 子程序
valid：(adj) 有效的
n：validity 有效性
vt：validate 验证，证实
n：validation
term：(vt) 把……称为，把……叫作
curved：(adj) 弯曲的
n：curve 曲线
n：curvature 曲率
contour：(n) 等值线
shaded：(adj) 填色的
ice melt：(n) 融冰
confine：(vt) 局限，限制
Intertropical Convergence Zone：热带辐合带
locus：(n) 地点，中心，核心，轨迹
拉丁语借词，多用于技术场合
pl：loci
~location：(n) 地点，地理位置
vt：locate：确定位置，定位

Notes

[1] measure 和 observe 意思接近。observe 可以指不用仪器而仅用眼睛留意观察，但海洋观测通常都是用仪器的。observe 还具有长期、一直密切注意、连续观测，或有组织地观测等含义。相比之下，measure 更倾向于指具体的某次测量。

[2] 此处表述并不严格。除高纬度海域外，河流注入较多的海域（如孟加拉湾）和高蒸发海域（如地中海和红海）盐度也可能对密度起较大作用。

[3] 此处的能量是指内能（inner energy），或称热含量（heat content）。热力学上，热量是内能的变化量。

[4] 热容（heat capacity）是物体容纳热量的能力，即温度变化 1℃时吸收或放出的热量，与质量有关。specific 译为“比”，表示单位质量的某属性，此处即比热容，是单位质量物质的热容，仅与物质的性质有关。

[5] 在标准大气压（standard atmosphere unit）下，淡水（盐度为 0）的冰点是 0℃，随着海水盐度升高，冰点降低。

[6] ease 即 easy 的名词，表示“简单性”。with which 引导定语从句修饰 the ease，其中 with 是与后面的动词 measure 搭配的。a variety of 表示各种各样的，wide 更强调其丰富程度。

[7] 颠倒温度计是电子时代之前应用最广的海温测量仪器。将它放置到观测目标深度后，船上的观测者通过特定装置使仪器颠倒，其中精巧的（ingenious）玻璃结构（glasswork）可以切断水银柱，从而保留了现场温度示数。

[8] 红外线（infrared）是一种电磁辐射（electromagnetic radiation），也叫热辐射（thermal radiation），因为任何温度大于绝对零度的物体都向外发射电磁波，电磁波的峰值频率与物体温度有关［黑体辐射定律（law of blackbody radiation）］。海洋温度对应的电磁波辐射频段是红外，因此称为红外辐射或长波辐射（longwave radiation）；而太阳辐射（solar radiation）的频段主要是可见光和紫外，因此又称短波辐射（shortwave radiation）。

[9] accuracy 与 precision 的区别见 "Hurricanes"（Text 5）。海面上存在温度较低的冷皮层（skin layer），而卫星遥感仅能反映皮层温度。若将温度计放入水中，则总有 0.5~1 m 的深度，不可能太贴近海面，反映的是一定深度层水体的温度，因此叫作块体温度（bulk temperature）。块体温度与皮层温度有差异，不能直接比较。卫星辐射计测量的即皮层温度（skin temperature）。

[10] quite 有"相当"和"绝对"两种含义，这里是后者。

[11] adiabatic 与 diabatic 是反义词，都是从希腊语进入英语的。其中 dia-词根表示"穿过"（"through"），-batic 表示"可通过"（"passable"），a-是否定前缀。以 a-作否定前缀构成的词本节还会遇到几个，请留意。绝热过程（adiabatic process）仅与压强变化，即垂向移动有关，而非绝热过程（diabatic process）还涉及热量交换，因此研究价值往往更高。除了海面上与大气的热量交换以及接收太阳辐射和发射长波辐射，海洋内部缺乏热源或冷源，因此，主要的非绝热过程是湍流（turbulence）导致的混合（mixing）。由于等位温面上的混合不涉及热量交换，因此，跨等位温面的混合才是非绝热过程。

[12] 相同热含量（heat content）的两个水团处在不同压强下则绝对温度不同，这是因为热含量与温度和比热容（heat capacity）成正比，而在高压下水团体积收缩，比热容降低，因此温度升高。经过绝热抬升至参考压强获得的位温即可排除压强对温度的影响。然而，在绝热湍流混合过程中会发生混合增密效应（cabbeling）或称混合收缩效应（contraction on mixing），导致体积减小、比热容降低，同样会使绝对温度升高。由于湍流混合时压强基本不变，位温也会与绝对温度一同升高，不能反映热含量。为此常利用湍流混合时焓（enthalpy）守恒的特点，将焓除以一个常数来定义温度，此常数可类比于比热容，随压强变化。经过与位温一样的绝热抬升过程，最终获得的温度称为保守温度（conservative temperature）。保守温度既与压强变化无关，又在湍流混合时守恒，因此总是能反映热含量。

[13] 这是引用参考文献的标准格式，即作者姓氏（年份）。

[14] result from 指示原因，而 result in 指示结果。即 A results from B 表示 A 是果，B 是因；而 A results in B 则相反。因为在海洋学研究范畴内盐量基本守恒，改变海面上层盐度的主因就是淡水收支（freshwater budget）。与温度一样，因为海洋内部没有淡水的源和汇（source and sink），盐度的改变主要由混合引起。

[15] 引用参考文献时，有两位作者时一般将二者都写出，即 A and B（年份）。超过两位时，根据格式规范的不同，一般省略写成第一作者 et al.，即某某等。et al. 是拉丁语借词，即“et alii”或“et alia”。其中 al 后面的点表示缩写，是单词的一部分，不能省略。学术界常用的其他拉丁语缩写词如 etc. = et cetera 表示“某某等”（用于非人事物），e.g. = exempli gratia 表示“例如”，i.e. = id est 表示“即”，ca. = circa 表示“约”（用于年份）。

[16] ppt（parts per thousand）= 1×10^{-3} 这个表示比例的“单位”并不能表明是根据质量还是体积来进行比较的。如果要明确指明是质量比例，可以写作 pptm，其中 m 表示“mass”；而体积比例可以写作 pptv，v 即“volume”。另一个常用的比例单位是 ppm，即 parts per million = 1×10^{-6}。例如，2023 年 5 月大气二氧化碳浓度为 424 ppm。

[17] 动词 refer 后面总是要跟 to，因此，此句不能说成“referred as chlorinity”。refer to as 在此处的意思是把……叫作。

[18] 此处 say 意为“例如”，是插入语。

[19] 标准海水最早由丹麦的哥本哈根水文研究所制定，氯度为 19.381‰，密封于玻璃安瓿瓶中，供各国海洋研究单位使用。后来由于各国的大量需要以及运输价格等原因，很多国家根据国际标准海水配制方法，自己生产出本国的标准海水，也就是所谓的副标准海水（substandard seawater）。使用标准海水定期标定仪器是很重要的，可以避免电子仪器出现偏差漂移（bias）。

[20] 比容即单位质量的体积，是密度的倒数。即使不考虑全球变暖导致的冰川和雪山融水，即在海水质量不变的情况下，体积的热膨胀即比容的上升也可导致海面上升。

[21] 位密即由位温带入状态方程算出的密度。位温和位密的参考面不一定是海面，特别是对深层海洋，用海面做参考面会引入较多的计算误差，因此，常改用某一较深层次做参考面。海水的压强常用分巴（deci-bar，dbar，1dbar = 1/10 bar）作单位，因为 1 dbar 近似等于 1 m 厚的海水的压强。深层海洋的位温和位密参考面常取为 4000 dbar。其他常用压强单位有：Pa（帕斯卡或帕）、hPa（百帕）、atm（标准大气压）、mmHg（毫米汞柱）等。压强单位巴（bar）较大，1 bar = 1000 hPa = 0.98 atm = 750 mmHg = 10 dbar。修饰单位的前缀：10^{-9} = nano（纳），10^{-6} = micro（微），10^{-3} = milli（毫），10^{-2} = centi（厘），10^{-1} = deci（分），10^{2} = hecto（百，即 hundred），10^{3} = kilo（千，即 thousand），10^{6} = mega（兆，即 million），10^{9} = giga（吉，即 billion），10^{12} = tera（太，即 trillion），10^{15} = peta（拍，即 quadrillion）。

[22] 地转平衡（geostrophic balance），即水平压强梯度力（pressure gradient force）与科氏力（Coriolis force）的平衡，其最终结果是地转流（geostrophic current），其地转调整（geostrophic adjustment）过程中会激发惯性重力波（inertial gravity waves）。热

成风关系（thermal wind relationship）将水平温度梯度（horizontal temperature gradient）与水平流速的垂向剪切（vertical shear of horizontal velocity,或 velocity shear）联系起来。不符合地转平衡的运动用形容词 ageostrophic 描述,这也是 a-作否定前缀的例子。

[23] 由于海水密度大部分情况下受温度控制,因此等位密度面近似于等位温面,故沿等位密/温面的混合是绝热混合,不需要额外能量输入,相比跨等位密/温面的混合更易发生。through stratification 即穿越层结,也就是穿越等密面,即 across isopycnal,或用形容词 diapycnal。dia-词根作“穿越”之意在注释[11]已提及,这里是另一个例子。由于海洋内部大尺度运动主要是沿等密面的,所以跨等密面的混合（diapycnal mixing）就成为影响海洋垂向物质和属性交换的重要过程,也是使热盐环流/翻转环流（thermohaline/overturning circulation）的上升部分完成闭合的重要过程。另外,熵（entropy, S）是热力学系统中不可用非化学能的量度,根据克劳修斯定律（Clausius theorem）,对于可逆过程有 $dS = Q/T$,因此绝热过程（$Q = 0$）熵守恒,等热含量面就是等熵面（isentropic surface）。根据注释[12],等热含量面严格等价于保守温度的等值面,近似等价于等位温面,在盐度作用较弱时又近似等价于等位密面。当然,事实上并不存在可逆过程,因此,用等熵面来代表等热含量面也是一种近似,只能用于远离上、下、侧边界且摩擦较弱的海洋内部区域。

[24] 表层海水的现场密度（in-situ density）的量值几乎总是在 1000~1030 kg/m^{3}范围内,6000 m 水深处现场密度约为 1060 kg/m^3,马里亚纳海沟（Mariana trench）内万米水深处可达 1100 kg/m^{3}左右。可见,相比于密度的绝对值,它的变化量仅占很小的一部分。为了更清晰地表达密度的变化,也为了书写方便,常使用密度的真实值减去 1000 得到的量来代替密度,这个量叫作密度超量,一般用符号 σ_t表示,因此也称 sigma-t density,但有时不做区分,直接称之为 density。

7. Turbulent Flow

Text

There are many opportunities to observe turbulent flows in our everyday surroundings, whether it be smoke from a chimney, water in a river or waterfall, or the buffeting of a strong wind. In observing a waterfall[1], we immediately see that the flow is unsteady, irregular, seemingly random and chaotic, and surely the motion of every eddy or droplet is unpredictable[2]. In the plume formed by a solid rocket motor, turbulent motions of many scales can be observed, from eddies and bulges comparable in size to the width of the plume, to the smallest scales the camera can resolve[3]. The features mentioned in these two examples are common to all turbulent flows. More detailed and careful observations can be made in laboratory experiments.

As implied by the above discussion, an essential feature of turbulent flows is that the fluid velocity field varies significantly and irregularly in both position and time. By observing the time history of the axial component of velocity measured on the centerline of a turbulent jet, it could be found that the velocity displays significant fluctuations, and that, far from being periodic, the time history exhibits variations on a wide range of timescales[4]. Very importantly, we observe that the velocity and its mean are in some sense "stable": huge variations of velocity are not observed; neither does it spend long periods of time near values different than the mean[5]. The profile of the mean velocity measured as a function of the cross-stream coordinate shows marked contrast to the instantaneous velocity: the mean velocity has a smooth profile, with no fine structure. Indeed, the shape of the profile is little different than that of a laminar jet.

In engineering applications turbulent flows are prevalent, but less easily seen. In the processing of liquids or gases with pumps, compressors, pipelines, etc., the flows are generally turbulent. Similarly, the flows around vehicles—e.g., airplanes, automobiles, ships, and submarines—are turbulent. The mixing of fuel and air in engines, boilers, and furnaces, and the mixing of the reactants in chemical reactors take place in turbulent flows.

An important characteristic of turbulence is its ability to transport and mix fluid much more effectively than a comparable laminar flow. This is well demonstrated by an experiment first reported by Osborne Reynolds (1883). Dye is steadily injected on the centerline of a long pipe in which water is flowing. As Reynolds (1894) later established, this flow is characterized by a single non-dimensional parameter[6], now known as the Reynolds number Re. In general, it is defined by $Re = UL/v$ where U and L are characteristic velocity and length scales of the flow, and v is the kinematic viscosity of the fluid[7]. For pipe flow, U and L are taken to be the area-averaged axial velocity and the pipe diameter, respectively. In Reynolds' pipe-flow experiment, if *Re* is less than about 2300, the flow is laminar—the fluid velocity does not change with time, and all streamlines are parallel to the axis of the pipe. In this (laminar) case, the dye injected on the centerline forms a long streak that increases in diameter only slightly with downstream distance[8]. If, on the other hand, *Re* exceeds about 4000, then the flow is turbulent. Close to the injector, the dye streak is jiggled about by the turbulent motion; it becomes progressively less distinct with downstream distance; and eventually mixing with the surrounding water reduces the peak dye concentration to the extent that it is no longer visible[9].

Turbulence is also effective at "mixing" the momentum of the fluid. As a consequence, on aircraft's wing and ships' hulls the wall shear stress (and hence the drag) is much larger than it would be if the flow were laminar. Similarly, compared with laminar flow, rates of heat and mass transfer at solid-fluid and liquid-gas interfaces, as well as at interfaces between stratified layers of liquid, are much enhanced in turbulent flows[10].

The major motivation for the study of turbulent flows is the combination of the three preceding observations: the vast majority of flows is turbulent; the transport and mixing of matter, momentum, and heat in flows is of great practical importance; and turbulence greatly enhances the rates of these processes.

Many different techniques have been used to address many different questions concerning turbulence and turbulent flows. The first step toward providing a categorization of these studies is to distinguish between small-scale turbulence and the large-scale motions in turbulent flows.

At high Reynolds number there is a separation of scales. The large-scale motions are strongly influenced by the geometry of the flow (i.e., by the boundary conditions), and they control the transport and mixing. The behavior of the small-scale motions, on the other hand, is determined almost entirely by the rate at which they receive energy from the large scales, and by the viscosity[11]. Hence these small-scale motions have a universal character, independent of the flow geometry. It is natural to ask what the characteristics of the small-scale motions are. Can they be predicted from the equations governing fluid motion? These are questions of *turbulence theory*, which are addressed in many books, e.g., Batchelor (1953), Monin and Yaglom (1975), Panchev (1971), Lesieur (1990), and McComb (1990).

Studies of *turbulent flows* can be divided into three categories.

(i) Discovery: experimental (or simulation) studies aimed at providing qualitative or quantitative information about particular flows.

(ii) Modelling: theoretical (or modelling) studies, aimed at developing tractable mathematical models that can accurately predict properties of turbulent flows.

(iii) Control: studies (usually involving both experimental and theoretical components) aimed at manipulating or controlling the flow or the turbulence in a beneficial way—for example, changing the boundary geometry to enhance mixing; or using active control to reduce drag.

The objective of studies in the first category is to develop an understanding for the important characteristics of turbulent flows, of the dominant physical processes, and how they are related to the equations of fluid motion.

For studies in the second category, that aim at developing tractable mathematical models, the word "tractable" is crucial. For fluid flows, be they laminar or turbulent, the governing laws are embodied in the Navier-Stokes (N-S) equations, which have been known for over a century[12]. Considering the diversity and complexity of fluid flows, it is quite remarkable that the relatively simple N-S equations describe them accurately and in complete detail. However, in the context of turbulent flows, their power is also their weakness: the equations describe every detail of the turbulent velocity field from the largest to the smallest length and time scales. The amount of information contained in the velocity field is vast, and as a consequence (in general) the direct approach of solving the N-S equations is impossible[13]. So, while the N-S equations accurately describe turbulent flows, they do not provide a tractable model for them.

For the high-Reynolds-number flows that are prevalent in applications, the natural alternative is to pursue a statistical approach. That is, to describe the turbulent flow, not in terms of the instantaneous velocity, but in terms of some statistics, the simplest being the mean velocity field. A model based on such statistics can lead to a tractable set of equations, because statistical fields vary smoothly (if at all) in position and time. Concepts and techniques used in the statistical representation of turbulent flow fields include: turbulent viscosity models, e.g., the k-ε model; Reynolds-stress models; models based on the probability density function (PDF) of velocity; and large-eddysimulations (LES).

Without turbulence and the mixing it causes, we would not have the same ocean that we do now, nor indeed the same climate. Turbulent mixing brings nutrients into the surface layer from below so that plankton can grow. Turbulence near the surface, driven by surface winds and cooling, transmits heat in and out of the ocean to create the reservoir of heat that governs climate[14]. It is turbulence that diffuses the permanent pycnocline separating the cold bottom waters of polar origin from the atmosphere-connected upper ocean, either directly, or indirectly by local mixing followed by distribution along isopycnals[15]. Turbulence in the bottom layer affects the deposition, resuspension and movement of sediments. Turbulence creates micro-environments for the small creatures that form the basis for life in the oceans.

Understanding turbulence and mixing in the ocean is important for many reasons. Most importantly perhaps, ocean models that can predict global circulation, climate change, pollutant dispersal and primary productivity, will provide reliable predictions only when we have the capability to quantify the subgrid-scale effects of turbulence[16].

Vocabulary

buffet：（vt）（风、浪等）反复敲打
unsteady：（adj）不稳定的
  ant：steady
    n：steadiness 稳定性
random：（adj）随机的
  syn：stochastic 随机的
chaotic：（adj）混乱的，混沌的
  n：chaos 混乱，混沌
unpredictable：（adj）不可预测的
plume：（n）羽状流
resolve：（vt）分辨，解析，辨析
  adj：resolvable
  n：resolution 分辨率
laboratory：（n）实验室
experiment：（n）实验
  adj：experimental
essential：（adj）必不可少的，至关重要的
axial：（adj）轴线的，轴向的
  n：axis 轴，轴线，坐标轴
    pl：axes
fluctuation：（n）起伏，波动，变动
  vi：fluctuate
periodic：（adj）周期性的
exhibit：（vt）表现出，显示出
timescale：（n）时间尺度
mean：（n）平均值，平均数；（adj）平均的
stable：（adj）稳定的
  n：stability 稳定性
  ant：unstable
    n：instability
profile：（n）剖面，切面，轮廓；（vt）显出……的轮廓（引申为测量）
as a function of：作为……的函数
marked：（adj）明显的，显著的
  adv：markedly
instantaneous：（adj）瞬时的，即时的
fine structure：（n）细结构
laminar：（adj）由薄片组成的，层流的
prevalent：（adj）盛行的，普遍的
reactant：（n）反应物
reactor：（n）反应器，反应堆
demonstrate：（vt）证实，演示，说明
dye：（n）染料
establish：（vt）证实，查明
non-dimensional parameter：无量纲参数
  dimension：（n）维，维度，量纲
  syn：dimensionless parameter
kinematic：（adj）运动学的
  n：kinematics 运动学
  ~dynamics 动力学
viscosity：（n）黏性，黏性，黏度
  adj：viscous 黏稠的，有黏性的
area-averaged：（adj）区域平均的
diameter：（n）直径
  ~radius /ˈreɪdiəs/ 半径
    pl：radii /ˈreɪdiaɪ/
streamline：（n）流线
  ~streamfunction 流函数
streak：（n）条纹
downstream：（adj/adv）下游的，在下游
  ant：upstream
progressive：（adj）逐渐的

distinct：（adj）明显的，可辨别的
momentum：（n）动量
drag：（n）拖曳力
interface：（n）界面
motivation：（n）动机，出发点
combination：（n）结合，共同效果
  vt：combine
address：（vt）解决（问题）
distinguish：（vt）区分
boundary condition：（n）边界条件
behavior：（n）行为
universal：（adj）通用的，普遍的
theory：（n）理论
  adj：theoretical
simulation：（n）模拟
  vt：simulate
  ~numerical simulation：数值模拟
qualitative：（adj）定性的
quantitative：（adj）定量的
modelling：（n）（用模型）模拟
  n：model 模型，模式
  vt：model 模拟
  ~numerical model（ling）数值模式（拟）
tractable：（adj）易于驾驭的，易于掌握的
mathematical：（adj）数学的
  n：mathematics
manipulate：（vt）操纵，操作
  n：manipulation
objective：（n）目标；（adj）客观的
  ant：subjective 主观的
crucial：（adj）至关重要的，关键的
law：（n）定律
Navier-Stokes equations：纳维－斯托克斯方程
diversity：（n）多样性
  adj：diverse
complexity：（n）复杂性
  adj：complex
context：（n）背景，上下文，语境
approach：（n）方法，方式；（vt）接近，着手处理
alternative：（n）另种选择，另种可能；（adj）其他可选的，另外的
  vt：alter 改变
pursue：（vt）追求，致力于，寻找
statistical：（adj）统计的
  syn：statistic（adj）统计的；（n）统计量
  n：statistics 统计学
in terms of：在……方面，与……有关
concept：（n）观念，概念，计划，设计
  adj：conceptual
  syn：conception 观点，看法，观念，概念
technique：（n）技巧，手段，方法
representation：（n）代表，描述，描写，表达，说明
  vt：represent
  adj：representative 有代表性的，典型的
probability density function：概率密度函数
  probability：（n）可能性，概率
    adj：probable 可能的
    ~possibility：（n）可能性，概率
      adj：possible 可能的
  probable 和 possible 的区别见注释
large-eddy simulation：（n）大涡模拟
plankton：（n）浮游生物
  adj：planktic/planktonic
  ~phytoplankton：浮游植物
  ~zooplankton：浮游动物
diffuse：（vt）扩散，弥散
  n：diffusion

n: diffusivity 扩散率
adj: diffusive 扩散性的
adj: diffusible 可扩散的
permanent: (adj) 永恒的,永久的,持久的
pycnocline: (n) 密度跃层
即垂向密度梯度大的层次
[-cline]: 跃层
~thermocline 温度跃层
~halocline 盐度跃层
deposition: (n) 沉积,释放
vt: deposit
~vorticity deposition 涡度释放
resuspension: (n) 再悬浮
suspension: (n) 悬浮
vt: suspend
micro-environment: 微环境
[micro-] 微小的,微观的
~microphysics 微物理
~microscope 显微镜
~microscopic: (adj) 微观的
ant: [macro-] 大的,宏观的
~macroscopic: (adj) 宏观的
dispersal: (n) 分散,分布,散布
syn: dispersion 分散,散布,弥散,频散
~dispersion relation 频散关系
vt: disperse
adj: dispersive 分散的,弥散的,频散的
primary productivity: 初级生产力
primary: 主要的,首要的
~secondary productivity 次级生产力
~primary production: 初级生产
sub-grid scale: (n) 次网格尺度
grid: (n) 网格

Notes

[1] in doing sth. 意为在做某事的时候,但并非表示时间,与 when 有细微的区别。介词 in 与动名词(动词的-ing 形式)相连,表示的意思有两重:① 表示结果,即"通过"或"因为"做某事而得到某结果,与 by 类似。例如,"In calling him a liar publicly you have put your life in danger."。② 表示做某事的方式,例如,"Take care in crossing the street.",表示过马路的方式要注意,而"Take care when crossing the street."强调过马路的时候要注意。有时这两重含义无法区分。此处应为第一重含义,即通过观察瀑布这一行为,我们会注意到一些现象。

[2] 这里 steady 指时间上稳定,regular 指空间上规则。random、chaotic、unpredictable 都表示随机的、无规则的、无法预测的。

[3] resolve(分辨、辨析)表示将两个不同物体区分开这一行为,而 resolution 表示区分不同物体的能力。显然,物体越小、观察仪器(人眼、相机或其他传感器)离物体的距离越远,区分不同物体的能力越低,同时也与仪器的能力有关。仪器所能分辨的最小尺度就是分辨率,尺寸小于分辨率的物体或距离小于分辨率的两个物体超出了仪器的分辨能力。

[4] time history 即某一属性随时间的变化历程,也可用 time series(时间序列)。由于存

在多时间尺度变化,这些时间尺度之间的相互叠加导致难以区分出单一的周期,所以说 far from periodic。

[5] 稳定性(stability)是指扰动可被保持在一定的限度内,且总有回到平衡位置的趋势。相反地,不稳定性(instability)则表示扰动会发展、增强。满足某些条件时流动可产生不同类型的不稳定性,如斜压不稳定(baroclinic instability)、对流不稳定(convective instability)、对称不稳定(symmetric instability)等。本部分中,Text 10、Text 12、Text 17 和 Text 20 都有提到不同类型的不稳定,学习相关章节的时候可以留意。

[6] 无量纲数(non-dimensional parameter 或 dimensionless parameter)是一类常用的指示参数,常表示某两个量或尺度的比例。常用的无量纲数除了雷诺数(Reynolds number)之外,还有罗斯贝数(Rossby number)、弗劳德数(Froude number)、理查森数(Richardson number)等。

[7] 运动学黏性系数(kinematic viscosity)是动力学黏性系数(dynamic viscosity)与密度的比,而动力学黏性系数是黏性应力[viscos stress,或称剪切应力(shear stress)]与流速剪切(shear velocity)的比。运动学黏性系数又叫作动量扩散系数(momentum diffusivity),动力学黏性系数又叫绝对黏性系数(absolute viscosity)或剪切黏性系数(shear viscosity)。

[8] 定语从句 a long streak that … 描述染色剂形成的长条随着向下游的流动仅仅稍微变粗。increase in diameter only slightly 部分的语序略难理解。

[9] 此句由三部分组成,分别描述接近染色剂注射器处、往下游一些的地方,以及最终充分混合的情况。最初染色剂长条变得弯弯曲曲,后来越来越难以辨认(即与周围水体发生了混合使染色剂被稀释),最终完全混合起来,不再有可见的条带。注意三个部分自身都是完整的句子,它们之间用了分号分隔,而不是逗号。这种有递进关系的并列子句都要用分号分隔,或者也可以改用句号将它们断开成单独的句子。

[10] 此段描述了扩散(diffusion)与耗散(dissipation)的关系。热量与物质通过混合被扩散开,导致原来集中在较小区域的热量或物质被分散到更大范围,其中物质扩散包括盐分、溶解氧(oxygen)、营养盐等各种物质。同时,动量也被扩散,但动量的扩散叫作耗散。扩散的能力或程度即扩散系数(diffusivity),各种物质和热量的扩散系数不同;耗散的能力或程度就是黏性系数(viscosity)。这也是上述运动学黏性系数也叫作动量扩散系数的原因。此处还提到了发生混合的三种界面:固-液界面、液-气界面,以及层化液体的不同层次界面。湍流可以引起强烈的混合,而层流仅依靠分子热运动(molecular thermal motion)来产生混合,即分子混合(molecular mixing),其效率比湍流混合(turbulent mixing)弱得多。拖曳(drag)是由于动量的耗散或黏性作用在物体上的阻力。

[11] 大尺度运动受边界条件(boundary condition)控制,反映了边界的影响向流体内区

的传递。小尺度运动由从大尺度得来的能量驱动，此即跨尺度能量传递（cross-scale energy transfer），或称能量串级或级联（cascade）。大尺度向小尺度传递能量，实际上就是指长距离、长时间的运动在过程中会被消耗，产生短距离、短时间的"碎片"运动，此种"正向"串级（forward cascade）的效率取决于大尺度流场的稳定性，而稳定性又与多种因素有关。事实上，海洋里有些情况下逆向串级（inverse cascade）也起主要作用，即小尺度向大尺度传递能量，物理上对应的是"碎片"运动经过整合产生较长时空距离的运动。

[12] be they laminar or turbulent 部分是插入语，用于解释 fluid flows，意思是不管它是层流还是湍流。注意此语法。

[13] 精确求解 N-S 方程组可谓世纪难题。因为此方程组包含所有尺度、所有信息，要精确求解要求有极高（厘米级或更高）的空间分辨率，使得连最小尺度的运动都能分辨，因此通常难以实际应用。随着计算机能力的提高，近年来有直接求解的应用实例，但都局限于很小的空间范围。如果分辨率不足以反映最小尺度的湍流运动，那么这部分运动对更大尺度运动的影响就必须要通过某种方式估算出来，这就是湍流参数化（parameterization），或下文所称的统计方法（statistical approach）。

[14] 海面附近的混合主要集中在混合层（mixed layer 或 mixing layer）内，主要的机制包括风的搅拌（stirring）、波浪的破碎（wave breaking），以及大气对表层海洋的冷却产生的对流混合（convective mixing）等。混合可将上层与大气相互作用导致的热量增益或损失储存到海洋内部，也可以将储存在海洋内部的正或负热量再次传递给大气。

[15] 混合层以下的海洋内部的混合通常较小，但十分重要，因为除了湍流混合外，海洋内部缺乏其他能导致跨等密面混合（diapycnal mixing）的机制。海洋是层化的，因此，海水在垂向上倾向于沿等密面运动，而跨等密面的运动则需要额外的能量。湍流混合可提供此种能量，使得水的属性和其中携带的物质在等密面上下发生交换，导致等密面变得更加平缓，同时将下层海水带到上层。此机制被认为是全球经向翻转环流（overturning circulating）的上升分支的重要机制。海洋内部产生湍流混合的物理过程主要是内潮（internal tide）与地形相互作用产生的高模态内波（high-mode internal wave，即尺度较小的内波）的破碎有关。

[16] 此处所指的 subgrid-scale effects of turbulence 即点明了湍流的小尺度特征，目前的海洋环流模式分辨率（即相邻网格点的距离）普遍是几十千米以上的量级，少数可达几千米甚至几百米量级，无法直接分辨湍流，因此，湍流的作用被称为次网格效应。在这样的海洋模式中要考虑湍流，必须对其进行参数化，而海洋内波产生的湍流是其中的重要研究对象。

8. Internal Waves

Text

The ocean is stably stratified almost everywhere. Therefore a water parcel that is displaced, say, upward, encounters water of lower density and falls back downward and vice versa[1]. This results in an oscillation, hence a wave. The restoring force is the buoyancy force, which is the product of gravity and the difference in density between the displaced water parcel and its neighbors at the same pressure[2]. Internal gravity waves are similar in this respect to the surface gravity waves on the strong air-sea density interface[3]. Because the stratification within the ocean is much weaker than that between the air and water, the restoring action is weaker and the waves have much lower frequencies than surface waves of comparable wavelengths. For the same reason, water particles can travel large distances up and down in internal waves: amplitudes of tens of meters are common for internal waves[4].

Internal waves are mostly generated by tides, which interact with topography and generate internal tides (baroclinic tides), and by the wind, which stirs the mixed layer and generates internal waves with frequencies close to the inertial frequency (associated with Earth's rotation)[5]. Following Gill (1982), we take two approaches to considering internal waves: (a) waves on an interface between two layers of different density and (b) waves in a continuously stratified ocean. These two types of waves have quite different behaviors.

An interfacial internal wave is illustrated in Figure Ⅰ-4. This kind of internal wave is strikingly similar to a surface gravity wave. It propagates horizontally and involves heaving up and down of the sharp interface between the two layers, whose densities are ρ_1 and ρ_2. The principal modification from surface gravity waves is that the density difference, $\Delta\rho = \rho_1 - \rho_2$, between the two layers is much smaller than the density difference between air and water. The phase and group speeds of the interfacial internal wave are like those of shallow water surface waves[6]:

$$c_p^2 \approx g\frac{\Delta\rho}{\rho}H_1 \equiv g' H_1$$

where H_1 is the mean thickness of the upper layer, ρ is the mean density and g' is called the "reduced gravity." It is assumed in this formula that the upper layer is much shallower than the deeper layer. If they are, otherwise, of comparable depth, then the factor H_1 becomes a more complicated combination of both layer depths.

In Figure Ⅰ-4, the wave is propagating to the right. The water at the node in the center of the diagram (zero between the crest and the trough) is moving downward. The horizontal velocities are highest at the crests and troughs. There is a convergence at the node behind the crest; if the wave has very large amplitude, it can produce a surface slick, which makes the wave observable from the sea surface[7]. Such observations are only possible on calm days because on windy days the very high surface wave activity obscures the slicks. Subsurface temperature fluctuations are also found due to internal waves heaving the thermocline up and down.

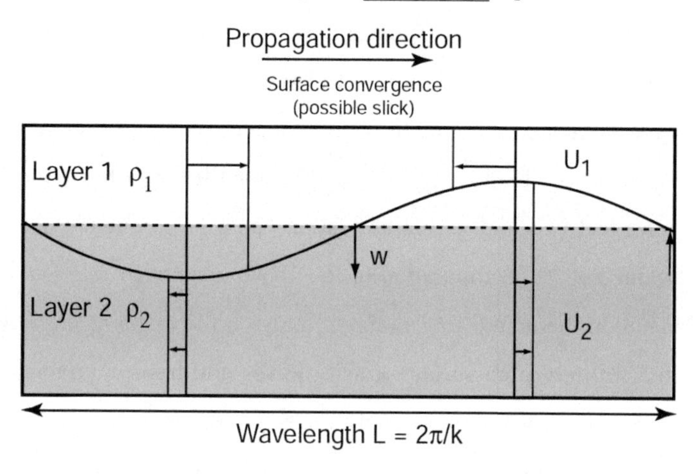

Figure Ⅰ-4: Schematic of a simple interfacial internal wave in a two-layer flow. [Source: Gill, 1982]

Now consider waves within a continuously stratified ocean (or atmosphere), ignoring the upper and lower boundaries. Vertical stratification is the most important external ocean property for characterizing these waves. The Brunt-Väisälä (buoyancy) frequency, defined as

$$N = \sqrt{\frac{g}{\rho}\frac{\partial \rho}{\partial z}}$$

is the maximum frequency for internal gravity waves[8]. The maximum frequency is higher for higher stratification (higher N). The wave periods range from several minutes in the well-stratified upper ocean, to hours in the weakly stratified deep ocean. Waves at the Brunt-Väisälä frequency propagate entirely horizontally, with water particles moving exactly vertically with maximum exposure to the stratification[9].

Because internal waves can have periods on the order of hours, low frequency internal gravity waves are influenced by Earth's rotation. The lowest frequency waves are pure inertial waves, whose frequency is equal to the Coriolis parameter, f[10]. These have particle motions that are entirely in the horizontal plane, with no vertical component that can feel the vertical stratification. The full range of internal wave frequencies, ω, is

$$f \leqslant \omega \leqslant N$$

Because f depends on latitude (0 at the equator and maximum at the poles), the allowable range of frequencies depends on latitude as well as on stratification.

The complete dispersion relation for internal waves in a continuously stratified flow is given here, without derivation, in terms of horizontal and vertical wavenumbers k, l, and m [11]:

$$\omega^2 = \frac{(k^2 + l^2)N^2 + m^2 f^2}{k^2 + l^2 + m^2}$$

This has been simplified by assuming that N has no variation and that f is constant (constant latitude). Even if more complicated stratification is included, this equation can still be a good approximation to the local behavior of the internal waves.

The internal wave frequencies from f to N are set entirely by the angle of the wave vector with the vertical (θ in Figure Ⅰ-5). As the wave vector tilts from horizontal toward the vertical, the water particles feel less and less stratification, and the frequency decreases until finally reaching its lowest value, f [12]. Manipulation of the above dispersion relation shows that the frequencies do not depend on the actual wavenumber, only on the angle of the wave vector with the horizontal. This differs entirely from surface gravity waves and from interfacial waves.

The group velocity (c_g) of internal gravity waves is exactly at right angles to the phase velocity (c_p). Thus the energy propagation direction, which is always given by the group velocity, is in the direction that the particles move[13]. And finally, the group velocity for both the highest frequency (N) and lowest frequency (f) internal waves is 0 in all directions[14].

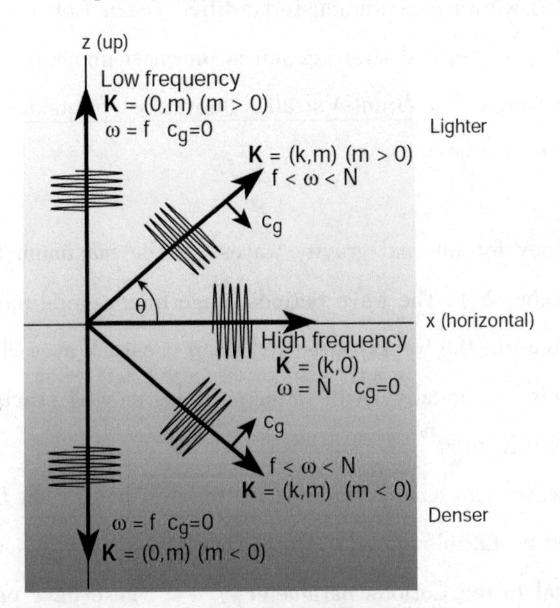

Figure Ⅰ-5: Schematic of properties of internal waves. The direction of phase propagation is given by the wavevector (k, m) (heavy arrows). The phase velocity (c_p) is in the direction of the wavevector. The group velocity (c_g) is exactly perpendicular to the wavevector (shorter, lighter arrows). [Source: Talley et al., 2011]

At near-inertial frequencies (close to f), downward group velocity from the mixed layer is accompanied by upward phase velocity, and the particles move in clockwise ellipses that are almost circular[15]. Because the Coriolis parameter, f, is 0 at the equator, internal waves of very low frequency can be found in the equatorial region, with periods of many days (10 days at 3 degrees latitude to infinite at the equator).

Internal waves within the water column (other than the interfacial waves) are primarily generated by winds that generate disturbances in the surface mixed layer and by the tides sloshing over bottom topography. Internal waves then propagate energy from the disturbances into the ocean interior (e.g., Polton et al., 2008). Nonlinear interactions between the internal waves generated at many different sites then spread the energy to internal waves at other frequencies[16].

Observed waves are usually analyzed by spectral analysis, including filters to remove frequencies that are not characteristic of internal waves[17]. Observed internal wave spectra are so similar from one place to another that it took several decades of work to begin to delineate variations in the spectrum due to local generation. The general form of the internal wave spectrum was introduced by Garrett and Munk (1972, 1975); their later modification is referred to as the GM spectrum (Munk, 1981), and remains widely used. Much of what is now known about internal wave distributions and generation has arisen from understanding the reasons for the nearly universal (empirical) spectral shape and from describing differences from this shape[18].

The energetic tides, which have very specific frequencies dictated by the moon and sun orbits, produce internal waves when they sweep over topography, if their frequency falls between f and N. This means that the propagation direction relative to the vertical of tidally generated internal waves can be precisely predicted because the direction is set, exactly, by the wave frequency.

Energy can pile up in internal waves, usually as the waves propagate toward shallow water near the coast, creating large localized disturbances called solitary waves or solitons. Internal solitons are associated with tides moving over banks or straits. Internal solitary waves have been observed in a number of locations. Acoustic backscattering from wave-generated turbulence was used to produce images of breaking internal solitary waves (e.g., Moum et al., 2003)[19].

Vocabulary

oscillation: (n) 振荡,振动,摆动,涛动
  vi: oscillate
restoring force: (n) 恢复力,回复力
buoyancy: (n) 浮性,浮力
  adj: buoyant 能浮起的,有浮力的,与浮力相关的
  ~buoy: (n) 浮子,浮标
product: (n) 乘积

internal wave：（n）内波
amplitude：（n）振幅
internal tide：（n）内潮
baroclinic：（adj）斜压的
 ~baroclinic tide 斜压潮
 n：baroclinicity 斜压性
 ant：barotropic（adj）正压的
 ~barotropic tide 正压潮
mixed layer：（n）混合层
inertial frequency：（n）惯性频率
be associated with：与……有关
interfacial internal wave：（n）界面内波
propagate：（vi）传播，传送
 n：propagation
heave：（vt/vi/n）起伏，抛举
phase speed/velocity：（n）相速度
 phase：（n）相位，位相
group speed/velocity：（n）群速度
reduced gravity：（n）约化重力
assume：（vt）假设，推测
 n：assumption
 adj：assumable
 adv：assumedly
node：（n）节点，交叉点
diagram：（n）图解，简图，示意图
slick：（n）光滑、平滑的条带
subsurface：（n）水下，次表层
thermocline：（n）温跃层
Brunt-Väisälä frequency/buoyancy
 frequency：（n）浮力频率，浮性频率
order：（n）量级
Coriolis parameter：（n）科氏参数
 ~Coriolis force：（n）科氏力
 ~Coriolis effect：（n）科氏效应
plane：（n）平面；（adj）平的，平面的

dispersion relation：（n）频散关系
 dispersion：（n）弥散，扩散，频散
derivation：（n）推导，推论
 vt：derive
simplify：（vt）简化，化简
 n：simplification
wave vector：（n）波矢量
clockwise：（adj/adv）顺时针的
 ant：anticlockwise/counterclockwise 逆时针的
ellipse：（n）椭圆
 adj：elliptic/elliptical
 syn：oval（n/adj）
infinite：（n）无穷，无限，无穷大；（adj）无穷的，无限的
 ~infinity（n）无限性，无穷性，无穷大
 ant：finite（adj）有限的
interior：（n）内部；（adj）内部的，里面的
 ant：exterior（n/adj）
nonlinear：（adj）非线性的
 n：nonlinearity
 ant：linear（adj）线性的
spectral analysis：（n）谱分析
 spectral：（adj）谱的
 n：spectrum 谱，波谱，光谱
 pl：spectra
delineate：（vt）详细描述，准确解释
empirical：（adj）经验的
specific：（adj）明确的，确定的，具体的
dictate：（vt）命令，规定，控制，支配，决定
localize：（vt）局部化，限制在某区域
solitary wave：（n）孤立波
 solitary：（adj）独立的，独自的，单个的
soliton：（n）孤立子，孤立波
bank：（n）堤岸，泥滩，沙洲

strait:(n) 海峡
acoustic:(adj) 声音的,声学的
backscatter:(vt/n) 后向散射
scatter:(vt/n) 散布,分散,散射
~scatterometer (n) 散射计

Notes

[1] 稳定层结 (stable stratification) 意指低密度在上、高密度在下的结构,即密度随深度的变化率为正。流体微团被扰动产生垂向位移后,由于浮力和重力的作用会在平衡位置附近做往复运动并产生波动。如果层结不稳定,则初始位移将被放大,不再回到平衡位置。

[2] 恢复力是任何波动必须具备的,也常被作为波动分类的依据。

[3] in this respect 意为"从这个角度讲"。

[4] 此处论证的思路是恢复力弱,所以需要更长时间才能完成一个周期,即频率低。也因此在流体微团的扰动被拉回之前已经走了较远的距离,即振幅大。同样的,波动有更多时间传播,因此水平方向的波长也较长。

[5] 此处提到了内波的两种主要来源:斜压潮 (baroclinic tide) 和近惯性波 (near-inertial wave)。斜压潮即正压潮过地形产生的垂向不一致潮,正压潮是天体引潮力直接作用在水体上产生的垂向一致潮。

[6] 相速度 (phase velocity) 和群速度 (group velocity) 是波动的两种最主要特征,前者表示相位 (phase) 即运动形式的传播速度,后者表示波群 [wave group,或波包 (wave package),或振幅包络线 (wave envelop)] 的传播速度,二者都是矢量。群速度代表能量的传播速度。相速度等于频率与波数的比,群速度等于频率对波数的导数,因此都可以通过频散关系 (dispersion relation) 确定。若相速度与频率无关(即频率正比于波数),则称为非频散波 (non-dispersive wave),否则为频散波 (dispersive wave)。非频散波相速度与群速度相等。

[7] 内波可引起表层海水辐聚辐散,改变海面粗糙度 (roughness),因此,可以从海面上观察到光滑和粗糙相间的条纹,反映水下的内波情况。

[8] 布伦特-维萨拉频率 (Brunt-Väisälä frequency) 又称为浮力频率 (buoyancy frequency) 或浮性频率,是衡量海洋层结强度的重要参量,也是内波频率的上限。一个与之紧密相关的物理量是静态稳定度 (static stability),$E = N^2/g$,由于只是平方和常数的关系,二者有时也不甚严格地混为一谈。

[9] 垂直运动的水质点在运动过程中经历的密度差最大,因此恢复力最强,即受到层结的影响最大,导致很快被拉回平衡位置,对应的周期短、频率高。斜向运动的水质点则由于恢复力弱,导致频率低。水平运动的水质点则完全感受不到层结的影响,仅在科氏力的作用下形成惯性振荡。可见,内波的传播方向与频率之间有对

应关系。

[10] 科氏参数（Coriolis parameter）仅与地球旋转速度和纬度有关,因此对于某特定地点可认为是常数。如果内波传播距离不是很远,也可近似认为科氏参数不变,即所谓的“f平面”近似（f-plane approximation）。

[11] 波动的频散关系（dispersion relation）即频率与波数的关系式,是决定波动特征的最重要因素。这里用 in terms of 表示将频散关系表示为波数的三个分量的函数。k、l、m 分别是波数矢量在 x、y、z 三个方向的分量。

[12] 波矢量（wave vector）的方向是波的传播方向,大小是波的传播速度(相速度)。对于包括内波在内的横波,波矢量与质点振动方向垂直。波的传播路径称为波射线(wave ray),波射线上各点的切线方向就是波矢量的方向,波射线的密集度则与波矢量大小成正比。

[13] 内波的群速度与相速度垂直,即与波矢量垂直,群速度与质点的振动方向在同一条线上。

[14] 最高频和最低频的内波群速度为0,其物理意义是这两种波的能量不传播。直观来讲,这两种波的振幅随着距离越远而越小。虽然波形(波峰和波谷)仍向外传播,但振幅越来越小,振幅最大值总是在波源地,即能量集中而不传播。相反,如果群速度不为0,则最大振幅位置也会传播。

[15] 此即近惯性振荡（near-inertial oscillation）。

[16] 此段描述了内波在上下边界附近生成,又向海洋内部传播的过程。不同来源的内波通过非线性相互作用（non-linear interaction）实现能量向其他频率的传递。另外,内波与海流（current）、中尺度涡（mesoscale eddy）等其他海水运动形式也存在相互作用,并能发生多普勒频移（Doppler shift）和能量的发散或汇聚等。

[17] 谱分析（spectral analysis）即通过傅里叶变换［Fourier transform,或傅里叶分解(Fourier decomposition)］将各频率混杂在一起的时间序列分解成不同频率的谐波(harmonic),将各谐波的能量(正比于振幅的平方)按照频率绘制成频谱（frequency spectrum）。同时,还可以将空间变化分解成不同波数的谐波,得到波数谱（wavenumber spectrum）。由于能量可用功率表示,又称能谱或功率谱（power spectrum）。采用滤波器［filter,或称数字滤波器（digital filter）］,可提取频谱的一部分而抑制另一部分,从而实现低通（lowpass,即低频通过）、高通（high-pass,即高频通过）或带通（bandpass,即中间频率通过）滤波。

[18] empirical 表示“经验的”,即没有通过理论分析,而是直接通过观测获取某关系式或曲线的定量形式。例如,海浪的波高与风速的关系,可通过大量观测拟合一条经验曲线,得出经验关系或经验公式。此处所述的经验谱形也是通过大量观测拟合得到的。

[19] 声学后向散射（acoustic backscattering）即海洋湍流造成的声波的散射,可通过发射声波,然后记录散射回来的声波强度进行观测。通过此种手段可以观测到海洋内部湍流较强的区域,由此发现了内孤立波破碎生成的剧烈湍流。

9.Ekman Transport and Pumping

Text

When the wind blows along the surface of the ocean, it causes both surface currents and waves. The quantitative details of how the stress of the wind is applied to the ocean surface are not completely understood. Energy is transferred from wind to ocean by some kind of turbulent process(es). More complete understanding requires a detailed examination and understanding of, not only the mean wind, current, and pressure fields, but of the variations of the wind, current, and pressure about the mean.

Several semiempirical observations are useful. One is that the surface current induced by a wind is approximately 3% of the wind. One might expect a 0.3 m/s surface current with a 10 m/s wind. The second is that the stress (τ) applied to the sea surface increases as the square of the wind speed according to

$$|\tau| = 0.002\ W^2$$

where W is the wind speed in meters per second and τ is the wind stress on the sea surface in Newtons per square meter. A 10 m/s wind causes a wind stress of approximately 0.2 N/m^2.

In fact, neither relationship is that simple. The "constant" 0.002 varies with wind speed and surface roughness[1]. It also depends on how far above the sea surface the wind is measured. For winds measured at "deck height", 5 to 10 m above the sea surface, the above equation is good to within a factor of 2 and probably considerably better than that[2].

If one assumes a wind blowing on a flat ocean surface with no horizontal pressure gradients, no acceleration, and no bottom friction, the only forces to be balanced in the equation of motion are the Coriolis force and the frictional term as represented by the wind stress[3]:

$$\frac{1}{\rho}\frac{\partial \tau_x}{\partial z} = -fv$$

$$\frac{1}{\rho}\frac{\partial \tau_y}{\partial z} = fu$$

If one now integrates between the surface and some depth Z below which the effect of the wind is no longer felt[4], one gets the following interesting steady state relationship:

$$\tau_x = -M_y f$$

$$\tau_y = M_x f$$

where the zonal and meridional mass transports are

$$M_x = \int_0^Z \rho u dz$$

$$M_y = \int_0^z \rho v \mathrm{d}z$$

The units of M_x and M_y are mass per unit time per unit length[5]. Note that a wind blowing from the north does not move the integrated water column to the south, but to the west in the Northern Hemisphere and to the east in the Southern Hemisphere. Again, the Coriolis term confounds our logic. Under the stress of the winds, water does not move downwind, but at right angles to the wind. For an observer facing downwind, the water moves to the right in the Northern Hemisphere and to the left in the Southern Hemisphere.

One can rewrite the equation of motion by replacing the wind-stress term with an eddy-viscosity term[6]:

$$A_z \frac{\partial^2 u}{\partial z^2} = -fv$$

$$A_z \frac{\partial^2 v}{\partial z^2} = fu$$

The solution of this equation gives the same results as before for the total wind-driven transport, but, in addition, it indicates the details of the velocity structure in the water column. The solution is a spiral in which the surface water moves at an angle of 45° to the wind (Figure I - 6)[7].

This kind of motion is called Ekman motion, after V. W. Ekman, who first examined the problem in 1902. Ekman's attention was called to the problem by Fridtjof Nansen, who, while frozen in the Arctic aboard the Fram, observed that when the wind blew, the ice appeared to move not downwind, but at an angle of 20°–40° to the right of the wind[8].

The ice-covered Arctic might be thought of as a special situation in which the basic assumptions aforementioned might be more likely to hold. However, the observational evidence in the open ocean for a wind-driven surface current moving at some angle to the right of the wind is good, as is the evidence for a mass transport of water to the right of the wind. Finally, one can often see some evidence of an Ekman spiral if one time averages the observed currents of the surface layer[9].

Upwelling is the term used in oceanography to describe the process by which deep water is brought to the surface. It has an importance well beyond its physical significance because the deep water carries remineralized nutrients to the surface layer, where they can be assimilated by phytoplankton[10]. Regions of upwelling are among the richest biological areas in the world.

Upwelling occurs wherever there is a divergent flow at the surface. Continuity requires upward vertical flow to replace the water lost by the surface divergence[11]. The most famous upwelling areas are those along certain coasts, where the wind drives the surface water offshore.

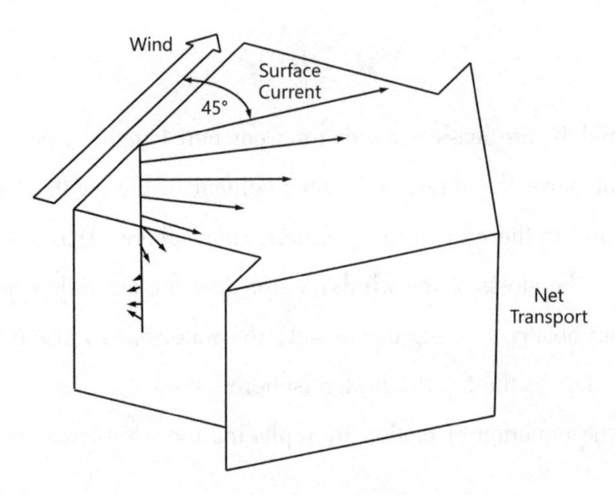

Figure Ⅰ-6: Water is set in motion by the wind. According to the Ekman relation, the effect of the Coriolis force is for each succeeding layer of water to move slightly to the right (in the Northern Hemisphere) of the layer above it. The result is the Ekman spiral as shown, with the net transport at 90° to right of the wind in the Northern Hemisphere. [Source: Knauss and Garfield, 2017]

According to Ekman theory, the effect of wind is to drive the water to the right of the wind in the Northern Hemisphere (left of the wind in the Southern Hemisphere). Thus maximum upwelling occurs when the wind is parallel to the shore, not offshore.

An examination of surface wind charts allows one to predict areas of coastal upwelling. These predictions can be verified by an examination of surface temperature charts. The cold coastal waters off the coasts of Peru and California can best be explained by coastal upwelling[12].

Upwelling can be found along the equator, where the easterly component of the trades causes an Ekman drift to the north in the Northern Hemisphere and to the south in the Southern Hemisphere. Where the easterly component of the trades is well developed, the surface temperature along the equator in the Central Atlantic and the Pacific is often 2℃ cooler than it is 160 km on either side of the equator[13]. On the equator, the sine of the latitude is zero, and one cannot expect the above equation of motion to hold, but it may apply within 100 km of the equator[14]. Given the strong vertical shear that exists along the equator because of the Equatorial Undercurrent, vertical mixing may also play a role in reducing the surface temperature along the equator, but both observation and theory suggest that the low surface temperature along the equator is at least in part the result of a divergence of the surface waters resulting from a poleward Ekman transport on either side of the equator.

Ekman upwelling is not confined to special situations, such as the equator or ocean boundaries. It can occur anywhere there is an adequate shear in the horizontal wind field[15]. A striking example is an intense cyclone or hurricane/typhoon. The effect of the cyclonic winds is to cause a divergence in the surface currents in the Ekman layer, which in turn requires upwelling of deeper water to maintain continuity. This process is often referred to as Ekman pumping, since Ekman transport pumps water to the surface and can cause a shoaling of the thermocline[16]. The depth of the Ekman layer is between 10 and 100 m, depending on the wind speed and the degree of stratification in the surface waters. The upwelling velocity is given as follows

$$w_e = \frac{1}{\rho f}\left(\frac{\partial \tau_y}{\partial x} - \frac{\partial \tau_x}{\partial y}\right)$$

where w_e is the vertical velocity at the bottom of the Ekman layer. A positive value means an upward velocity.

For a hurricane with 50 m/s winds, the upwelling velocity from the above equation would be on the order of 10^{-3} m/s, several orders of magnitude larger than what is generally accepted as a typical vertical velocity in the ocean. It could also be demonstrated that water can be pumped down as well as up. The anticyclonic winds cause a convergence in the Ekman layer, downwelling, and a deepening of the thermocline.

Vocabulary

examination: (n) 检查,调查,考察
  vt: examine
semiempirical: (adj) 半经验的
  [semi-]: 半
square: (n) 平方
roughness: (n) 粗糙度
deck: (n) 甲板
pressure gradient: (n) 压强梯度
  pressure gradient force 压强梯度力
  gradient: (n) 梯度
acceleration: (n) 加速,加快,加速度
  vt: accelerate
equation of motion: (n) 运动方程
term: (n)(数学表达式的)项
integrate: (vt) 合并,使结合,求积分
  n: integration 积分法,求积分
  n: integral (n) 积分;(adj) 积分的
  ~differentiate (vt) 求微分
  n: differentiation 微分法,求微分
  n: differential (n) 微分;(adj) 微分的
steady state: (n) 稳定态
zonal: (adj) 纬向的,沿纬线方向的,东西向的
meridional: (adj) 经向的,沿经线方向的,子午向的,南北向的
confound: (vt) 使糊涂,使疑惑
logic: (n) 逻辑,逻辑推理,原理
  adj: logical 逻辑的,符合逻辑的
downwind: (adj/adv) 顺风的/地,向下风向的/地
  ant: upwind (adj/adv) 逆风的/地,向上风向的/地
solution: (n)(方程的)解,(问题的)答案或解决方案
indicate: (vt) 表明,指出,指示
  n: indication

n: indicator 标志,迹象,指标

adj: indicative 能表明的

aforementioned: (adj) 上述的

likely: (adj) 很可能的

n: likelihood 可能性

evidence: (n) 证据,根据

adj: evident 显然的,明显的

Ekman spiral: (n) 埃克曼螺旋

net: (adj) 净的,总的

oceanography: (n) 海洋学

remineralized: (adj) 再矿化的

assimilate: (vt) 吸收,领会,同化

n: assimilation

~data assimilation 数据同化

biological: (adj) 生物的,生物学的

n: biology 生物学

divergent: (adj) 辐散的

continuity: (n) 连续性

~equation of continuity 连续性方程

chart: (n) 曲线图,海图,地图,图表,表格;(vt) 制图,绘图

verify: (vt) 证明,核实,证实

n: verification

easterly: (n) 东风(从东方来的风)

Ekman drift: (n) 埃克曼漂流

hold: (vi)(公式、原理等)成立

Equatorial Undercurrent: (n) 赤道潜流

adequate: (adj) 足够的,满足需要的

Ekman pumping: (n) 埃克曼泵压

~Ekman suction 埃克曼抽吸

shoal: (vi) 变浅;(n) 浅水处,浅滩

degree: (n) 程度

as follows: 如下

order of magnitude: (n) 量级

typical: (adj) 典型的,有代表性的

Notes

[1] 此公式实际上争议很大。更一般的形式应为 $\tau = \rho_a C_D W^2$,其中, ρ_a 是空气密度,可视为常数。C_D 为拖曳系数 (drag coefficient),实际上它与多种因素有关,包括风速和海面粗糙度,以及气泡 (bubbles) 和飞沫 (spray) 等复杂因素。拖曳系数的具体形式仍无定论,且在高风速情况下认识很少,争议很大。

[2] be good to within a factor of 2 表示此公式的精度,即其估计值在真实值的0.5~2倍之间。这一精度实际上相当粗糙。

[3] 埃克曼理论的假设:无限广阔海洋(没有水位堆积导致的压强梯度力)、均质海洋(没有密度变化导致的压强梯度力)、无限深海(没有底摩擦)、稳定风场吹拂足够长时间[流场达到稳定态 (steady state),忽略调整过程,即局地加速项 $\partial u/\partial t$ 消失]。

[4] 此深度称为摩擦深度 (frictional depth),或埃克曼层厚度 (Ekman layer thickness),通常用于表示风应力的作用能够传递到的最大深度,记为 D,定义为流速 u 降低至表层流速 u_0 的 $e^{-\pi}$ 倍(即4.3%)的深度。推导可知 $D = \pi\sqrt{2A/f}$,其中,A 是动力学黏性系数 (dynamic viscosity),f 是科氏参数,可见摩擦深度与风速无关。摩擦层底处的流速 $u_D = u_0\, e^{-\pi}$, $u_0 = \dfrac{\tau}{\rho\sqrt{Af}}$,其中,τ 为风应力,与风速有关。可见摩擦层底处的

流速虽比例固定,但绝对值仍与风速有关。因此,若按传统将风的作用深度定义为摩擦深度,则确实与风速无关;但若将风的作用深度定义为埃克曼流流速超过某固定阈值的深度,则其实际上与风速有关,我们认为后一种定义更符合"更大的风搅动更深的水"的直观认知。

[5] 此处表述不够严谨。mass per unit time per unit length 是量纲(dimension),而不是单位(unit)。同一量纲可以有不同单位,例如,质量(mass)的单位可以是千克(kilogram)也可以是磅(pound),长度(length)的单位可以是厘米(centimeter)也可以是英寸(inch)等。

[6] 此处的 eddy viscosity 是指湍黏性(turbulent viscosity)。eddy 一词在不同的语境下可以指代不同对象。描述湍流时,指湍流形成的涡或扰动,因此译作"涡动"或"扰动"。描述中尺度现象时,指环形的流场,即中尺度涡。后者的尺度(几十至几百千米)比前者(厘米量级)大得多。湍黏性力的表达式是湍黏性系数与流速在垂直方向的二阶导数的乘积。这来源于雷诺应力(Reynolds stress)的处理方式,即将湍黏性类比于分子黏性(molecular viscosity)的牛顿黏性定律,得到 $\tau_x = A_z \partial u / \partial z$,然后求任意流体微团上下界面处的湍应力的微分,表示二者的合力,即得到了二阶导数。

[7] 此即著名的埃克曼螺旋(Ekman spiral),是不同层次上流速矢量的末端连线构成的一个三维的螺旋,其在水平面上的投影则是一个二维的螺旋线,通常叫作埃克曼螺旋。

[8] 挪威科学家 Fridtjof Nansen 带领的科考团队于 1893—1896 年乘坐"前进号"(挪威语 Fram,或译为"弗莱姆号")进行北极科考,期间发现了后来被称为埃克曼流的现象。表层流偏转角度达不到理论给出的 45°的原因有多种,包括存在压强梯度、底摩擦,或风不够稳定导致科氏力来不及起作用,等等。反映了实际风生漂流的复杂性。

[9] 时间平均(time average)操作可在一定程度上消除风的高频变化带来的影响,提取相对稳定风场的作用,因此更能体现埃克曼流。

[10] an importance well beyond its physical significance 意思是它的重要性远远不仅限于物理方面。下文讲述了上升流带来营养盐(矿物质),从而促进海洋初级生产力。

[11] 上层辐散导致水减少,因此诱导出上升流。此过程可以通过连续性方程(equation of continuity)来理解,即密度的局地变化与散度(divergence)相平衡,直观理解是如果一个固定的立方体网格内发生了净的进水,那么密度一定增加,反之亦然。如果认为水是不可压缩的(incompressible),则密度不变,所以散度必须为 0。由此,水平方向的辐散必然要有垂向的上升流,否则此网格就变成了空的,违背了流体的连续性。

[12] 秘鲁和加利福尼亚分别处在东南和东北信风(trade wind)作用下,而海岸线基本是南北走向的,因此,风有沿岸分量。两地沿岸风方向相反,但埃克曼输运相对风向的偏转方向在南北半球也相反,因此,恰好都有离岸的埃克曼输运,导致上升流。

[13] 在信风纬向分量作用下,赤道两侧埃克曼输运反向,产生辐散和上升。

[14] 科氏力不存在时任意流体微团上下界面处的湍流黏性力反向平衡,而海面处的湍流黏性力与风应力平衡,由此得出流速随深度线性衰减且方向不偏转。

[15] 具体来讲,此处水平风场的剪切(shear in the horizontal wind field)是指风应力旋度(wind stress curl)。旋度(curl)实际上就是涡度(vorticity),但通常仅用在风场上,表达式为 curl $\tau = \dfrac{\partial \tau_y}{\partial x} - \dfrac{\partial \tau_x}{\partial y}$。注意此处 τ_x 对 y 求导而 τ_y 对 x 求导。若是 $\dfrac{\partial \tau_x}{\partial x} + \dfrac{\partial \tau_y}{\partial y}$,则为风应力的散度(divergence)。涡度又用 $\nabla \times \tau$ 表示,散度用 $\nabla \cdot \tau$ 表示,其中,∇读作 nabla。埃克曼理论将风应力旋度与埃克曼输运的散度联系起来。

[16] 这里将气旋式风应力旋度产生上升流(upwelling)的过程叫作 Ekman pumping,属于不是很严格的说法。严格来讲,此过程应该叫作 Ekman suction(埃克曼抽吸),而不是 pumping。Ekman pumping 应该指反气旋式风应力旋度产生下降流(downwelling)的过程,即埃克曼泵压。但是英文语境里常用 Ekman pumping,中文语境里用埃克曼抽吸这个说法模糊指代抽吸和泵压两个过程,或者单独指代上升流过程(可能因为上升流的生态学意义更强),因此造成中文的"抽吸"与英文的"pumping"对应的情况。

10. Geostrophic Wind and Thermal Wind

Text

In atmospheric science, geostrophic flow is the theoretical wind that would result from an exact balance between the Coriolis force and the pressure gradient force. This condition is called geostrophic equilibrium or geostrophic balance (aka geostrophy)[1]. The geostrophic wind is directed parallel to isobars, i.e., lines of constant pressure at a given height. This balance seldom holds exactly in nature. The true wind almost always differs from the geostrophic wind due to other forces such as friction from the ground. Thus, the actual wind would equal the geostrophic wind only if there were no friction (e.g., above the atmospheric boundary layer)[2]. Despite this, much of the atmosphere outside the tropics is close to geostrophic flow much of the time and it is a valuable first approximation. Geostrophic flow may be either barotropic or baroclinic. Geostrophic flow may also be thought of as a zero-frequency inertial wave[3].

A useful heuristic is to imagine air starting from rest, experiencing a force directed from areas of high pressure toward areas of low pressure, called the pressure gradient force. If the air began to move in response to that force, however, the Coriolis "force" would deflect it, to the right of the motion in the northern hemisphere or to the left in the southern hemisphere. As the air accelerated, the deflection would increase until the Coriolis force's strength and direction balanced the pressure gradient force, a state called geostrophic balance[4]. At this point, the flow is no longer moving from high to low pressure, but instead moves along isobars. Geostrophic balance helps to explain why, in the northern hemisphere, low-pressure systems (or cyclones) spin counterclockwise and high-pressure systems (or anticyclones) spin clockwise, and the opposite in the southern hemisphere.

The effect of friction, between the air and the land, breaks the geostrophic balance. Friction slows the flow, lessening the effect of the Coriolis force. As a result, the pressure gradient force has a greater effect and the air still moves from high pressure to low pressure, though with great deflection. This explains why high-pressure system winds radiate out from the center of the system, while low-pressure systems have winds that spiral inwards[5].

The neglect of frictional effects is usually a good approximation for the synoptic scale instantaneous flow in the midlatitude mid-troposphere[6]. Although ageostrophic terms are relatively small, they are essential for the time evolution of the flow and, in particular, are necessary for the growth and decay of storms. Quasigeostrophic theories are used to model flows in the atmosphere more widely. These theories allow for a divergence to take place and for weather systems to then develop[7].

Flow of ocean water is also largely geostrophic. Just as multiple weather balloons that measure pressure as a function of height in the atmosphere are used to map the atmospheric pressure field and infer the geostrophic wind, oceanographers can infer ocean currents from measurements of the sea surface height (by combined satellite altimetry and gravimetry) or from vertical profiles of seawater density taken by ships or autonomous buoys[8]. The major currents of the world's oceans, such as the Gulf Stream, the Kuroshio Current, the Agulhas Current, and the Antarctic Circumpolar Current, are all approximately in geostrophic balance and are examples of geostrophic currents.

In atmospheric and oceanic sciences, the thermal wind is the vector difference between the geostrophic wind or current at upper altitudes minus that at lower altitudes. Oceanographers tend to keep the nomenclature of "thermal wind" rather than using "thermal current". It is the hypothetical vertical wind shear that would exist if the winds obeyed geostrophic balance in the horizontal, while pressure obeys hydrostatic balance in the vertical[9]. The combination of these two force balances is called thermal wind balance.

Since the geostrophic wind at a given pressure level flows along geopotential height contours on a map, and the geopotential thickness of a pressure layer is proportional to virtual temperature, it follows that the thermal wind flows along thickness or temperature contours, its magnitude a function of horizontal temperature gradient. Also called baroclinic flow, the thermal wind varies with height in proportion to the horizontal temperature gradient[10].

The term thermal wind is often considered a misnomer, since it really describes the change in wind with height, rather than the wind itself. However, one can view the thermal wind as a geostrophic wind that varies with height, so that the term wind seems appropriate. In the early years of meteorology, when data was scarce, the wind field could be estimated using the thermal wind relation and knowledge of a surface wind speed and direction as well as thermodynamic soundings aloft. In this way, the thermal wind relation acts to define the wind itself, rather than just its shear.

If a component of the geostrophic wind is parallel to the temperature gradient, the thermal wind will cause the geostrophic wind to rotate with height. If geostrophic wind blows from cold air to warm air (cold advection) the geostrophic wind will turn counterclockwise with height (for the northern hemisphere), a phenomenon known as wind backing[11]. Otherwise, if geostrophic wind blows from warm air to cold air (warm advection) the wind will turn clockwise with height, also known as wind veering. Wind backing and veering allow an estimation of the horizontal temperature gradient with data from an atmospheric sounding.

A horizontal temperature gradient exists while moving North-South along a meridian because curvature of the Earth allows for more solar heating at the equator than at the poles. This creates a westerly geostrophic wind pattern to form in the mid-latitudes. Because thermal wind causes an increase in wind velocity with height, the westerly pattern increases in intensity up until the tropopause, creating a strong wind current known as the jet stream. The Northern and Southern Hemispheres exhibit similar jet stream patterns in the mid-latitudes.

The strongest part of jet streams should be in proximity where temperature gradients are the largest. Due to land masses in the northern hemisphere, largest temperature contrasts are observed on the east coast of North America (boundary between Canadian cold air mass and the Gulf Stream/warmer Atlantic) and Eurasia (boundary between the boreal winter monsoon/Siberian cold air mass and the warm Pacific). Therefore, the strongest boreal winter jet streams are observed over east coast of North America and Eurasia[12]. Since stronger vertical shear promotes baroclinic instability, the most rapid development of extratropical cyclones (so-called bomb cyclones) is also observed along the east coast of North America and Eurasia[13]. The lack of land masses in the Southern Hemisphere leads to a more constant jet with longitude (i.e. a more zonally symmetric jet).

Vocabulary

aka: also known as 又叫作,亦称为
geostrophy: (n) 地转特性,地转状态
isobar: (n) 等压面,等压线
boundary layer: (n) 边界层
inertial wave: (n) 惯性波
heuristic: (adj/n) 启发式的(方法或步骤)
response: (n) 响应
  in response to 对……做出响应
  vi: respond
lessen: (vt/vi) (使)降低、减弱、缩小
synoptic: (adj) 天气的,天气学的
midlatitude: (n/adj) 中纬度(的)
  syn: midlatitudinal
  mid 与 latitude 之间可加连词符
troposphere: (n) 对流层
ageostrophic: (adj) 非地转的
evolution: (n) 演化过程
  vi: evolve 演化
growth: (n) 生长,成长
  vi: grow
decay: (n/vi) 消亡,衰减,衰退
quasigeostrophic: (adj) 准地转的
  [quasi-]: 准……
largely: (adv) 很大程度上,主要地
infer: (vt) 推断,推论
sea surface height: (n) 海面高度
altimetry: (n) 高度计,测高仪
gravimetry: (n) 重力计,测重仪
autonomous: (adj) 自治的,自动的
Kuroshio Current: (n) 黑潮(流)
  ~Kuroshio Extension: (n) 黑潮延伸体
Agulhas Current: (n) 厄加勒斯流
  ~Agulhas Return Current 厄加勒斯回流
Antarctic Circumpolar Current: (n) 南极绕

极流

[circum-]：环绕

thermal wind：(n) 热成风

vector：(n) 矢量

minus：(prep) 减去；(adj) 负的；(n) 减法

nomenclature：(n) 术语，命名法

hypothetical：(adj) 假设的，假象的

n：hypothesis 假设，假说

pl：hypotheses

vt：hypothesize / hypothesis

geopotential height：(n) 位势高度

geopotential：(n/adj) 重力位势 (的)

proportional：(adj) 成比例的

n：proportion 比例

virtual temperature：(n) 虚温

virtual：(adj) 虚的，虚拟的(不存在但通过计算呈现出来的)；(adj) 实际上的，事实上的

misnomer：(n) 错误或不恰当的名字

data：(n pl) 数据(通常用此复数形式)

sing：datum

sounding：(n) 水深测量，探空测量多指用绳索、杆或回声等方式测量，气象学探空测量则用探空气球 (sounding balloon)

vt：sound 测量，探测

advection：(n) 平流

vt：advect

wind backing：(n) 风的逆转

wind veering：(n) 风的顺转

tropopause：(n) 对流层顶

jet stream：(n) 急流，射流

proximity：(n)(时间、空间、关系的)靠近，接近

in (close) proximity to：离……很近

syn：in vicinity of

land mass：(n) 陆地

Eurasia：(n) 欧亚大陆

boreal：(n) 北的，北方的，北半球的

ant：austral

monsoon：(n) 季风

winter monsoon：(n) 冬季风

Siberian：(adj) 西伯利亚的

promote：(vt) 促进

bomb cyclone：(n) 炸弹气旋

=explosive cyclone 爆发性气旋

lack：(n/vt) 没有，缺乏，欠缺

symmetric：(adj) 对称的

=symmetrical

ant：asymmetric 不对称的

~antisymmetric 反对称的

~axisymmetric 轴对称的

~centrosymmetric 中心对称的

Notes

[1] 此句用 would 引导了虚拟语气，表示严格的平衡实际上并不存在。aka 是 also known as 的缩写，已经可以当成一个单独的词来使用。类似的情况还有 asap (as soon as possible)、FYI (for your information)，但这种缩写词较少用在正式的科技英语场合。

[2] 如果存在除科氏力和压强梯度力外的任何其他力，都可以打破地转平衡，产生非地转运动 (ageostrophic motion)。一般认为摩擦力仅在大气边界层，即直接与地面或海

面接触的一层内才重要,而在边界层之上的自由大气(free atmosphere)中仅存在空气之间的摩擦力,其量级小得多,可以忽略。例外情况也是存在的,如湍流混合特别强时。

[3] 地转平衡是在水平面上定义的,仅考虑水平科氏力和压强梯度力的水平分量。至于不同层次的地转风是同向[正压(barotropic)]还是不同向[斜压(baroclinic)],则依赖于各层上压强的分布情况。由于地转流总是环绕压强的高值或低值旋转,可以认为是受到科氏力这个惯性力控制的波(inertial wave)或惯性振荡(inertial oscillation),而且它不随时间变化,因此频率为0,且不随空间传播,因此波数也为0。当然这是针对科氏参数不变(即f平面)的情况而言的。

[4] 科氏力总是正比于流速,因此没有流动的时候没有科氏力,刚开始流动的时候科氏力小,速度越快则科氏力越大。

[5] 此段点出了地转流的另一个重要特征——无辐散(non-divergent)。而非地转运动的特征则是有辐散(divergent)。实际的运动可以分解成地转的或旋转的(rotational)分量,以及无旋的(irrotational)或有辐散的(divergent)分量。

[6] synoptic scale 是天气尺度,指时间尺度在2~8 d(有人认为2~10 d)的运动形式,包括各种常见天气现象如台风、寒潮(cold wave)、热带外气旋(extratropical cyclone,或译温带气旋)、爆发性气旋(explosive cyclogenesis 或 bomb cyclone 或 bombogenesis)等。

[7] 准地转(quasigeostrophic)理论放松了地转平衡,允许非地转的辐散运动存在,但假定仍以旋转运动为主,辐散分量很小。

[8] 卫星高度计可以测量卫星到海面的距离,而重力卫星可以测量地球的重力场,从而得出卫星到大地水准面(geoid)的距离,二者结合可以得到绝对动力高度(absolute dynamic topography),是海面高度(sea surface height)的表示方式之一。

[9] 热成风平衡或热成风关系是由水平方向的地转平衡与垂直方向的静力平衡(hydrostatic balance)结合推导出来的。静力平衡即重力与浮力的平衡,而浮力就是垂直方向的压强梯度力。垂直方向的压强梯度力又反比于两个等压面之间的距离(厚度)。本句第一个 obey 用了过去分词 obeyed,因为这是 would 引导的虚拟语气。第二个用了一般现在式 obeys,因为是 while 引导的并列从句。

[10] 此段涉及几个重要概念:

(1)气象学中通常用位势高度(geopotential height)代替几何高度(geometric height),位势高度等于重力位势(geopotential)除以一个常数,即 $z=\frac{\Phi}{g_0}=\frac{1}{g_0}\int_0^z g(Z)\,\mathrm{d}Z$,$g_0=9.8\ \mathrm{m/s^2}$,$Z$是几何高度。等位势面(即等位势高度面)处处与重力垂直,是水平面。以位势高度为垂直坐标轴的坐标系叫做z坐标,以压强为垂直坐标轴的坐标系叫作 p 坐标。

(2) 等位势面上的压强与等压面上的位势高度是等效的,即 z 坐标与 p 坐标的高低关系相同。以下图为例,相同位势高度的两个点 A 和 B,若$p_A>p_B$,任意一个倾斜的等压面 $p=p_0$与两个点所在的垂直线的交点的 z 坐标分别为 z_{0A}和 z_{0B},则 $z_{0A}>z_{0B}$。因此,地转风在 z 坐标下沿等压线且正比于压强的水平梯度 ∇p;在 p 坐标下则沿等位势线且正比于位势高度的水平梯度 ∇z。

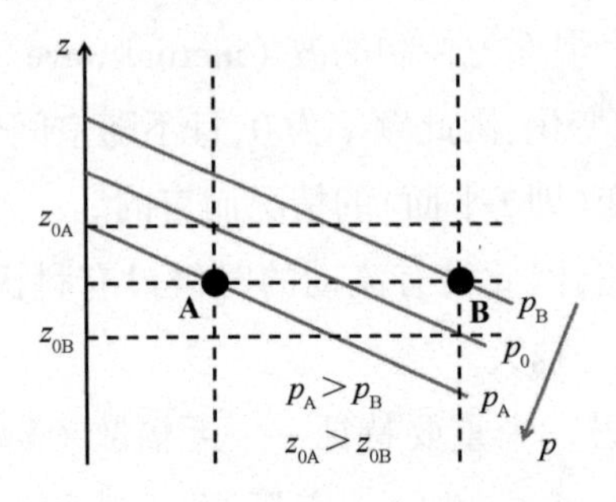

(3) 两个等压面之间叫作一个压强层(pressure layer),由于压强就是上方空气柱的总重量,那么压强层一旦确定则内部的空气质量不变。因此压强层的厚度越大则密度越小,由理想气体方程(ideal gas law)得知层内的温度就越高,所以等厚度线就是等温线。此处用虚温(virtual temperature)表示温度,虚温是把水汽(water vapor)的影响考虑进来之后的温度,又称密度温度(density temperature)。上述关系的数学表达即 $z_2-z_1=\frac{R\overline{T}_v}{g}\ln\left(\frac{p_1}{p_2}\right)$,其中 z_1、z_2 分别是压强层下界面和上界面的位势高度,p_1 和 p_2 是对应的压强,R 是气体常数(specific gas constant),$\overline{T}_v$ 是层内的平均虚温。这就是所谓的"气压—位势高度公式"(hypsometric equation,或译为测高公式)或"厚度公式"(thickness equation)。

(4) 热成风表示的是压强层上下两个面的地转风之差,由第(2)条得知它正比于两个面上的位势高度水平梯度之差($\nabla z_2-\nabla z_1$),后者等于位势高度之差的水平梯度[$\nabla(z_2-z_1)$,因为梯度算子是线性的],因此,根据测高公式,热成风也就正比于温度的水平梯度 $\nabla\overline{T}_v$。热成风和温度的关系与地转风和压强或位势高度之间的关系类似。

[11] 注意此处说的是地转风平行于温度梯度的情况,即垂直于等温线,与热成风垂直,这表明地转风随高度发生旋转。平行于温度梯度,即发生高低温区之间的跨温度流动。这样就会出现高温空气到达原来的低温区或反之,即温度平流(advection),数学表达式是 $u\cdot\nabla T$。有风、有温度梯度,且风有平行于温度梯度的分量,是使温度平流不为 0 的条件。

[12] 由于相同时间南、北半球季节相反,在谈论季节时,通常需要指明是针对哪个半球,用类似 boreal winter、austral summer 来表达。或者直接说明月份,如 DJF(12 月至

翌年2月)、JJA(6—8月)等。

[13] 水平温度梯度除了决定热成风,还是斜压不稳定性(baroclinicity instability)的一个重要成分。因此,西风急流最强处也是最不稳定处。由斜压不稳定性控制的斜压涡旋(baroclinic eddy)是中纬度大气的重要特征。

11. Mesoscale Eddy

Text

Mesoscale eddies are energetic, swirling, time-dependent circulations about 100 km in width, found almost everywhere in the ocean. Several modern observational techniques are used to profile these "cells" of current, and to describe briefly their impact on the physical, chemical, biological, and geophysical aspects of the ocean[1]. The ocean is turbulent. Viewed either with a microscope or from an orbiting satellite, the movements of ocean water shift and meander, and eddying motions are almost everywhere[2]. These unsteady currents give the ocean a rich "texture" (Figure Ⅰ-7). If you stir a bathtub filled with ordinary water, it will quickly be populated with eddies: whirling, unstable circulations that are chaotically unpredictable[3].

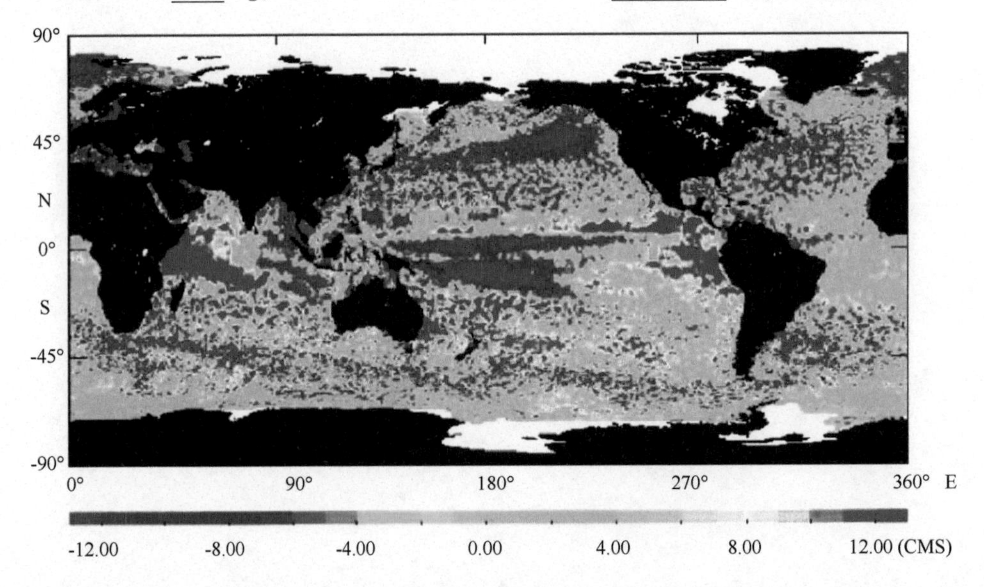

Figure Ⅰ-7: Sea surface elevation (cm) from the Topex-Poseidon and ERS satellites, 25 March 1998. The mean sea surface elevation for this time of year has been subtracted, so that only "anomalies" from normal conditions are shown. The speckled pattern shows mesoscale eddies almost everywhere. In addition, there are larger-scale patterns associated with El Nino (where the Equatorial Pacific has more level sea surface than normal) and large bands of high and low sea surface at middle latitudes. These may be associated with climate variability. [Source: Rhines, 2011]

One may think about the fineness of the pattern of fluid motion in analogy to the resolution of an image on a computer screen. In a bathtub, the fluid eddies have scales from about 1 mm to 1 m, hence spanning a thousandfold range of sizes. In the oceans, the smallest circulations are also a few millimeters in size, but the largest are of the order of 10,000 km in diameter: this represents a range in scale of about 10^{10} between the smallest and the largest[4]. There is thus room for many sizes of motion, each with a distinct dynamical nature: from tiny eddies that strongly feel viscosity, to "mesoscale eddies" that strongly feel the Earth's rotation, to great "gyres" of circulation filling entire oceans that feel also the curvature of the Earth[5]. At scales in between are also numerous types of wave motion.

Mesoscale eddies are whirling and localized yet they densely populate the ocean. Typically 100 km across, their size varies with latitude and other factors of their environment: energy level, bottom topography, and the nature of their generation[6]. Eddies need not necessarily be round, with circular streamlines. They are often generated by unstable meandering of an intense current like the Gulf Stream. In this case the waving deflection of the Stream is itself a form of latent eddy[7], which may eventually grow and "break" to form a circular eddy (as an ocean surface wave grows and "breaks" at a beach).

Eddies are important because they have so much kinetic energy, and because they can transport momentum and trace water properties[8]. They have deep "roots" that often reach 5 km or more downward, carrying energy and momentum to the seafloor[9]. They are responsible for the irreversible mixing of waters with different properties[10]. Mesoscale eddies are typically as energetic as the concentrated currents that give birth to them. They may owe their existence to several sources other than meandering of strong currents: for example, direct generation by winds or cooling at the sea surface; flow over a rough seafloor or past islands and coastal promontories; or generation by mixing or waves of smaller scale.

Before describing the "physics" of mesoscale eddies, we should discuss their "geography." A satellite image of the surface of the global ocean can be assembled from many orbits, as the Earth turns below[11]. A particularly basic measurement is that of the height of the sea surface. If ordinary waves are averaged out, we are left with a surface smooth to the eye yet varying by a meter or so relative to the "geoid," which determines the gravitational horizon[12]. Small variations in height of the sea surface correspond to small variations in pressure in the ocean below. Lines of constant pressure (isobars) are approximate lines of flow, or streamlines for horizontal circulation[13].

If one subtracts from this field the time-averaged sea surface height the result (Figure I-7) is a dramatic display of time-varying mesoscale eddies: they are nearly everywhere. By looking at animations of this field, one can see that many of the features move westward[14]. Mesoscale eddies (as currents at the ocean surface) are particularly apparent in Figure I-7 along the paths of intense, major ocean currents. These delineate the Antarctic Circumpolar Current round Antarctica, which has a "sawtooth" form, flowing south-eastward across the South Indian and Pacific Oceans, and jogging northward where it encounters major seafloor ridges or gaps. In each subtropical ocean there are western boundary currents like the Gulf Stream and Kuroshio, which are marked by time-dependent energy after they leave the coasts and flow eastward and poleward. The jet-like equatorial currents show fine-scale energy that is more related to meandering than to separated, circular eddies. The westward flow in the low subtropical latitudes develops eddies in mid-ocean. Altimetry measurement has a large "footprint" that misses eddies smaller than about 50 km in diameter. From direct measurements in the ocean we know that the texture of the circulation includes mesoscale eddies smaller than this, particularly at high latitudes[15].

Eddies formed by convection can be seen over most of the Earth, but they are particularly energetic in the cold, high latitudes[16]. Eddies formed directly by winds blowing on the sea surface are thought to occur widely, and yet the large size of wind patterns is not well-matched to the small, roughly 50 km diameter of mesoscale eddies. However, near ocean boundaries, wind forcing can have demonstrable effect on eddies. Larger-scale eddies more in tune with wind forcing take on the characteristics of Rossby waves[17]. Eddies formed by flow over an irregular seafloor are common, and can be identified in tracks of floats and drifters. These range across the spectrum of turbulent sizes, all the way to the grand scale setting the path of the Antarctic Circumpolar Current. Eddies formed by flow past an irregular coastline are seen widely. When fluid flows past a cylindrical island, it sheds a regular pattern of eddies with alternating rotation direction. This is known as a Karman vortex street. The interesting thing is that, when the same experiment is done in a laboratory, the regularity of the vortex street disappears as the flow is made stronger or the cylinder is made larger[18]. At the much greater scales of oceanic flow it is at first surprising that the turbulence regime is not encountered. The likely reason is that fluid motions restricted to two dimensions cannot fragment their energy into a full state of turbulence as readily as can a fluid with full freedom to move in all three dimensions[19]. Thus, your kitchen sink looks more turbulent than does a much larger ocean basin.

Eddies are crucial to the transport of heat, momentum, trace chemicals, biological communities, and the oxygen and nutrients relating to life in the ocean. They are also active in air-sea interaction, both through response to weather and in shaping the patterns of warmth that drive the entire atmospheric circulation[20]. As a member of the huge family of turbulent motions, eddies contribute to the stirring and mixing of the oceans, to the creation of its basic, layered density field, and to its general circulation. The fundamental physics of eddies is expressed in terms of its potential vorticity, which is a traceike property that "moves with the fluid." The distribution of potential vorticity can be turned into knowledge of the currents and fluid density variations[21]. The smallness and great energy of mesoscale eddies, the great thermal and chemical capacity of the oceans, and the slowness of the circulation conspire to challenge computer models, but rapidly increasing computer power is producing ever better representations of the ocean's fabric[22].

Vocabulary

mesoscale：(adj/n) 中尺度(的)

[meso-]：中

time-dependent：(adj) 时变的，与时间有关的

cell：(n) 单元

geophysical：(adj) 地球物理的

n：geophysics 地球物理学

texture：(n) 结构，构造，纹理，肌理

ordinary：(adj) 普通的，平常的

whirl：(vi/vt)(使)急转、飞转、回旋

chaotically：(adv) 混乱地

adj：chaotic

n：chaos 混乱，混沌

elevation：(n) 高度，海拔，高处

vt：elevate 举起，抬升，提高

fineness：(n) 精致/精细的程度、状态

adj：fine

in analogy to：与……相似/可类比

analogy：(n) 相似性，可类比性，可类比物

adj：analogous

resolution：(n)(模式、仪器等的)分辨率

thousandfold：(adj/n) 千倍(的)

[-fold]：倍，重，由若干部分组成

nature：(n) 本质，特征，实质

latent：(adj)(性质、状态)潜在的，隐藏的

~latent heat flux：(n) 潜热通量

kinetic energy：(n) 动能

kinetic：(adj) 与运动有关的，运动造成的

~potential energy：(n) 势能，位能

~internal energy：(n) 内能

trace：(vt) 追踪，追溯，追究；(n) 痕迹，踪迹

n：tracer 追踪者，示踪物

irreversible：(adj) 不可逆的

ant：reversible

promontory：(n) 岬角，海岬，海角

geography：(n) 地理，地理学

assemble：(vi/vt) 聚集，集合，组合，组装

n：assembly 聚会，集合，会议

geoid：(n) 大地水准面

correspond：(vi) 相似，相符，一致，对应

subtract：(vt) 减，减去

time-varying：(adj) 时变的

animation：(n) 动画

sawtooth：(adj) 有锯齿的，锯齿般的

syn：zigzag (n/adj/adv/vi) 锯齿状线条(的/地)

jog：(vi) 慢跑，缓慢移动

footprint：(n) 足迹，脚印，印记，影响，覆盖区，覆盖范围

convection：(n) 对流

vt/vi：convect 对流，通过对流传递(热、物质)

forcing：(n) 强迫(作用、因子)

demonstrable：(adj) 明显的，可论证的

in tune with：与……协调一致

tune：(n) 曲调；旋律 (vt) 调节，调整

Rossby wave：(n) 罗斯贝波

identify：(vt) 指认，认出，识别

drifter：(n) 漂浮物

grand：(adj) 宏伟的，宏大的

cylindrical：(adj) 圆柱体的

= cylindric

n：cylinder 圆柱体

shed：(vt) 射出，发出(光，涡旋等)

Karman vortex street：(n) 卡门涡街

vortex：(n) 涡旋，漩涡

pl：vortexes / vortices

regime：(n) 体制，体系，组织方法

restrict：(vt) 限制，控制，约束

fragment：(vt/vi)(使)裂成碎片，分裂；(n) 碎片，小块，片段

readily：(adv) 轻而易举地，容易地

trace chemical：(n) 痕量化学物质

community：(n) 社群，群落

air-sea interaction：(n) 海气相互作用

warmth：(n) 温暖，温热

general circulation：(n) 大洋环流，大气环流

potential vorticity：(n) 位涡

vorticity：(n) 涡度

conspire：(vi) 合作，协同，共同导致

fabric：(n) 基础结构，构造

Notes

[1] 此处 profile 作动词，意为“取剖面/廓线”，即测量的意思。describe briefly 也可以调整顺序为 briefly describe。

[2] eddy 作动词表示产生涡旋，而它的现在分词 eddying 被当成形容词使用时表示能产生或正在产生涡旋的。海洋运动基本上总是能产生或正在产生涡旋，但用于模拟海洋的数值模式由于受到分辨率的限制不一定能，因此用 eddying 修饰 model 或 simulation 可以表达“能做到分辨和模拟涡旋的”意思。

[3] populate 表示产生很多(涡旋)以充满(容器)。涡旋是不稳定的流，而且是混乱不可预测的。这里讲的浴缸里的涡旋尺度很小，实际上是湍流的涡旋，而不是本文后面主要讲的中尺度涡旋，因此为了区分，在中文里可叫作“涡流”。详见“Turbulent Flow”(Text 7) 的讲解。

[4] 英文里的幂次的读法：2 次方(平方)读作 squared，如 x^2 读作 x squared；3 次方(立方)读作 cubed；更高次方读作 to the power of，如 x^n 读作 x to the power of n。

[5] 此句对海洋不同尺度的描述十分深入。“tiny eddies”强烈感受到黏性或摩擦，但因为尺度很小，感受不到地转的效应，不需要考虑科氏力。从概念上看，这对应着低埃克曼数 (Ekman number，衡量黏性与旋转的相对重要性) 的情况。“mesoscale eddies”强烈感受到地球的旋转，也就是需要考虑科氏力，对应着低罗斯贝数。罗斯贝数用于衡量惯性与旋转的相对重要性，见“Submesoscale Processes”(Text 12) 注释[7]。而“great gyres”还能感受到地球的曲率 (curvature)，也就是必须考虑科氏力随纬度的变化(即 β 效应)，此时运动的水平尺度与地球半径的比不再是一个很小的数。

[6] 中尺度涡的大小受到罗斯贝变形半径 (Rossby radius of deformation) 的控制，而罗斯贝变形半径随纬度增大而减小。

[7] 由于大弯曲（large meander）还没有形成首尾相接的流环，因此还不是涡旋，而只是“潜在的涡旋（latent eddy）”。不过，有些语境里 eddy 一词更多表示类似“扰动（perturbation）”的意思，不一定指环形的涡旋，此时大弯曲就可能被包含在内。

[8] 涡动能（eddy kinetic energy）占总动能（total kinetic energy）的绝大部分。一般认为大部分涡旋都能将其中的水“捕获（trap）”在内部，并带着这些水移动，因此可以造成物质和属性的输运。这种涡旋称为 coherent eddy，其中 coherent 意为“一体的，整体的，结合的，连贯的”，表示涡旋内的水质点一直保持在一起。最近上述认识被挑战，有研究认为传统方法识别出的很多涡旋并不是 coherent eddy。

[9] 厚度在千米量级的中尺度涡，水平尺度是几百千米。因此，中尺度涡其实是扁平形状的，不是瘦高形状。这与整个海洋的“薄层流体”特征一致。

[10] 由于不同属性的水体混合后就无法再分开，这里用了“irreversible”（不可逆的）这个词。有研究表明，中尺度涡内部可以捕获内波，造成混合增强。

[11] 卫星运行一圈仅扫描了地球上的一条带（swath）。由于地球同时在旋转，经过多圈扫描，卫星可以基本覆盖整个地球。

[12] 中尺度涡的海面高度异常最大是几十厘米到 1 m 之间，水平尺度是几百千米，因此它的海面高度起伏用肉眼是无法看到的，现实中总是被海浪、潮汐等引起的海面起伏覆盖。大地水准面是衡量重力位势的参考面，也就是这里用比喻手法说的“gravitational horizon”。

[13] 流线（streamline）是流函数（stream function）的等值线，流函数的差代表了通过两点之间的流量大小，切线方向是流速的方向。流函数可以是任意符合此要求的物理量，而海洋和大气领域常用来表示流函数的量是压强。在海洋表层，假定密度在垂向上是均匀的，则海面高度也可以用作流函数。流函数只有在无辐散（non-divergent）情况下才能定义，而地转流恰好满足此要求。

[14] 中尺度涡与罗斯贝波（Rossby wave）一样，理论上都向西传播，但有时会出现偏转，甚至东传，如在背景东向平流很强时。罗斯贝波作为一种线性波动，它只传递运动形式而不传递物质；而中尺度涡通常具有较强的非线性，或可理解为非线性的罗斯贝孤立子（Rossby soliton），它可以携带水体运动。事实上，从海面高度场上得出的信号很难区分罗斯贝波和中尺度涡，因为他们都能引起相似的海面扰动，且二者常常相互吸引、重合、共同传播。通常认为罗斯贝波在传播过程中可逐渐非线性化转换为中尺度涡。

[15] 卫星的刈幅是固定的，决定了卫星高度计资料的分辨率在 25 km 左右，因此无法分辨小于 50 km 的中尺度涡。高纬度罗斯贝变形半径减小，涡旋较小，因此卫星高度计无能为力。现场的浮标（buoy）、潜标（submerged buoy）等手段可以观测到更小的涡旋。

[16] 对流直接生成的涡旋，即由于海面异常降温，使水体变重下沉，导致浮力损失（buoyancy loss）。下沉的水进入混合层以下，并沿等密面下沉，即潜沉（subduction），形成次表层涡旋（subsurface eddy），进而汇聚成为模态水（mode wa-

ter)。研究表明,模态水的潜沉很大一部分是由中尺度涡边缘处而来的。

[17] 由于风应力旋度导致的埃克曼抽吸可引起海面和温跃层起伏,此种信号在β效应的配合下产生向西传播的罗斯贝波,与中尺度涡类似,见注释[14]。

[18] 卡门涡街指在流体中安置圆柱形的阻流体,在特定条件下会出现不稳定的边界层分离,在阻流体下游的两侧会产生两道非对称排列的旋涡,其中一侧的旋涡顺时针方向转动,另一旋涡则反方向旋转,这两排旋涡相互交错排列,各个旋涡和对面两个旋涡的中间点对齐,如街道两边的街灯一般,故名涡街。该现象由匈牙利裔美国空气动力学家西奥多·冯·卡门(Theodore von Kármán)于1911年最先从理论上阐明而得名。卡门涡街的出现与否取决于雷诺数(Reynolds number,衡量惯性与黏性的相对重要性)。雷诺数超过47时才能出现卡门涡街,但大于188.5后变得不对称,而大于10^5后流体进入湍流状态,涡街消失。此处所讲的在实验室里使流增强或使圆柱增大,对应的就是增大雷诺数的情况。

[19] 海洋中经常能观察到卡门涡街现象,而卡门涡街是一种有序状态,因此才说"surprising"。这里的解释实际上是说,二维流动限制了流涡在垂向的拉伸,导致湍流不能自我维持。

[20] 已有研究证明中尺度涡与周围海洋之间的热量差异可以使大气底边界层发生响应,改变风速、云量、降雨等,这种影响可以到达大气边界层以上,甚至影响大气风暴活动,从而将影响传播至整个对流层和很远的水平距离。

[21] 位涡(potential vorticity)是海洋流动的重要特征量。在正压流体中定义为$q=\frac{\zeta+f}{H}$,其中,ζ是相对涡度(relative vorticity);f是行星涡度(planetary vorticity),也即科氏参数(Coriolis parameter);H是水深。在层结流体中定义为$q=\frac{\zeta+f}{g}N^2$[即埃特尔位涡(Ertel's potential vorticity)],其中,N是浮力频率。由于位涡既包含涡度这样的动力学量,又包含浮力频率这样的热力学量,它将海洋流动的动力学和热力学视角结合起来。位涡只能通过非绝热加热或冷却(diabatic heating/cooling)和摩擦过程改变,绝热、无黏运动中位涡守恒(conserve)。对于海洋,除了海面与大气之间有非绝热交换之外,海洋内部没有热源。因此在混合不是很强烈的情况下,可认为位涡是守恒的,这是一般规律,不仅限于中尺度涡。中尺度涡捕获水体并携带运动,该水体在运动过程中依旧保持位涡守恒。

[22] 要能很好地模拟中尺度涡,模式的分辨率至少需要达到10 km左右,这叫作eddy-resolving。如果分辨率是几十千米,则仅能分辨部分较大的涡旋,这叫作eddy-permitting。不同学者定义的此种标准可能不同,而且依赖于关注的区域[高纬度涡旋小,因为罗斯贝变形半径(Rossby radius of deformation)小]。目前eddy-resolving的数值模拟已经不太难实现。

12. Submesoscale Processes

Text

Introduction

Submesoscale currents (SMCs) are intermediate-scale flow structures crucial for understanding the intricate dynamics of the ocean. With horizontal scales of 0.1 to 10 kilometers, these currents manifest as density fronts, filaments, topographic wakes, and coherent vortices, occurring both at the surface and within the ocean's interior[1]. Dynamically distinct from mesoscale eddies and microscale turbulence, SMCs play a pivotal role in mediating energy transfer across scales and facilitating vertical material exchange between the surface and deeper ocean layers. The generation mechanisms of submesoscale processes in the ocean are closely tied to the interaction of various forces and scales of motion. These processes typically emerge at spatial scales of 0.1–10 km and time scales of hours to days, acting as a bridge between mesoscale eddies and smaller turbulent scales. Some key generation mechanisms include: frontogenesis[2], mixed-layer baroclinic instabilities[3], symmetric instability[4] and so on, owing to the forced and unforced processes.

Studying SMCs has historically been challenging due to their size—too small for many satellite remote sensing tools yet too large for traditional shipboard instruments. The advent of high-resolution numerical simulations in the 2000s, coupled with the increasing use of high-resolution two-dimensional surface imagery, marked a turning point (Figure Ⅰ-8)[5]. These advancements revealed organized material patterns associated with SMCs, sparking new theoretical work and targeted in-situ measurements[6]. In recent years, interest in SMCs has grown due to their potential impact on climate models and predictions. They are now recognized as integral to initializing weather and climate models and for parameterizing subgrid-scale processes in ocean general circulation models. Furthermore, SMCs are critical for understanding the transport and dispersion of pollutants, nutrients, and heat within the ocean.

Due to their transient and intermittent nature, studying submesoscale processes remains a significant challenge. Ongoing research aims to refine numerical models, develop innovative observation tools, and unravel the interactions between SMCs and other scales of ocean motion. Recognized as vital components of the global ocean system, SMCs are a focal point of oceanographic research, promising to deepen our understanding of the ocean processes.

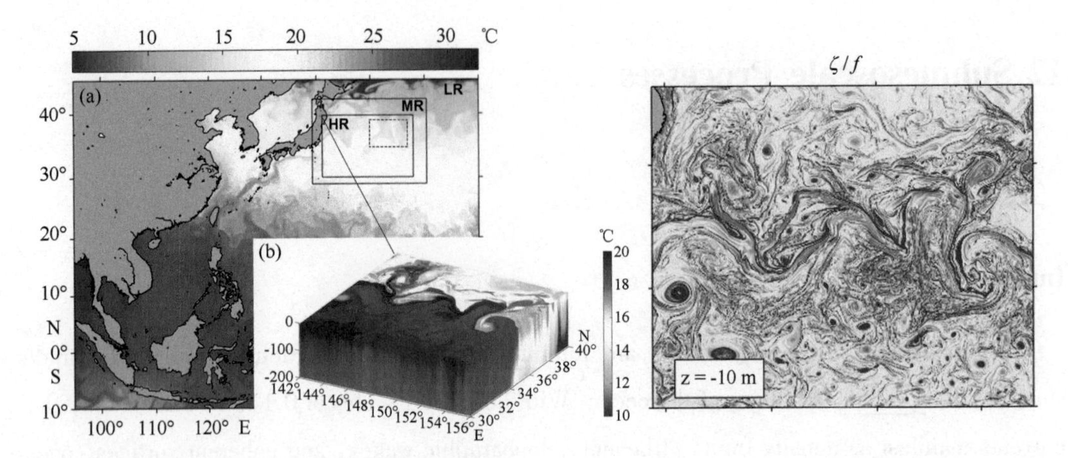

Figure Ⅰ-8: Submesoscale features in high-resolution simulations. Shown are (left) temperature and (right) Rossby number defined as relative vorticity divided by planetary vorticity[7]. The regions marked as LR, MR, and HR denote the three-level nested simulation domain[8].

Role in Energy Cascade[9]

Previous research on the oceanic energy cascade has predominantly focused on large-scale climatological forcing, the mesoscale as the primary reservoir of kinetic energy with a dominant inversecascade, and the forward cascade associated with internal waves and microstructure turbulence. However, recent studies have emphasized the critical role of submesoscale processes and their instabilities as a dynamic pathway for energy transfer between large-scale motions and dissipation scales. These processes are thought to facilitate downscale energy transfer through mechanisms such as barotropic kinetic energy conversion or baroclinic pathways linked to potential energy transfer. For instance, surface-trapped modes can easily drive submesoscale flows characterized by Rossby number and Richardson numbers on the order 1 via various submesoscale instabilities[10]. These include mixed layer instability, symmetric instability—drawing energy from geostrophic shear production, lateral shear instability, barotropic energy conversion, and centrifugal instability.

Ecological Impacts and Mechanisms

Submesoscale currents play a vital role in transporting nutrients from the deep ocean into the euphotic zone, where they support phytoplankton growth[11]. In nutrient-depleted regions such as the subtropical gyres, these upward nutrient fluxes can locally enhance phytoplankton growth rates and biomass[12]. While the uplift of nutrient-rich layers in the core of cyclonic eddies (eddy pumping) is a well-known mechanism, submesoscale processes significantly contribute to vertical tracer exchange. The uplift of isopycnals into the sunlit zone within the eddy core enhances light availability for photosynthesis and creates sloping isopycnal surfaces, serving as a crucial vertical pathway for material transport. Compared to diapycnal mixing, isopycnal submesoscale stirring is the dominant mechanism for disrupting eddy coherence, dispersing phytoplankton and nutrients on timescales of hours to days. This process efficiently delivers nutrients into the euphotic zone, sustaining high phytoplankton biomass in the upper eddy core[13]. Feedback mechanisms, such as the influence of phytoplankton concentration on light penetration, further modulate primary production. Such interactions exemplify the coupling of physical and biological processes in the ocean[14]. Given the widespread presence of mesoscale eddies, these findings likely apply to many other ocean regions, highlighting the global importance of mesoscale and submesoscale dynamics in shaping marine ecosystems. Furthermore, these processes play a key role in supporting marine biodiversity, emphasizing their broader ecological significance.

Effects on Vertical Heat Transport

The upper ocean heat budget strongly influences sea surface temperature and air-sea heat exchange, thereby shaping the climate system[15]. While mesoscale eddies play a critical role in lateral (primarily poleward) heat transport, recent studies highlight the greater efficiency of submesoscale processes in driving vertical heat transport within and below the mixed layer. This enhanced vertical transport arises from intensified vertical flows, typically ranging from 10 to 100 m day^{-1}. In energetic regions such as the Kuroshio Extension, the Gulf Stream, and the Antarctic Circumpolar Current, submesoscale processes are widespread and vigorous. Frontogenesis, resulting from the convergence of background flows, triggers ageostrophic secondary circulations as fronts intensify. These circulations promote upwelling on the warmer side of fronts and downwelling on the colder side, yielding a net upward vertical heat transport[16]. High-resolution numerical simulations reveal submesoscale-induced vertical heat fluxes reaching up to 100 W m^{-2}, far exceeding those driven by mesoscale processes. Below the mixed layer, large vertical velocities are primarily generated by unbalanced motions. Potential drivers of this enhanced vertical velocity include frontogenesis, along-isopycnal flows, and high-mode internal gravity waves[17]. These mechanisms, linked to ageostrophic frontal dynamics near regions like the Kuroshio, play a significant role in facilitating vertical heat fluxes between the mixed layer and the ocean interior.

Research Interests and Future Challenges

SMCs are challenging to observe using traditional methods. Emerging technologies and approaches are crucial for bridging this observational gap. Key priorities include the expanded use of autonomous platforms, such as gliders, and wave-powered vehicles, which can provide high-resolution vertical and horizontal observations. These platforms are particularly effective in capturing the three-dimensional structure of submesoscale processes. Drifting sensor arrays, including clusters of Lagrangian drifters, are also being increasingly deployed to track horizontal dispersion and material transport at submesoscale resolutions[18].

Remote sensing advances, including improved satellite-derived sea surface temperature, salinity, and velocity fields, are essential for detecting submesoscale features. Novel methods, such as synthetic aperture radar, high-resolution optical sensors and Surface Water and Ocean Topography (SWOT) mission[19], can reveal surface signatures of submesoscale processes like fronts and filaments. However, combining surface remote sensing with in-situ measurements remains a major challenge due to the difficulty of resolving subsurface dynamics from surface observations.

Another focus is on developing integrated observational frameworks that combine multiple platforms and data sources. These frameworks, supported by machine learning and data assimilation techniques[20], can enhance the resolution and coverage of submesoscale observations while improving the interpretation of complex datasets.

Continued efforts in high-resolution modeling will also inform observational strategies by identifying key regions, timescales, and variables to target. Enhanced observations are expected to improve our understanding of submesoscale processes' role in energy transfer, biogeochemical cycling, and climate-relevant exchanges, ultimately advancing global ocean modeling and ecosystem predictions.

Vocabulary

submesoscale:(n) 亚中尺度

intermediate:(adj) 中间的,过渡的

intricate:(adj) 错综复杂的,复杂精细的,盘根错节的

manifest:(vt) 表明,显示,证实;(adj) 显然的,清楚的,明显的

front:(n) 锋面

filament:(n) 丝,纤维

wake:(n)(船、飞机、山脉等的)尾流,尾迹

coherent:(adj) 有条理的,连贯的;说话条理清晰的,易于理解的;团结一致的,凝聚的;(波)相干的,相参的;黏着的,粘连的

microscale:(adj) 微尺度

pivotal:(adj) 关键的,起中心作用的

mediate:(vt) 连接,沟通,传递,传达,引起,形成

facilitate:(vt) 使(行动、过程)更容易,使便利

mechanism:(n) 机制,机理,机器,机械装置

emerge:(vi) 出现,浮现,出来

frontogenesis:(n) 锋生(作用)
remote sensing:(n) 遥感
  remote:(adj) 远程的,遥远的
high-resolution:(adj) 高分辨率的
numerical:(adj) 数值的
coupled:(adj) 耦合的
  vt:couple
  n:coupling
  n:coupler 耦合器
reveal:(vt) 揭示,展现,使显露
target:(vt) 把……作为目标,瞄准;(n) 目标,靶子
  adj:targeted 有目标的,定向的
initialize:(vt) 初始化
  n:initialization
critical:(adj) 严重的,关键的,决定性的,临界的
intermittent:(adj) 间歇的,断断续续的,不持续的
ongoing:(adj) 持续的,正在进行的
innovative:(adj)(观念、产品等)革新的,创新的
  n:innovation 革新,变革,创新,新产品,新观念,新方法
unravel:(vt) 弄清,解决,阐明
focal point:(n) 焦点
  focal:(adj) 焦点的,最关注的
Rossby number:(n) 罗斯贝数
relative vorticity:(n) 相对涡度
planetary vorticity:(n) 行星涡度
nested:(adj) 嵌套的
  vt:nest
domain:(n) 区域,范围,领域
cascade:(n)(一个引发下一个的)连续阶段,系列过程,串级,级联;(vt/vi)(水)倾泻,流注,大量落下,垂下,连续传递,传授,使(装置,物品)串联
predominately:(adv) 占优势地,主导地,支配性地
  vi:predominate
inverse cascade:(n) 逆向串级/级联
forward cascade:(n) 正向串级/级联
microstructure:(n) 细结构
pathway:(n) 途径,道路,通道
downscale:(adj) 降尺度的;(vi) 降尺度(指向小尺度转移或使能分辨的最小尺度降低)
  n:downscaling 降尺度
surface-trapped mode:表面捕获模态
Richardson number:(n) 理查森数
lateral:(adj) 侧面的,侧向的
euphotic zone:(n) 真光层,透光层
  euphotic:(adj) 透光的
deplete:(vt) 使减少,使衰竭
  adj:depleted
biomass:(n) 生物量
uplift:(n/vt) 提起,举起,上升
photosynthesis:(n) 光合作用
  [photo-]:光,照相
  synthesis:(n) 综合,合成
    pl:syntheses
    adj:synthetic 合成的,人造的
feedback:(n) 反馈
  vi:feed back (to)
penetration:(n) 穿透,刺穿
  vt/vi:penetrate
modulate:(vt) 调节,调整,调制
  n:modulation
exemplify:(vt) 是……的例证,作为……的例证

highlight：(vt) 使显著，使突出，强调；(n) 最显著的部分
biodiversity (n.) 生物多样性(维持着生态环境平衡的大量各种生物的共存)
budget：(n) 收支，方程中各项的平衡
secondary circulation：次级环流
unbalanced motion：非平衡运动
glider：(n)(水下)滑翔机
sensor：(n) 传感器
array：(n) 数组，阵列
cluster：(n) 一丛，一簇，一串，集群
Lagrangian：(adj) 拉格朗日的
~Eulerian：(adj) 欧拉的
~Laplacian：(adj) 拉普拉斯的
~Gaussian：(adj) 高斯的
synthetic aperture radar：合成孔径雷达
aperture：(n) 开孔，缝隙
optical：(adj) 光学的
n：optics 光学
mission：(n) 任务，使命，可特指一次航天或飞行任务
signature：(n) 识别标志，鲜明特征，影响印记
framework：(n)(系统、概念或文章的)基本框架，体系
machine learning：(n) 机器学习
interpretation：(n) 解释，说明，阐释
vt：interpret
dataset：(n) 数据集
可拆分写作 data set
biogeochemical：(adj) 生物地球化学的
n：biogeochemistry
ultimately：(adv) 最终，最后
adj：ultimate

Notes

[1] 亚中尺度过程 (submesoscale processes) 和亚中尺度流 (submesoscale currents) 是相同的概念，其中"亚"字也可译为"次"字。所谓"中间尺度"(intermediate-scale)表示其介于中尺度(如涡旋)和小尺度(如朗缪尔环流和湍流)之间。亚中尺度过程的水平尺度特别是其上限有不同的定义，此处采取的定义为 100 m 至 10 km。根据不同的研究目的，可见有些作者将亚中尺度的上限定为中尺度的下限 (~50 km)；或定义为罗斯贝变形半径 (Rossby radius of deformation)，即不再固定，而是与纬度和海洋层结有关。亚中尺度过程的时间尺度通常为几小时至几天，其上限同样没有固定标准，通常取为 2~10 d。这里列举了亚中尺度过程的多种形式，即锋面 (front)、丝 (filament)、尾流(wake)、涡 (vortex)。

[2] 锋生过程 (frontogenesis) 是指温度、盐度或密度等物理属性的水平梯度在一定时间内增强，以致产生锋面 (front) 的过程。它通常发生在大尺度流场中，由于流体动力学过程，这些梯度被拉伸或压缩，从而形成锋面或增强现有的锋面。海洋中，在西边界流(如湾流、黑潮)或环流中的锋区，流场的强剪切和水团交汇会导致密度梯度增强。在中尺度涡的边缘，涡旋流动也会通过形变场(strain field)触发锋生。

[3] 混合层斜压不稳定 (mixed-layer baroclinic instabilities) 是亚中尺度过程中的重要机

制,发生在海洋表层混合层(mixed layer)内的强密度梯度区域。这种不稳定性驱动了能量从大尺度向小尺度的传递,并导致显著的垂直和水平物质输送。锋面区域或其他密度梯度较强的区域容易触发斜压不稳定性,不稳定性产生的扰动能引发强烈的横向和垂向速度场。通过混合层斜压不稳定性,势能(potential energy,由密度梯度储存)转化为动能(kinetic energy,以流动形式释放)。垂直输送过程会在混合层内显著改变密度场。

[4] 对称不稳定性(symmetric instability)是海洋动力学中的一种重要不稳定机制,通常发生在锋区或其他存在强水平密度梯度和速度剪切的区域。这种不稳定性是亚中尺度动力过程的关键之一,能够有效驱动垂直和横向的混合和物质输送。对称不稳定性能够有效地在混合层或锋区增强垂直混合,将深层的营养盐输送至表层,同时将表层热量带入更深处。在水平密度梯度较强的区域,它通过横向的对流运动增强物质的扩散,促进水团之间的交换。对称不稳定将背景流场的动能转化为小尺度湍流耗散,是正向能量串级的重要驱动力。

[5] 针对不同历史阶段、不同研究目的,高分辨率(high-resolution)一词可以有不同的含义,即对“高”的定义不是一成不变的。显然,此处指为了模拟或观测亚中尺度过程而必须达到的分辨率,至少需要千米至百米量级。

[6] 高分辨率模式和观测可通过观察结构化的物质分布形态(organized material pattern)来获知亚中尺度过程的存在。targeted in-situ measurement 也可作 targeted observation,译为“目标观测”,即事先确定关键观测点位的方法,而观测点位的确定需要通过数值模式确定,找到最可能出现关键过程之处,或者最能提高模式的预报效果之处。

[7] 罗斯贝数(Rossby number) $Ro=\frac{U}{Lf}$,其中,U 是运动的特征流速,L 是运动的特征尺度,f 是科氏参数。此数表示惯性与科氏力的比,大尺度和中尺度运动的 $Ro \ll 1$,即运动明显受到科氏力的作用;而亚中尺度的 Ro 接近 1 甚至更大,表示非地转运动变得明显。罗斯贝数的另一种定义是 $Ro=\frac{\zeta}{f}$,即相对涡度(relative vorticity)与行星涡度(planetary vorticity)的比,物理意义相同。

[8] 数值模式为了在关心的区域获得最高的分辨率,同时保证为此区域尽量提供合理的边界条件(boundary condition),常采用多级嵌套(multi-level nesting)的策略,分辨率从最外层的大区域开始向研究区域逐级递增。此过程又可叫作降尺度(downscaling)。

[9] 关于能量串级或级联(energy cascade)的解释见“Turbulent Flow”(Text 7)注释[11]。在中尺度及以上,正向串级和逆向串级都常见;而在亚中尺度及以下,能量的正向串级显著增强,即中尺度能量通过亚中尺度过程传递给湍流,最终耗散,因此,

可认为亚中尺度过程是中尺度能量耗散的途径。目前尚不明确亚中尺度能量是否能逆向串级反馈给中尺度过程。

[10] 理查森数（Richardson number）$R_i = \frac{g}{\rho}\frac{\partial\rho/\partial z}{(\partial u/\partial z)^2}$，又称梯度理查森数（gradient Richardson number），即层结与剪切的比。当密度差与密度的绝对值相比很小时（即布辛尼斯克近似，Boussinesq approximation），往往用约化重力 $g' = g\frac{\Delta\rho}{\rho}$ 代替 $\frac{g}{\rho}$，此时的理查森数叫作密度理论理查森数（densitometric Richardson number）。由于层结越强则越能阻止不稳定的发展，而剪切对不稳定起驱动作用，因此，理查森数衡量此两种作用的相对重要性。大尺度和中尺度运动的理查森数远大于 1，而亚中尺度的理查森数接近 1 甚至更小。

[11] 浮游植物的生长（phytoplankton growth）是海洋生态系统中的基本生物过程，涉及浮游植物通过光合作用（photosynthesis）将无机物质（如二氧化碳、氮、磷等）转化为有机物质的能力，即初级生产力（primary productivity）。这一过程对海洋生物化学循环（biogeochemical cycle）、食物链动态（food chain dynamics）以及全球碳循环（global carbon cycle）至关重要。亚中尺度过程通过增强垂直输送或局部混合改变真光层内的营养盐分布，从而影响浮游植物的生长。

[12] 副热带流圈是缺乏营养盐的区域（nutrient-depleted region），因为此区域是辐聚下沉的区域，而营养盐自海表、陆地、河流等进入海洋后向海底沉积聚集，故此区域上层有光照的区域缺乏营养盐，导致初级生产力低，成为所谓的“海洋沙漠”（ocean desert）。通量即单位时间内穿过单位面积的某物理量的值，因此，可用该平面的法向流速与此物理量的乘积表示。如穿过南北走向的一个单位面积（即东西向）的热通量为 $c_p uT$（c_p 是比热容），动量通量为 uu。通量的瞬时值通常意义不大，而更关注一段时间内的平均通量，如 $c_p\,\overline{u'T'}$，其中撇号（prime）表示去掉平均值之后的扰动值，横线（bar）表示平均。此处所述为向上的营养盐通量，即所关注的物理量为营养盐含量，参考面为单位面积的水平面。亚中尺度过程引起的强烈的垂向运动和混合都能加强营养盐的垂向输送，从而在向上输送处形成条带或丝状的高初级生产力区。

[13] 涡旋泵（eddy pumping）是指气旋式中尺度涡引起的等密面上拱（isopycnal doming）导致下层的部分营养盐随之抬升进入真光层，促进浮游植物的光合作用。同时，由于海洋水团主要沿等密面运动，因此，涡旋中心处等密面的上拱意味着物质输运出现了垂向分量。而根据此段原文，亚中尺度过程引起强烈的沿等密面混合，借此在中尺度涡的基础上进一步促进营养物质的垂向输运。

[14] 反馈机制（feedback mechanism）是指受影响的一方反过来影响对方的过程，即形成相互作用的过程。浮游植物浓度显然受光照的穿透特性（light penetration）的影

响,而它又可以通过影响海水的透明度(seawater transparency)和水色(water color)影响光照衰减曲线(light extinction/attenuation curve),从而反过来影响海水温度。

[15] 上层海洋热收支(heat budget)即温度方程中的各项,它们的大小和分布直接影响海洋温度。海洋中的热量通过垂直混合输送到深海,进而影响海洋的温度结构。通过亚中尺度过程引起的垂直热量输送,热量能够更迅速地从海洋表层(尤其是热带地区)传输到深层,并能影响热盐环流,进而影响大尺度的热量和物质交换,这在调节全球气候中起着关键作用。另外,亚中尺度过程的强烈垂直热量输送也会影响海洋与大气之间的热量交换。海洋表层的热量直接影响大气的温度、湿度和风力等气候变量。在热带和极地地区,亚中尺度过程对热量的快速垂直输送可能会增强或减弱当地的海气热量通量,激发更多的极端热通量事件,从而影响大气对流事件等的发生。

[16] 锋生过程引起的非地转次级环流(ageostrophic secondary circulation)是指在锋面即强水平温度梯度形成过程中,由于局部的非地转力(即与地转平衡不一致的力)的作用产生的次级环流,即叠加在主要环流基础上的弱一些的环流。此次级环流在锋面的暖侧产生上升流,而在冷侧产生下降流,二者都对应向上的热输运。

[17] 强烈的垂向运动是亚中尺度过程的特征之一,导致强烈的垂向热交换。此段指出垂向运动主要是由非平衡过程(unbalanced motions)引起的,其中的“平衡”特指水平方向的地转平衡(geostrophic balance)和垂直方向的静力平衡(hydrostatic balance)。事实上,静力平衡本身就是通过在垂向动量方程中忽略垂向速度得到的。此种运动形式包括亚中尺度锋生过程和高模态内波(high-mode internal gravity wave)等。内波的模态数(mode number)即在垂向水柱内波的个数,模态数高则内波波长短。高模态内波的尺度与亚中尺度存在重合,但由于机制不同,通常不把它算为亚中尺度过程。

[18] 此处提到几种先进的观测仪器,即水下滑翔机(glider 或 underwater glider)、波浪能无人船[wave-powered(unmanned/autonomous)vehicle]、拉格朗日漂流浮标群(clusters of Lagrangian drifters)等。

[19] Surface Water and Ocean Topography(SWOT)任务是由美国国家航空航天局(NASA)、法国国家空间研究中心(CNES)、加拿大空间局(CSA)和英国空间局(UK Space Agency)共同实施的一个全球性海洋研究卫星任务。SWOT 任务的主要目的是通过高分辨率的海面高度(sea surface height)测量,提供关于海洋动力学、海洋表层流动、海洋热量输送等方面的全新数据,从而为海洋研究提供重要的支持。SWOT 卫星于 2022 年 12 月发射,搭载的仪器对星下点(nadir)两侧 50 km 宽的两个平行观测带(swath)进行扫描测量,两个观测带之间有 20 km 的空缺。SWOT 的空间分辨率可达 2 km,星下点相邻两次经过同一位置的最长时间(即采样

周期 sampling period)为 21 d,刈幅相邻两次覆盖同一位置的平均时间,即重现时间(revisit time)为 11 d。

[20] 机器学习（machine learning）是人工智能（artifitial intelligence）领域的一个分支,专注于构建和实现算法（algorithm),使机器能够从数据中学习并做出预测。机器学习使用各种方法,包括回归分析（regression analysis)、聚类分析（cluster analysis）等传统统计方法,以及随机森林（random forest)、深度学习（deep learning）等新兴方法。深度学习的基础是深度神经网络（deep neural network),而卷积神经网络（convolutional neural network）是深度神经网络的一种。数据同化（data assimilation）是一种在数值模型的动态运行过程中融合新的观测数据的方法,可以纠正模式误差,使模式运行结果更贴近观测。见“Numerical Modelling”（Text 16）注释[22]。

13. Thermohaline Circulation

Text

The thermohaline circulation is that part of the ocean circulation induced by deep-reaching convection driven by surface buoyancy loss in polar latitudes[1]. The phrase "thermohaline circulation" is widely used but not precisely defined (see Wunsch, 2002). It means different things to different people. Perhaps its most literal interpretation is the circulation of heat and salt in the ocean and thus involves both wind-driven and buoyancy-driven circulation. Here, however, we adopt its more common, narrow usage, to mean the circulation induced by polar convection[2].

Deep convection in the ocean is highly localized in space and only occurs in a few key locations; in particular, the northern North Atlantic Ocean and around Antarctica. However, the response of the ocean to this localized forcing is global in scale. Giant patterns of meridional overturning circulation are set up that cross the equator and connect the hemispheres together (Figure I -9). Unlike the faster wind-driven circulation, which is confined to the top kilometer or so, the thermohaline circulation plays a major role in setting properties of the abyssal ocean[3]. Both wind-driven and buoyancy-driven circulations play an important role in meridional ocean heat transport. However, because of the very long timescales and very weak currents involved, the thermohaline circulation is much less well observed or understood.

Deep-reaching convection often creates very deep mixed layers that can ventilate the abyss[4]. Although deep convection is itself driven by surface buoyancy loss, there is no direct relationship between the pattern of air-sea buoyancy forcing and the pattern of mixed layer depth. This is because the strength of the underlying stratification plays an important role in "preconditioning" the ocean for convection[5]. The deepest mixed layers are seen in the polar regions of the winter hemisphere and are particularly deep in the Labrador and Greenland Seas of the North Atlantic, where they can often reach depths well in excess of 1 km. Here the ambient stratification of the ocean is sufficiently weak and the forcing sufficiently strong to trigger deep-reaching convection and bring fluid from great depth into contact with the surface[6]. Note that deep mixed layers are notably absent in the North Pacific Ocean. Waters at the surface of the North Pacific are relatively fresh and remain buoyant even when cooled. Deep mixed layers over wide areas of the southern oceans in winter are also observed, but they are considerably shallower than their counterparts in the northern North Atlantic[7].

Evidence from observations of mixed layer depth and interior tracer distributions suggest that convection reaches down into the abyssal ocean only in the Atlantic (in the Labrador and Green-

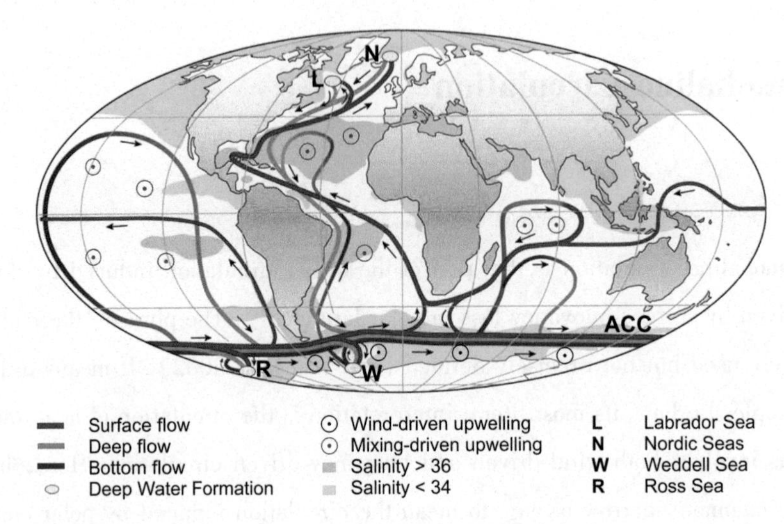

Figure Ⅰ-9: Strongly simplified sketch of the global overturning circulation system. In the Atlantic, warm and saline waters flow northward all the way from the Southern Ocean into the Labrador and Nordic Seas. By contrast, there is no deepwater formation in the North Pacific, and its surface waters are fresher. Deep waters formed in the Southern Ocean become denser and thus spread in deeper levels than those from the North Atlantic. Note the small, localized deepwater formation areas in comparison with the widespread zones of mixing-driven upwelling. Wind-driven upwelling occurs along the Antarctic Circumpolar Current. [Source: Kuhlbrodt et al., 2007]

land Seas) and also in the Ross Sea and Weddell Sea[8]. These sites, despite their small areal extent, have global significance in setting and maintaining the properties of the abyss. They are thought to play a major role in climate variability[9]. Observations suggest that there are certain common features and conditions that predispose these regions to deep-reaching convection. First, there is strong atmospheric forcing because of thermal and/or haline surface fluxes. Thus open ocean regions adjacent to boundaries are favored, where cold, dry winds from land or ice surfaces blow over water inducing large sensible and latent heat and moisture fluxes[10]. Second, the stratification beneath the surface mixed layer is weak, made weak perhaps by previous convection. And third, the weakly stratified underlying waters are brought up toward the surface so that they can be readily and directly exposed to buoyancy loss from the surface. This latter condition is favored by cyclonic circulation associated with density surfaces, which "dome up" to the surface, drawn upward by Ekman suction over subpolar gyres. In places where deep convection is occurring, weak vertical buoyancy gradients are observed and isopycnals dome up toward the surface[11].

The time-mean abyssal flow in the ocean is so weak that it cannot be measured directly. However abyssal circulation, and the convective processes forcing it, leaves its signature in the distribution of water properties, from which much can be inferred[12]. Water masses modified by deep convection are tagged with temperature and salinity values characteristic of their formation region, together with other tracers, such as tritium from the atomic weapon tests of the 1960s and chlorofluorocarbons (CFCs) from industrial and household use. Tracers can be tracked far from their formation region, revealing interior pathways through the ocean[13].

From the zonal-average sections of temperature and salinity across the Atlantic Ocean, we see three distinct layers of deep and abyssal ocean water, fed from different sources. Sliding down from the surface in the Southern Ocean to depths of 1 km is Antarctic Intermediate Water (AAIW)[14], with low salinity (34.4 psu) and slightly lower temperature than water immediately above and below. This water appears to originate from about 55°S in the circumpolar ocean. At a depth of 2 km or so—indeed, filling most of the Atlantic basin—is North Atlantic Deep Water (NADW)[15], with high salinity (34.9 psu) originating in high northern latitudes, but identifiable as far south as 40°S and beyond. At the very bottom of the ocean is the Antarctic Bottom Water (AABW), less saline but colder (and denser) than NADW. Together, these give us a picture of a multilayered pattern of localized sinking and horizontal spreading of the dense water.

Another useful tracer of the circulation is dissolved oxygen. Surface waters are near saturation in oxygen content (in fact, they are slightly super-saturated). As the water leaves the surface (the source of oxygen), its oxygen content is slowly used up by biological activity. Hence, oxygen content gives us a sort of clock by which we can get a feel for the "age" of the water (i.e., the time since it left the surface); the lower the content, the "older" the water. In the Atlantic, water in the deep ocean shows a progressive aging from north to south, implying that the dominant source is in the far north. However, the water is generally "young" (oxygen saturation > 60%) everywhere except at depths shallower than 1 km in low latitudes, where "old" water is (we infer) slowly upwelling from below. That water is relatively young near the Antarctic coast, around 40°–50°S and, especially, in high northern latitudes, is evidence that surface waters are being mixed down in these regions of the Atlantic[15]. In contrast, the Pacific Ocean cross-section shows young water only near the Antarctic. Deep water in high northern latitudes has very low oxygen content, from which we infer that there is no sinking of surface waters in the North Pacific.

Water whose properties are set (oceanographers use the term "formed") at the source regions of deep-reaching convection must spread out before slowly upwelling back to the surface to complete the circuit of mass flow. Estimates of the strength of the major source, NADW, are about 14 Sv. Using this source rate we can make several estimates of parameters indicative of the strength of the circulation. The area of the Atlantic Ocean is about 1014 m^2. The depth of the ocean ventilated by the surface sources is perhaps 3 km. So one estimate of the time scale of the overturning circulation is t = ocean volume/volume flux = 1014 $m^2 \times 3\times10^3$ m / 1.4×10^7 $m^3 \cdot s^{-1} \approx 700$ y. The net horizontal flow velocity in the deep ocean must be about v = volume flux / depth / width = 1.4×10^7 m^3 s^{-1}/ 3×10^3 m × 5×10^6 m ≈ 10^{-3} m · s^{-1}. If compensating upwelling occupies almost all of the ocean basin, the upwelling velocity must be about w = volume flux / area of ocean = 1.4×10^7 $m^3 \cdot s^{-1}$/ 10^{14} $m^2 \approx 4$ m · y^{-1}, several tens of times smaller than typical Ekman pumping rates driven by the wind[17].

Thus the interior abyssal circulation is very, very weak, so weak that it is all but impossible to observe directly. Indeed progress in deducing the likely pattern of large-scale abyssal circulation has stemmed as much from the application of theory as direct observation[18].

Vocabulary

literal:(adj) 按字面最常用或最基本意义理解的

adopt:(vt) 采用,采纳

giant:(adj)(体积、力量)巨大的

meridional overturning circulation:经向翻转流

abyssal:(adj) 深海的,海底的

n:abyss

~hadal:(adj) 超深渊的

ventilate:(vt) 使通风,使与空气接触

n:ventilation

underlying:(adj) 位于……之下的

vt:underlie

ant:overlying

precondition:(vt) 使按某方式发生,制约,预处理;(n) 先决条件,前提

in excess of:超过

ambient:(adj) 环境的,周围的,背景的

trigger:(vt) 引发,引起,激发;(n) 启动装置

notably:(adv) 尤其,明显地,显著地

adj:notable

counterpart:(n) 对应的人或物

areal:(adj) 面积上的

n:area

variability:(n) 可变性,易变性,变率

predispose:(vt) 使……倾向于某种状态或行为

favor/favour:(vt) 赞同,喜爱,偏爱,有利于,有助于

sensible:(adj) 能被感知到的

sensible heat flux:感热通量

latter:(adj)(两者中的)后者

ant:former

water mass:(n) 水团

tag:(vt) 给……加上标签

tritium:(n) 氚

chlorofluorocarbon:(n) 氟氯烃

originate:(vi) 源自,来自,起源于

saturation:(n) 饱和状态,饱和度

vt:saturate 使浸透,使饱和

super-saturated:(adj) 过饱和的

cross-section:(n) 截面,断面,切面

Sv:流量单位 Sverdrup, 1 Sv = 10^6 m^3/s

compensate:(vt) 补偿

all but:(adv) 几乎,接近

= almost

deduce：(vt) 推论，演绎出(事实、结论)

Notes

[1] 由于浮力与水体和环境的密度差成正比，密度增大对应浮力损失，而密度增大可由降温和增盐两个过程引起。高纬度海域降温过程和通过结冰析盐造成的增盐过程都较为明显，因此易发生浮力损失。

[2] 此处是说"热盐环流"，字面上可以理解为热量和盐分的流动，但一般指由于热量和盐分变化引起的浮力损失导致的深对流，以及由此驱动出的环流，包括深对流和深层水平流，以及将深层水翻到上层的机制。现代观点认为，单纯密度变化不足以驱动出全球范围的闭合环流，还是需要配合风生环流和垂向混合过程。因此，目前有些人通常更倾向于用"翻转环流 (overturning circulation)"来代替"热盐环流"。

[3] abyssal 或 abyss 一般指几千米甚至更深的"深渊层"，通常比 deep 深，比 bottom 浅，但没有明确的划分标准。hadal 一词通常指 6000 m 以深的超深渊层，主要指海沟 (hadal trench)。

[4] 由于混合层内的混合十分强烈，整层水体性质几乎均匀，原本远离海面的海水有机会被带至表层与大气接触，即所谓的"通风 (ventilation)"。

[5] buoyancy forcing 意为浮力强迫，也就是大气通过海-气之间的热交换和淡水交换影响海水温度、盐度、密度和浮力的过程，对应的通量就是浮力通量 (buoyancy flux)。与之相对的，大气还能通过风的摩擦和搅拌作用影响海洋的动量，即 momentum forcing/flux。此句的意思是说，仅有较强的浮力强迫不一定能产生深对流，还要看下层的层结状况是否允许表层经过浮力损失的水体下沉。

[6] 冬季在大气的冷却作用下，海洋表层密度增大、损失浮力，发生对流混合，且因环境密度随深度增大，水团能下沉到的最深深度应该是与其密度一致处，即对流混合深度与大气冷却的强弱和海洋层结的强弱有关。到了春季，表层增温产生一个薄的暖水层，混合层变浅 (detrainment)，将下面混合均匀的水与大气隔离，这些水在重力作用下沿等密面下沉并离开源地，下个冬季不再被混合。这就是潜沉 (subduction) 过程，也是一种模态水的生成机制 (mechanism of mode water formation)，主要发生在副热带海域。另一种生成机制发生在混合层深度在水平方向上变化较强之处，此时若有强的背景平流将混合层较深处的水输运至混合层较浅的一侧，则这部分水同样与大气隔离而在重力作用下潜沉。潜沉发生处的海面密度就是与海面接触(即通风 ventilate 或露头 outcrop)的那个等密面的密度，模态水即沿此等密面潜沉。此等密面与海面接触之处连成的线叫作露头线 (outcrop line)，通过观察露头线上何处的示踪参数[tracer，如温度、盐度、位涡(potential vorticity)等]与在此等密面上发现的模态水最接近，可以揭示模态水的来源。当然，此方法假设所用示踪参数在模态水移

动过程中是守恒的（conservative）或近似守恒的。与潜沉相反，浮露（obduction）是模态水沿等密面向上运动，并因为混合层的加深（entrainment）而被卷至混合层内的过程，主要发生在副极地海域的晚秋和早冬季节，或者由背景平流被输运至混合层较深的一侧。

[7] 此处强调的深混合层特指由于深对流（deep convection）引起的深混合层。太平洋缺少深对流是共识，而文中所述南大洋的深对流比北大西洋浅则属于一家之言。

[8] 此处的 Greenland Seas 具体应该是北欧海（Nordic Sea）。南极洲周围除了大西洋扇区的威德尔海（Weddel Sea），太平洋扇区的罗斯海（Ross Sea）也是深层水的源区。

[9] variability 意为变化的性质和能力，可译为“变化”或“变率”。但“变化”容易与 change 混淆，后者常表示 climate change 这样的长期单向变化，因此用“变率”更好些。variability 一定要指明尺度，例如，seasonal variability、interannual variability、climate variability，分别表示在季节、年际、气候（十几年至几十年）尺度上往复变化的程度。因此，从数学上来讲 variability 接近标准差（standard deviation）或方差（variance）的概念。深层水的源区在气候变率中的作用至少包括将热量和二氧化碳带至深层，从而缓和全球变暖。同时由于全球变暖使得上层变热、冰川融化淡水增多，因此上层层结加强，阻碍深对流。

[10] 海气之间的热通量包括四个部分：感热（sensible）、潜热（latent）、长波辐射（longwave 或 thermal）、短波辐射（shortwave 或 solar），其中前两个是由湍流造成的，因此又叫湍流通量（turbulent flux）；后两个是辐射造成的，因此又叫辐射通量（radiative flux）。淡水通量主要是通过蒸发（evaporation）和降雨（precipitation），这里用了 moisture flux 这个说法。

[11] dome 此处做动词，表示向上拱。同时它也可以做名词，意为拱形结构。等密面的形状通常有如下几种：拱形（dome）、碗形（bowl）、凸透镜形（convex）、凹透镜形（concave），其中后两者合称透镜型（lens）。由于凸透镜形等密面意味着包含一团性质近乎均匀的水团，其关注度更高也更易观测，因此，常用 lens 特指凸透镜形。

[12] 由于没有直接观测，而仅从水体属性的分布中“推测”深层流，此处十分恰当地用了 inferred，而没有用 found、known 等词。

[13] 温度和盐度是最常用的“示踪”参数（tracer），其次是位涡，而氚和氟氯烃也被海洋学家巧妙地用来追踪水团。不管用何种示踪参数或示踪物质，其中隐含两个基本假设。一是此种 tracer 是守恒的，即海洋内部不存在它的源或汇，因此，从海面上获得后在海洋内部只存在输运和再分布，而没有增减。二是跨等密面的混合（diapycnal mixing）很弱，因此，在特定等密面上观察 tracer 的分布可知其输运路径，否则需要在三维空间进行追踪，难度增大。如此，等密面上的 tracer 输运仅需要对抗沿等密面的混合（isopycnal mixing），也就是在混合过程使得 tracer 分布均匀、浓度减弱为 0 之前先将其输送一段距离，从而为我们观察输送路径提供可能。因此，示

踪法不适合于混合很强、输运较弱的情况。

[14] AAIW是一个典型的水团名称。世界大洋中其他主要水团名称还有:Antarctic Bottom Water、Circumpolar Deep Water、Indian Deep Water、Labrador Sea Water、Mediterranean Water、Mediterranean Overflow Water、North Pacific Subtropical Mode Water、North Pacific Central Mode Water、Pacific Deep Water等。

[15] 此句用破折号插入了一个补充句,表示实际上NADW不仅存在于2 km深度,而是填满整个北大西洋。

[16] 此句语序也许不易理解。开头的that引导的从句作主语,中间是插入语,后面is是谓语,evidence是宾语,其后的that又引导一个定语从句修饰evidence。主语从句不太多见,开头的that容易被误会为代词,指代前面的某个对象。

[17] 与深对流不同,补偿上升过程的确应该分布在很大的范围上,因此速度很弱。但海洋中存在一些混合较强的"热点(hotspot)"区域,这些区域的混合和因此导致的深层水上翻应该比本文用全球海洋面积带入估算的量级大得多。这些热点区域多在海底地形梯度较大处,即海底山脉较多处,如大洋中脊区域。

[18] 在此方面做出奠基性贡献的是亨利·斯托梅尔(Henry Stommel),他在泰勒-普劳德曼定理(Tayler-Proudman theorem)基础上得出的深海环流理论预言了深层西边界流的存在。

14. Sea Ice

Text

Sea ice, any form of ice found at sea that originated from the freezing of sea water[1], has historically been among the least-studied of all the phenomena that have a significant effect on the surface heat balance of the Earth. Fortunately, this neglect has recently lessened as the result of improvements in observational and operational capabilities in the polar ocean areas.

About 7% of the surface of the Earth is covered by sea ice during some time of the year. In the northern hemisphere the area covered by sea ice varies between 8×10^6 and 15×10^6 km^2, with the smaller number representing the area of multiyear (MY) ice remaining at the end of summer. In summer this corresponds roughly to the contiguous area of the United States and to twice that area in winter, or to between 5% and 10% of the surface of the northern hemisphere ocean[2]. At maximum extent, the ice extends down the western side of the major ocean basins, following the pattern of cold currents and reaching the Gulf of St. Lawrence (Atlantic) and the Okhotsk Sea off the north coast of Japan (Pacific)[3]. The most southerly site in the northern hemisphere where an extensive sea ice cover forms is the Gulf of Bo Hai, located off the east coast of China at 40°N[4]. At the end of the summer the perennial MY ice pack of the Arctic is primarily confined to the central Arctic Ocean with minor extensions into the Canadian Arctic Archipelago and along the east coast of Greenland.

In the southern hemisphere the sea ice area varies between 3×10^6 and 20×10^6 km^2, covering between 1.5% and 10% of the ocean surface. The amount of MY ice in the Antarctic is appreciably less than in the Arctic, even though the total area affected by sea ice in the Antarctic is approximately a third larger than in the Arctic. These differences are largely caused by differences in the spatial distributions of land and ocean. The Arctic Ocean is effectively landlocked to the south, with only one major exit located between Greenland and Svalbard. The Southern Ocean, on the other hand, is essentially completely unbounded to the north, allowing unrestricted drift of the ice in that direction, resulting in the melting of nearly all of the previous season's growth.

In addition to its considerable extent, there are good reasons to be concerned with the health and behavior of the world's sea ice covers. Sea ice serves as an insulative lid on the surface of the polar oceans. This suppresses the exchange of heat between the cold polar air above the ice and the relatively warm sea water below the ice[5]. Not only is the ice itself a good insulator, but it provides a surface that supports a snow cover that is also an excellent insulator. In addition, when the sea ice forms with its attendant snow cover, it changes the surface albedo, i.e., the reflection coefficient for visible radiation, from 0.15 for open water to 0.85 for newly formed snow[6], leading to a 70% decrease in the amount of incoming short-wave solar radiation that is absorbed. As a result, there are inherent positive feedbacks associated with the existence of a sea ice cover[7]. For instance, a climatic warming will presumably reduce both the extent and the thickness of the sea ice. Both of these changes will, in turn, result in increases in the temperature of the atmosphere and of the ocean, which will further reduce ice thickness and extent. It is this positive feedback that is a major factor in producing the unusually large increases in Arctic temperatures that are forecast by numerical models simulating the effect of the accumulation of greenhouse gases[8].

The presence of an ice cover limits also the flux of moisture. This effect is revealed by the common presence of linear, local clouds associated with individual leads (cracks in the sea ice that are covered with either open water or thinner ice)[9]. In fact, sea ice exerts a significant influence on the radiative energy balance of the complete atmosphere-sea ice-ocean system. For instance, as the ice thickness increases in the range between 0 and 70 cm, there is an increase in the radiation absorption in the ice and a decrease in the ocean. There is also a decrease in the radiation adsorption by the total atmosphere-ice-ocean system. It is also known that the upper 10 cm of the ice can absorb over 50% of the total solar radiation, with all the ultraviolet and infrared radiation absorbed in the upper 50 cm of the ice; only visible radiation penetrates into the lower portions of thicker ice and into the upper ocean beneath the ice[10]. Significant changes in the extent and/or thickness of sea ice would result in major changes in the climatology of the polar regions.

However, there are even less obvious but perhaps equally important air-ice and ice-ocean interactions. Sea ice drastically reduces wave-induced mixing in the upper ocean, thereby favoring the existence of a 25–50 m thick, low-salinity surface layer in the Arctic Ocean that forms as the result of desalination processes associated with ice formation and the influx of fresh water from the great rivers of northern Siberia. This stable, low-density surface layer prevents the heat contained in the comparatively warm (temperatures of up to +3℃) but more saline denser water beneath the surface layer from affecting the ice cover[11]. As sea ice rejects roughly two-thirds of the salt initially present in the sea water from which the ice forms (called brine rejection), the freezing process is equivalent to distillation, producing both a low-salinity component (the ice layer itself) and a high-salinity component (the rejected brine)[12]. Both of these components play important geophysical roles. Over shallow shelf seas, the rejected brine, which is dense, cold, and rich in CO_2, sinks to the bottom, ultimately feeding the deep-water and the bottom-water layers of the world ocean. Such processes are particularly effective in regions where large polynyas exist (semi-permanent open water and thin-ice areas at sites where climatically much thicker ice would be anticipated)[13].

Because ice is a thermal insulator, the thicker the ice, the slower it grows, other conditions being equal. As sea ice either ablates or stops growing during the summer, there is a maximum thickness of first-year (FY) ice that can form during a specific year. The exact value is, of course, dependent upon the local climate and oceanographic conditions, reaching values of slightly over 2 m in the Arctic and as much as almost 3 m at certain Antarctic sites. It is also clear that during the winter the heat flux from areas of open water into the polar atmosphere is significantly greater than the flux through even thin ice and is as much as 200 times greater than the flux through MY ice. This means that, even if open water and thin ice areas comprise less than 1~2% of the winter ice pack, lead areas must still be considered in order to obtain realistic estimates of ocean-atmosphere thermal interactions[14].

If an ice floe survives a summer, during the second winter the thickness of the additional ice that is added is less than the thickness of nearby FY ice for two reasons: it starts to freeze later and it grows slower. Nevertheless, by the end of the winter, the second-year ice will be thicker than the nearby FY ice[15]. Assuming that the above process is repeated in subsequent years, an amount of ice is ablated away each summer (largely from the upper ice surface) and an amount is added each winter (largely on the lower ice surface)[16]. As the year pass, the ice melted on top each summer remains the same (assuming no change in the climate over the ice), while the ice forming on the bottom becomes less and less as a result of the increased insulating effect of the thickening overlying ice. Ultimately, a rough equilibrium is reached, with the thickness of the ice added in the winter becoming equal to the ice ablated in the summer. Such steady-state MY ice floes can be layer cakes of ten or more annual layers with total thicknesses in the range 3.5-4.5 m.

Historical data based on direct observations of sea ice extent are rare, although significant long-term records do exist for a few regions such as Iceland. In monitoring the health of the world's sea ice covers the use of satellite remote sensing is essential because of the vast remote areas that must be surveyed. Unfortunately, the satellite record is very short. If data from only microwave remote sensing systems are considered, because of their all-weather capabilities, the record is even shorter, starting in 1973[17]. The imagery shows that there are definitely large seasonal, interannual and regional variations in ice extent. For instance, a decrease in ice extent in the Kara and Barents Seas contrasts with an increase in the Baffin Bay/Davis Strait region and out-of-phase fluctuations occur between the Bering and the Okhotsk Seas. The most recent study, which examined passive microwave data, concludes that the areal extent of Arctic sea ice has decreased by 12.6% per decade. In addition, record minimum areas of Arctic sea ice have been observed once in a few years. Up to now, the minimum winter-max Arctic sea ice area was found in March 7, 2017, with the value of 8.2×10^6 km^2[18].

Off the Antarctic the situation is not as clear. One study has suggested a major retreat in maximum sea ice extent over the last century based on comparisons of current satellite data with the earlier positions of whaling ships reportedly operating along the ice edge. As it is very difficult to access exactly where the ice edge is located on the basis of only ship-board observations, this claim has met with some skepticism. An examination of the satellite observations indicates a very slight increase in areal extent since 1973[19].

Vocabulary

operational:(adj) 日常业务的

~operational forecast 业务化预报

multiyear ice:(n) 多年冰

contiguous:(adj) 接壤的,共边的,连续的

extensive:(adj) 广阔的,广大的,广泛的

perennial:(adj) 长期的,永久的,持久的

ice pack:(n) 流冰群,浮冰群

archipelago:(n) 群岛,列岛

appreciably:(adv) 可察觉的,可观的

adj: appreciable

spatial:(adj) 空间的,空间上的

unbounded:(adj) 无边的,无限的

insulative:(adj) 隔热的,绝缘的,隔音的

n: insulator 隔热材料,绝缘材料,隔音材料

lid:(n) 盖子

suppress:(vt) 镇压,抑制,阻止

attendant:(adj) 伴随的

albedo:(n) 反射率,反照率

reflection:(n) 反射,反映

vt: reflect

coefficient:(n) 系数

visible radiation:(n) 可见光辐射

incoming:(adj) 入射的

ant: outgoing 出射的

inherent:(adj) 内在的,固有的,本性的

presumably:(adv) 据推测,大概,可能

adj: presumable

vt: presume 推测,假设

numerical:(adj) 数值的

greenhouse gas:(n) 温室气体

lead:(n) 水道,冰间水道

climatology:(n) 气候学,气候态

drastically:(adv) 猛烈地,激烈地

adj: drastic

thereby:(adv) 因此,由此,从而

desalination:(n) 脱盐,淡化

influx:(n) 涌入,流入,注入

comparatively:(adv) 比较而言地,相对而言地

adj: comparative

brine rejection:(n) 结冰析盐(现象、过程)

brine:(n) 盐水,卤水

distillation:(n) 蒸馏过程

vt: distill

CO_2 = carbon dioxide:(n) 二氧化碳

polynya:(n) 冰间湖

anticipate:(vt) 期望,预料,认为……很有

可能
ablate：(vt/vi)(使)脱落，切除，毁坏
first-year ice：(n) 一年冰
floe：(n) 浮冰块
subsequent：(adj) 随后的，后续的，紧接的
annual：(n) 每年的，一年一度的，全年的
monitor：(vt) 监视，监测，监控
remote sensing：(n) 遥感
survey：(vt) 审视，调查，测量，勘测，测绘；(n) 测量，测量部门
microwave：(n) 微波
all-weather：(adj) 全天候的，适应各种天气的
imagery：(n)(总称)图像
definitely：(adv) 肯定地，当然，明确地
adj：definite
out-of-phase：(adj) 异相位的，不同步的去掉连词符可做副词短语
ant：in-phase (in phase) 同相位的，同步的
passive：(adj) 被动的
~passive microwave sensing 被动微波遥感
ant：active 主动的
record：(n)(有记录以来的)最高、最好纪录
retreat：(n/vi) 撤退，衰退，退缩
reportedly：(adv) 据传，据报道
ship-board：(adj) 船载的
skepticism：(n) 怀疑，怀疑态度
adj：skeptic

Notes

[1] 这是海冰的广义定义，即海洋中见到的冰都是海冰，包括海水冻结的冰、河流带来的陆地冰、冰川断裂入海形成的冰山等。当然海水冻结的冰是其最主要成分，也是其狭义定义。

[2] 此句语序需要梳理清楚。correspond 后面分别用了三个 to 表示对应的三个对象，其中第一个直接在 correspond 后面，第二个用 and 引出一个分句，第三个则跟在逗号和 or 之后。twice that area 指的是前述的美国陆地面积的两倍。

[3] 此处 down 不是向下的意思，而是“沿着”。所谓沿着寒流，只是表示空间对应关系，并不代表海冰是由这些寒流从北冰洋带来的。虽然北冰洋海冰的确随着东格陵兰流等海流南下，但文中提到的圣劳伦斯湾和鄂霍次克海海冰以局地生成为主。从白令海峡 (Bering Straight) 进入太平洋的海冰很少。

[4] “在……外海”用的介词是 off。

[5] 此处提到极地海洋相对于大气更暖这一事实，其原因是北大西洋与北冰洋之间的通道较为宽阔，加上北美东岸的地形，使得北大西洋流的暖水从东格陵兰海和北欧海进入北冰洋。而大气缺少这样强烈的经向热输送，极地涡旋 (polar vortex) 将冷空气相对稳定地禁锢在极地。

[6] 这里的描述不是很准确，因为反照率 (albedo) 一般不是仅针对可见光的，而是针对

太阳辐射的全频段。若确要针对某个频段,则叫作光谱反照率(spectral albedo)。这里给的反照率数值也未必准确,开阔海洋的反照率一般在0.05~0.1范围内,典型值是0.06;无雪海冰的反照率在0.5~0.7范围内;新雪的反照率是0.8~0.85。

[7] positive feedback 即正反馈,即A的增大使B增大,B增大后又使A更大,如此不断强化,形成愈演愈烈之势,直至别的原因阻断继续发展。相反的是负反馈(negative feedback),即A的增大使B增大,B增大后使A减小,A减小后使B减小,B减小后又使A增大,形成周期性的循环往复。

[8] 这一正反馈导致了“北极放大现象(Arctic Amplification)”,即文中说的unusually large increases in Arctic temperatures。实际上,北极地区的增温速度是全球平均增速的两倍且仍在加速,而且不仅如文中所说可以在数值模式中看到,观测资料也已经证实。北极放大现象的成因有不同的说法,本文所述的海冰-反照率反馈(ice-albedo feedback)是其中接受度最高的一种。

[9] 请注意lead一词的读法。它最常见的意思就是动词“引导”“领先”,读作/liːd/,也可以做名词表示“表率”“优势”“主角”。如果读作/led/,则表示金属铅。此处还是读作/liːd/,是名词,专指冰的缝隙处露出的狭长海水通道,可译为“冰间水道”,与冰间湖(polynya)类似。由于海水裸露,海洋向大气放热,而此过程通过感热通量、潜热通量、长波辐射三个途径共同进行,其中潜热应该占主。潜热通量即由蒸发的水蒸气携带热量进入大气,并于对流层中上部降温冷凝后释放,同时形成云。因此冰间水道上空常见与之同样细长形的线状云,即“linear cloud”。除了冰间水道处之外,这种线状云在自然界中并不常见,而更多为人所知的是飞机飞过后形成的线状“飞机云”,英文叫作contrail或condensation track,字面意思是凝结尾迹。

[10] 总结起来,此段阐明随着海冰的生长增厚,它吸收的辐射越来越多,而透过海冰进入海水的辐射则越来越少。未在表面上被反射的那15%~20%的辐射进入海冰,海冰越厚自然就吸收越多的辐射。10 cm厚的海冰可吸收全部辐射的50%;50 cm厚的海冰则吸收全部的紫外辐射和红外辐射,仅可见光透过;而70 cm厚的海冰则连可见光也全部吸收。

[11] 结冰时的脱盐过程(desalination,本段即将介绍)提供了低盐水的来源(与淡水输入一起),而生成的海冰起到了消浪作用,从而减小混合加强层结,保护了自己免受下层暖水的侵袭。请注意把握此处的逻辑关系,特别是thereby、as the result of、associated with和prevent from等逻辑连接词。

[12] 实际上,若给予充足的时间,海水结冰的过程将排出所有的盐分,即结冰析盐过程(brine rejection)。只是由于结冰过程往往较快,有一些海水泡被包裹在冰里面留下来,因而被阻断与外界的热交换,不能再结冰。结冰(液变固)与蒸馏(液变汽)这样的相变过程都只涉及水,不携带其中的盐分,因此能将盐分分离出来。结成的冰可能包括一些较软的小冰碴,旋即又再次融化,与海冰一起形成覆盖在上的淡水

层,析出的盐则形成咸水层。

[13] 由于接近大气,这些海水能接触到二氧化碳;又由于温度较低,其溶解度(solubility)高。因此,这些海水的下沉是海洋储碳(carbon storage)的通道之一。然而,只有源源不断的结冰析盐才能供应充足的高盐高密水,但正常情况下一旦结冰就阻断了海气热交换,切断了继续结冰的条件。只有冰间湖(polynya)满足此条件,因为这里刚生成的新冰就会被风吹走,使海洋继续裸露,结冰析盐继续进行。这里的风往往是从陆地(南极大陆或格陵兰岛)上吹下来的冷风。

[14] 此处的几个数字再次体现了海冰的隔热效果。因此,如冰间水道这样的狭窄区域却对海气热交换有着极大的作用。这对数值模式提出了巨大的挑战,因为模式的分辨率往往不够分辨冰间水道,却又不得不考虑它的效应,因此,必须借助参数化(parameterization)来考虑这些次网格效应(sub-grid scale effect),但目前尚没有很好的参数化方案(scheme)。

[15] 在已经存在的冰的基础上增长的冰(即两年冰)与重新生成的冰(即一年冰)相比,两年冰的新增厚度小于一年冰的厚度,这还是由于冰的隔热作用。但毕竟还是有新增,使得两年冰比一年冰厚,这都是不言而喻的。

[16] 融冰从上表面开始,因为极昼(polar day)期间接受太阳辐射。结冰从下表面开始,因为水在冰的下面。

[17] 卫星遥感可通过监测反照率来探知海冰覆盖范围。微波的波长长,可以衍射(diffract)绕过云里的水滴,因此穿透性强,适应各种天气(all-weather)。其他波长较短的波段则无法透视云下的海面情况,只能用于晴空(clear sky)情况。

[18] 此处我们用 NASA 的最新数据代替了原文的旧数据。看这样的年际或年代际变化数据时,需要注意区分是冬季还是夏季海冰,是海冰面积(extent 或 area)还是体积(volume)。不管何种数据,北极海冰衰退(Arctic sea ice decline)是毋庸置疑的。

[19] 20 世纪 80 年代以来南极海冰面积增加而不是减少是目前的共识,普遍认为的原因是南极陆地冰川(glacier)和冰架(ice shelf)融化断裂进入海洋的量增加了。但是,2014 年南极海冰覆盖面积到达最大值后出现了断崖式下降,至今仍未恢复,2023 年达到历史最低,原因尚不明确。

15. Sea Level Rise

Text

Between 1901 and 2018, the average global sea level rose by 15–25 cm, or an average of 1–2 mm per year. This rate accelerated to 4.62 mm/yr for the decade 2013 – 2022. Climate change due to human activities is the main cause. Between 1993 and 2018, thermal expansion of water accounted for 42% of sea level rise (hereafter SLR)[1]. Melting temperate glaciers accounted for 21%, with Greenland accounting for 15% and Antarctica 8%[2]. SLR lags changes in the Earth's temperature. So SLR will continue to accelerate between now and 2050 in response to warming that is already happening. What happens after that will depend on what happens with human greenhouse gas emissions. SLR may slow down between 2050 and 2100 if there are deep cuts in emissions. It could then reach a little over 30 cm from now by 2100. With high emissions it may accelerate. It could rise by 1 m or even 2 m by then. In the long run, SLR would amount to 2–3 m over the next 2000 years if warming amounts to 1.5℃. It would be 19–22 metres if warming peaks at 5℃.

Rising seas ultimately impact every coastal and island population on Earth. This can be through flooding, higher storm surges, king tides, and tsunamis[3]. These have many knock-on effects. They lead to loss of coastal ecosystems like mangroves. Crop production falls because of salinization of irrigation water. And damage to ports disrupts ocean trade. The SLR projected by 2050 will expose places currently inhabited by tens of millions of people to annual flooding[4]. Without a sharp reduction in greenhouse gas emissions, this may increase to hundreds of millions in the latter decades of the century. Areas not directly exposed to rising sea levels could be affected by large scale migrations and economic disruption.

SLR is not uniform around the globe[5]. Some land masses are moving up or down as a consequence of subsidence (land sinking or settling) or post-glacial rebound (land rising due to the loss of weight from ice melt)[6]. Therefore, local relative SLR may be higher or lower than the global average. Gravitational effects of changing ice masses also add to differences in the distribution of sea water around the globe.

When a glacier or an ice sheet melts, the loss of mass reduces its gravitational pull. In some places near current and former glaciers and ice sheets, this has caused local water levels to drop, even as the water levels will increase more than average further away from the ice sheet. Consequently, ice loss in Greenland has a different fingerprint on regional sea level than the equivalent loss in Antarctica. On the other hand, the Atlantic is warming at a faster pace than the

Pacific. This has consequences for Europe and the US East Coast, which receives a SLR 3-4 times the global average. The downturn of the Atlantic meridional overturning circulation (AMOC) has been also tied to extreme regional SLR on the US Northeast Coast.

There are two ways of modeling SLR and making future projections. In one approach, scientists use process-based modeling, where all relevant and well-understood physical processes are included in a global physical model. An ice-sheet model is used to calculate the contributions of ice sheets and a general circulation model is used to compute the rising ocean temperature and its expansion. While some of the relevant processes may be insufficiently understood, this approach can predict non-linearities and long delays in the response, which studies of the recent past will miss[7].

In the other approach, scientists employ semi-empirical techniques using historical geological data to determine likely sea level responses to a warming world, in addition to some basic physical modeling[8]. These semi-empirical sea level models rely on statistical techniques, using relationships between observed past contributions to global mean sea level and global mean temperature. This type of modeling was partially motivated by most physical models in previous Intergovernmental Panel on Climate Change (IPCC) literature assessments having underestimated the amount of SLR compared to observations of the 20th century[9].

The IPCC provides multiple plausible scenarios of 21st century SLR in each report, starting from the IPCC First Assessment Report in 1990. The differences between scenarios are primarily due to the uncertainty about future greenhouse gas emissions, which are subject to hard to predict political action, as well as economic developments. The scenarios used in the 2013-2014 Fifth Assessment Report (AR5) were called Representative Concentration Pathways, or RCPs[10]. An estimate for sea level rise is given with each RCP, presented as a range with a lower and upper limit, to reflect the unknowns. The RCP2.6 pathway would see greenhouse gas emissions kept low enough to meet the Paris climate agreement goal of limiting warming by 2100 to 2℃[11]. Estimated SLR by 2100 for RCP2.6 was about 44 cm (the range given was as 28-61 cm). For RCP8.5 the sea level would rise between 52 and 98 cm.

Sea level changes can be driven by variations in the amount of water in the oceans, by changes in the volume of that water, or by varying land elevation compared to the sea surface. Over a consistent time period, assessments can source contributions to SLR and provide early indications of change in trajectory, which helps to inform adaptation plans[12]. The different techniques used to measure changes in sea level do not measure exactly the same level. Tide gauges can only measure relative sea level, whilst satellites can also measure absolute sea level changes. To get precise measurements for sea level, researchers studying the ice and the oceans on our planet factor in ongoing deformations of the solid Earth, in particular due to landmasses still rising from past ice masses retreating, and also the Earth's gravity and rotation[13].

The three main reasons warming causes global sea level to rise are the expansion of oceans due to heating, along with water inflow from melting ice sheets and glaciers. SLR since the start of the 20th century has been dominated by retreat of glaciers and expansion of the ocean, but the contributions of the two large ice sheets (Greenland and Antarctica) are expected to increase in the 21st century. The ice sheets store most of the land ice (~99.5%), with a sea-level equivalent (SLE) of 7.4 m for Greenland and 58.3 m for Antarctica[14].

The oceans store more than 90% of the extra heat added to Earth's climate system by climate change and act as a buffer against its effects. When the ocean gains heat, the water expands and sea level rises. The amount of expansion varies with both water temperature and pressure. For each degree, warmer water and water under great pressure (due to depth) expand more than cooler water and water under less pressure. Consequently cold Arctic Ocean water will expand less than warm tropical water. Because different climate models present slightly different patterns of ocean heating, their predictions do not agree fully on the contribution of ocean heating to SLR. Heat gets transported into deeper parts of the ocean by winds and currents, and some of it reaches depths of more than 2,000 m[15].

Each year about 8 mm of precipitation (liquid equivalent) falls on the ice sheets in Antarctica and Greenland, mostly as snow, which accumulates and over time forms glacial ice. Some of the snow is blown away by wind or disappears from the ice sheet by melt or by sublimation (directly changing into water vapor). The rest of the snow slowly changes into ice. This ice can flow to the edges of the ice sheet and return to the ocean by melting at the edge or in the form of icebergs[16]. If precipitation, surface processes and ice loss at the edge balance each other, sea level remains the same. However, scientists have found that ice is being lost, and at an accelerating rate.

Sea ice loss contributes very slightly to global sea level rise. If the meltwater from ice floating in the ocean was exactly the same as sea water then, according to Archimedes'principle, no rise would occur. However melted sea ice contains less dissolved salt than sea water and is therefore less dense, with a slightly greater volume per unit of mass. If all floating ice shelves and icebergs were to melt sea level would only rise by about 4 cm[17].

Vocabulary

thermal expansion：热膨胀

account for：解释，说明，提供，构成（数量/比例）

hereafter：（adv）自此，此后

temperate：（adj）温带的，（气候）温和的

king tide：（n）国王大潮（天文极大潮）

knock-on:(adj) 附带效果,间接影响

salinization:(n) 盐化

vt: salinize

irrigation:(n) 灌溉,浇灌

vt: irrigate

project:(vt) 预计,推断,推测

n: projection

uniform:(adj) 完全一样的,不变的,均一的

subsidence:(n)(土地)下陷,下沉

vi: subside

post-glacial rebound:(n) 冰(期)后回弹

rebound:(n/vi) 弹回,跳回,反弹

ice sheet:(n) 冰原,冰盖,冰冠

process-based:(adj) 基于过程的

employ:(vt) 用,利用,使用

Intergovernmental Panel on Climate Change: 政府间气候变化委员会

panel:(n) 小组,专题讨论小组,委员会

literature:(n) 文献资料,文学作品

assessment:(n) 评估,评价,估计

vt: assess

plausible:(adj) 似乎有道理的,似乎可能的

scenario:(n) 假想中的事态发展情况,情景

uncertainty:(n) 不确定性

Representative Concentration Pathway: 代表性/典型浓度路径

Paris climate agreement: 巴黎气候协定

consistent:(adj) 一致的,一贯的,符合的,协调的,不矛盾的,连续的,持续的

trajectory:(n) 轨迹,路径

adaption:(n) 适应,调整(的方法)

= adaptation

tide gauge:(n) 潮位计

gauge:(n) 量具,量器;(vt) 测量

deformation:(n)(受压而)变形

vt: deform

inflow:(n) 入流

ant: outflow 出流

sea-level equivalent:(n) 海平面当量

sublimation:(n) 升华

meltwater:(n) 冰川融化水

可以拆成两个单词

Archimede's principle:(n) 阿基米德定理

~Archimedean:(adj) 阿基米德的

Archimedean force: 阿基米德力(即浮力)

Notes

[1] 热膨胀是在质量不变的情况下的体积变化,即单位质量的体积(也就是比容 specific volume)变大,或单位体积的质量(也就是密度)变小。比容与密度是倒数关系。此部分海平面变化与质量无关,而仅与水分子之间的距离增大有关,因此,术语叫作 steric sea level rise/change。steric 是形容词,意为与分子的空间排列有关的。中文术语则称作"比容海平面"。实际上比容海平面还包括由盐度变化引起的部分,因为盐度降低同样能减小密度、增大比容。细分起来,比容海平面可分为热比容(thermosteric)和盐比容(halosteric)两部分。目前,通常认为盐比容部分贡献较小,因此

本文没有提到,但存在很大的不确定性。

[2] 冰川融水(meltwater)直接增加海洋的水量,因此,直接叫作 seawater mass change。实际上,冰川融水除了改变质量,还可以改变温度和盐度,特别是通过降低盐度而引起盐比容海平面变化。实际上此处给出的 21%的数字已经过时,目前的共识是冰川融化对海面上升的贡献可达 2/3。

[3] 要注意的是,与这里提到的几种现象相比,气候变化导致的海面上升非常缓慢,前面给出的每年上升速率仅为毫米量级,而这几种灾害现象可在几天甚至几小时内引起数米甚至数十米的海面起伏。因此,海面上升的威胁与这些短时间的剧烈灾害不同,它的作用更多体现在缓慢提高“平均”海平面,使得剧烈灾害发生时有一个相对更高的背景水位,借此使灾害变得更严重。当然,海面上升的缓慢性使人类有更多时间去应对,比如通过加高堤防等方式,但也容易被忽视。此处提到的 king tide 直译为“国王大潮”,是指天文潮的极大潮位,是有规律、可预测的。此词汇发源于澳大利亚和新西兰,近年来在英语国家逐渐流行。

[4] project 作为动词意思是预测、预估,是指在已有的变化曲线的基础上,将之延伸至未知部分,可以理解为外推。除了时间上的外推,还可用在其他情境,比如将已有观测数据的中低风速下的风速-浪高关系外推至没有观测数据的高风速情况等。这是一种较为主观的外推,与 forecast 或 predict 是不同的。

[5] 全球平均的海平面变化可用术语 eustatic sea level change 表示,与之相对的局地海平面变化则用 local 表示。

[6] 陆地沉降(subsidence of land)的原因有多种,包括自然原因和人为原因,其中后者包括地下水流失、建筑物负重、地下资源开采等。冰后回弹(post-glacial rebound)即冰盖消失后陆地的反弹,例如,加拿大东部 7000~8000 a 前消失的冰川导致当前仍有 1 cm/a 左右的回弹。

[7] 虽然海洋-大气-海冰耦合模式(coupled model)早已有之,但包含陆地冰川在内的耦合模式尚不多见。原因包括陆地冰川模式本身的不确定性,以及海洋模式处理淡水输入的方法问题等。若这样的模式足够成熟,确实可以用于模拟长时间、非线性的海平面问题,是基于观测资料的短期研究无法比拟的。

[8] 这可算作古海洋学(paleoceanography)或古气候学(paleoclimatology)的一部分,即利用与现在情况类似的地质时期发生的事情,类比、推测将来的可能变化。

[9] 此句结构稍难理解,若在 having 前加逗号,或将 having 改成 that have 可能更容易明白 having 后面的部分是对 most physical models 的形容。也就是说,由于 IPCC 报告里的物理模型低估了 20 世纪海面变化,才催生出了此种经验模拟方法。实际上,物理模型对海面变化的低估很大程度上是由于冰川融水的速率、位置等信息缺失,因而无法考虑冰川融水造成的。

[10] 由于无法知道未来温室气体的排放情况会是怎样的,只能假定一些“情景(scenar-

ios)”,在此基础上进行预测(projection)。每个情景用一个“代表性浓度路径(representative concentration pathway)”表示。这里的 representative 强调只是诸多可能性中的一种,concentration 表示以温室气体浓度为目标,而 pathway 则不仅仅指一个浓度,而是达到这个浓度的过程。考虑的温室气体包括 CO_2、CH_4[甲烷(methane)]、N_2O[一氧化二氮(nitrous oxide)]等。

[11] IPCC AR5 定义了 4 个 RCP:2.6、4.5、6、8.5。AR6 又增加了 1.9、3.4、7,以及“共享社会经济路径(shared socioeconomic pathway, SSP) 1-5”。RCP 后面的数字表示到 2100 年时的辐射强迫(radiative forcing)值,即大气层顶(top of atmosphere)的净辐射收益与 1750 年情况的差,单位是 W/m^2。

[12] 此句初看似乎语焉不详。consistent time period 指一段足够长的连续时间,assessments 指基于观测的分析评估,source 是动词“溯源”,change in trajectory 指发展趋势的变化,adaption plans 指应对变化做出的调整计划。provide early indications of change in trajectory 即当海面上升的发展趋势发生变化(如明显加速)时给出预警。climate change adaption/adaptation 是目前气候变化领域一个重要的方向,但更多涉及社会学、人类学、经济学、政策等领域。IPCC 给出的定义是:“In human systems, as the process of adjustment to actual or expected climate and its effects in order to moderate harm or take advantage of beneficial opportunities. In natural systems, adaptation is the process of adjustment to actual climate and its effects; human intervention may facilitate this.”

[13] factor in … 表示将……因素包括进来。这里包括(考虑)进来的因素是地球的变形(deformation)、重力场的改变、地球自转(速度)的改变。landmasses still rising from past ice masses retreating 部分特意安排了前后对称的结构(landmasses rising、ice masses retreating),但将 retreating 放在 from 和 past 之间可能更易懂。前面已经提过的地球重力场的改变可引起质量分布变化(变形)和自转加速。有观点认为两极冰川融化,地壳隆起,地球变得更“瘦高”,因此地转加速。

[14] sea-level equivalent(海平面当量)是近年来出现的一个名词,表示将某范围的陆地冰全部融化后可导致的全球平均海面升高幅度。准确的估计依赖于对南极和格陵兰岛冰川体积的准确测量。

[15] 目前 2000 m 以深的温度观测数据极其匮乏,因此,我们对深海在气候变化情况下的响应几乎一无所知。

[16] 由积雪形成冰川的时间通常需要许多年,甚至几百年。冰川的移动速度差异很大,从静止不动到 30 m/d 都有,平均速度 1 m/d。

[17] volume per unit of mass 即 specific volume(比容)。注意最后一句的虚拟语气,用了 were 而不是 are。而且用了 were to melt,而不是 melted,此处 to 表示将来的动作。

16. Numerical Modelling

Text

Numerical ocean modeling is simply a special case of computational fluid dynamics (CFD) which, in its most general form, involves solutions of a coupled set of nonlinear partial differential equations (PDEs) governing the time-dependent behavior of properties of a fluid flowing in three-dimensional space and acted upon by various forces[1]. Therefore, the governing equations are invariably one form or another of Navier-Stokes equations, supplemented by conservation equations for relevant scalar properties such as fluid temperature and salinity with appropriate source and sink terms[2]. More appropriately, since most practical situations involve high Reynolds number turbulent flows, ensemble-averaged Reynolds-type equations derived from Navier-Stokes equations are almost always used instead[3]. Here lies one of the principal conceptual difficulties in CFD: how to parameterize the unknown turbulent stress and diffusion terms that result, in other words, the turbulence closure problem[4]. The task of finding the optimum solution to the resulting PDEs in the most efficient manner possible requires intimate knowledge of the nature of the flow, careful attention to the available computer resources, and ingeniously designed simplifications that can be made without adversely affecting the solutions sought[5].

The principal differences between CFD in different fields of science, such as aerodynamics, geophysical fluid dynamics, and astrofluid dynamics, are related to the differences in the underlying equations[6]. The fluid can be considered to be incompressible in most geophysical fluid flows. Mach number M, the ratio of typical flow velocity to the speed of sound, is small (less than 0.002 in the oceans and 0.2 in the atmosphere). However, this simplification is more than counterbalanced by the fact that the fluid is stratified (density is not constant) and hence under the action of Archimedean (buoyancy) forces in a gravitational field[7]. Because of the planetary rotation, the equations are usually written in a rotating (non-inertial) frame of reference, giving rise to fictitious body forces that nevertheless have a dominant effect on the flow[8].

In contrast to aerodynamical CFD, techniques used in the oceans and the atmosphere, especially the former, tend to be rather conservative. Highly advanced CFD techniques such as 3D unstructured grids, adaptive grids (where the computational grid is fitted to the flow and changed as needed during the computation) are not as easily implemented. This is simply because one often does not have the luxury of verification data under controlled conditions to assess the efficacy of the schemes. The flows also tend to be more complex and less well understood[9].

Numerical computations involve two types of errors: round-off errors and truncation errors.

The former result from the inability of a computer to represent a floating-point number to infinite precision. Thus floating point numbers are inevitably rounded off and represented only approximately in arithmetic operations[10]. Round-off errors are unavoidable in any numerical operation on a digital computer, and in sequential operations on an operand they tend to accumulate. One has very little control on the magnitude of the cumulative round-off error.

Truncation errors arise from the need for discretization in computing on a digital computer. A variable continuous in space can only be represented on a digital computer at pre-selected discrete points in space. The ordinary and partial derivatives of a variable at such discrete points have to be represented in terms of the values of the variable at that grid point and its neighbors[11]. This is done using Taylor series expansions. The value of a function $y(x)$ at the neighboring grid point $j+1$ ($x = x_0 + h$, where h is the grid spacing) can be expressed as an infinite series[12]:

$$y_{j+1} = y_j + \left.\frac{dy}{dx}\right|_j h + \left.\frac{d^2y}{dx^2}\right|_j \left.\frac{d^2y}{dx^2}\right|_j \frac{h^2}{2!} + \cdots$$

Expansions such as these can be used to derive expressions for any order derivative by truncating the expansions suitably. For example, the first derivative at grid point j ($x = x_0$) can be written as

$$\left.\frac{dy}{dx}\right|_j = \frac{y_{j+1} - y_j}{h} + \varepsilon$$

$$\varepsilon = \frac{h}{2!} \left.\frac{d^2y}{dx^2}\right|_j + \cdots = O(h)$$

The derivative is now represented in terms of finite differences between the values of the variable at two adjacent grid points. ε is the truncation error, and in this case, it is of order h. This finite difference approximation is therefore called a first order accurate scheme. An n-th order accurate scheme would have a truncation error of $O(h^n)$ [13].

Truncation error is also cumulative. In theory, the truncation error is under the control of the modeler, since reducing the grid size (equivalently, increasing the model resolution) or going to a higher order scheme reduces the truncation error. However, this is not practical since doubling the resolution in a numerical model requires an order of magnitude increase in computing time and manyfold increase in memory requirements. Even if it were possible to make h arbitrarily small to minimize truncation errors, on a finite precision computer, it causes increased round-off errors, since the same calculation will now involve a correspondingly larger number of arithmetic operations. In most cases, the choice of the model resolution is dictated by available computer resources, and often it is impractical to perform a convergence test by carrying out calculations at even two different model resolutions[14].

Ocean models make a large demand on computer resources: CPU time, core memory, and disk storage[15]. Because ocean eddies are much smaller than weather systems, the resolution needed is therefore much finer. Fine resolution also forces one to take smaller time steps in explicit models, as per the CFL condition which requires that $C = \frac{u\Delta t}{\Delta x} \leqslant C_{max}$ (1D case), C being the Courant Number, u the magnitude of velocity, Δt the time step, and Δx the grid spacing[16]. For explicit free surface models, the time step is limited by the speed of the fast-moving surface gravity waves, and one has to take a large number of small time steps to integrate over a simulation or forecast period[17] (mode splitting helps alleviate this problem)[18]. For implicit ocean models, which filter out these gravity modes, the CPU time requirement is governed by the rate of convergence of the iterative method[19].

Taking advantages of high-performance computer architecture, explicit model codes are usually readily vectorizable and parallelizable, and generally need few additional arrays to store the auxiliary variables that may be needed to speed up the computations. In contrast, the vectorization/parallelization of implicit codes is usually a nontrivial problem, and for some schemes that have been used up to now on serial machines, not at all feasible[20]. The extra work resulting from the iterative or matrix inversion solution can often increase the total CPU time so that it is comparable to, or even exceeds, that for explicit codes, especially on vector/parallel computers. In addition, there are almost always extra arrays needed during this stage of the computations. The successive overrelaxation and conjugate gradient methods are two techniques that are well suited to vectorization and parallelization in implicit codes[21].

Disk storage requirements for storing ocean model results are often in tens to hundreds of gigabytes, and depend on the length of the simulation, and how often and how many variables are required to be stored for later analyses. Disk storage and postprocessing requirements often constrain the temporal resolution and the details of the analyses carried out on the results of an ocean model.

Data-assimilative ocean models require even more resources than the free-running ones, the additional memory and CPU time requirements depending very much on the method of assimilation. It is not unusual for assimilation to more than double the CPU time requirements, even for simple optimal interpolation (OI)-type schemes. Methods such as Kalman filters and adjoint methods are even more demanding. Generally, data assimilation on massively parallel computers requires considerable investment of time and effort for efficient implementation[22].

Vocabulary

computational fluid dynamics：计算流体力学
computational：（adj）计算的
partial differential equation：偏微分方程
differential：（adj/n）微分（的）
~differential calculus：微分学
~integral calculus：积分学
~calculus：（n）微积分
~ordinary differential equation：常微分方程
governing equations：（n）控制方程组
invariably：（adv）一律地，总是，始终
adj：invariable
supplement：（vt/n）补充（物），增补（物）
adj：supplemental
scalar：（adj/n）纯量（的），标量（的），无向量（的）
ensemble /ɒnˈsɒmbl/：（n）整体，全体
parameterize：（vt）参数化
n：parameterization
turbulence closure problem：湍（流）封闭问题
optimum：（adj）最佳的，最优的
ingeniously：（adv）灵巧的，创造性的，巧妙的
adj：ingenious
adversely：（adv）有害地，不利地，破坏性地
adj：adverse
aerodynamics：（n）空气动力学
［aero-］：空气
geophysical fluid dynamics：地球物理流体力学
astrofluid dynamics：天体物理流体力学
=astrophysical fluid dynamics
［astro-］：天文的，宇宙的
~astrophysics：（n）天体物理学
Mach number：（n）马赫数
counterbalance：（vt）使平衡，中和，抵消；（n）平衡力，抗衡力，抵消因子
planetary：（adj）行星的
~planetary wave：行星（尺度）波动
frame of reference：参照系
fictitious：（adj）假的，虚构的
body force：（n）体积力
ant：surface force 表面力
conservative：（adj）保守的，不激进的，不先进的，守恒的
unstructured：（adj）无结构的，非结构化的
adaptive：（adj）适应性的，自适应的
implement：（vt）实现，履行；（n）工具，器具
n：implementation 一种实现，实现方法
efficacy：（n）功效，效力，效能
scheme：（n）方案，规划，计划
round-off error：（n）舍入误差
round：（vt）把（数）四舍五入，使凑整
truncation error：（n）截断误差
truncation：（n）截断
vt：truncate
floating-point number：（n）浮点数
inevitably：（adv）不可避免地，必然地
adj：inevitable
arithmetic：（adj/n）算术（的）
~algebra：（n）代数（学）
~linear algebra：线性代数

sequential：（adj）连续的，接连的，顺序的
operand：（n）操作数，运算对象
[-and]：对象
~analysand：（n）分析对象
discretization：（n）离散，离散化
vt：discretize
variable：（n）变量，变元
ordinary derivative：（n）常微分/导数
derivative：（n）导数，微商
partial derivative：（n）偏微分/导数
Taylor series expansion：泰勒级数展开
grid spacing：（n）网格距
finite difference：有限差分（法）
~finite volume：有限体积（法）
~finite element：有限元（法）
manyfold：（adj）多倍的
memory：（n）（计算机的）内存
arbitrarily：（adv）任意地
adj：arbitrary
convergence test：收敛性测试
convergence：（n）收敛性，汇集，相交
core：（n）（计算机的）核
time step：（n）时间步长
explicit：（adj）明确的，显式的
as per：（prep）按照，根据
CFL condition：（n）CFL 条件
Courant number：库朗数
free surface：自由表面
ant：rigid lid 刚盖
mode splitting：模态分解
alleviate：（vt）避免，避开
implicit：（adj）隐含的，隐式的
mode：（n）波形，振荡模，振荡型，模态
iterative：（adj）迭代的
vt：iterate 迭代，重复
n：iteration
high-performance：（adj）高性能的
performance：（n）（机器的）性能
architecture：（n）架构，构架，结构
code：（n）代码
vectorizable：（adj）可向量化的
vt：vectorize 向量化
n：vectorization
parallelizable：（adj）可并行化的
vt：parallelize 并行化
n：parallelization
auxiliary：（adj/n）辅助（的）
nontrivial：（adj）非平凡的，不简单的
ant：trival（adj）琐碎的，无价值的，不重要的，（数学）平凡的
serial：（adj）顺序的，连续的，串行的
feasible：（adj）可行的，办得到的
matrix：（n）矩阵
inversion：（n）（矩阵的）逆，求逆
adj：inverse 相反的，倒转的，反向的，逆的
successive：（adj）连续的，接连的，依次的
overrelaxation：（n）超松弛
relaxation：（n）松弛，弛豫，重新平衡
conjugate：（adj）（复数、角、弧等）共轭的
gigabyte：（n）10^9个字节
byte：（n）（计算机的）字节
postprocessing：（n）后处理
temporal：（adj）时间上的
data-assimilative：（adj）（数值模式）同化观测数据的，受观测数据约束的
=assimilative
n：assimilation
free-running：（adj）（数值模式）自由运行的
optimal interpolation：（n）最优插值

optimal：(adj) 最佳的，最优的
interpolation：(n) 插值
　vt：interpolate
Kalman filter：(n) 卡曼滤波
adjoint method：(n) 伴随方法
　adjoint：(adj/n) 伴随矩阵(的)
demanding：(adj) 要求高的，难度大的，费时费力的

Notes

[1] 计算流体力学就是用数值的方法解一组描述流体运动的偏微分方程，得到的结果不是某变量的解析（analytical）的数学表达式，而是此变量在某时某地的具体数值。此处 coupled(耦合的)意指流体运动的各维度相互之间有关系。nonlinear(非线性的)则主要指方程中的平流项（advection term）$\boldsymbol{u}\cdot\nabla\boldsymbol{u}$。acted upon by various forces 意为被各种力（various forces）作用（act upon）。下文讲到守恒方程，其实运动方程也可以看作动量的守恒方程，各种力就是动量的源（source）和汇（sink）。

[2] 这里虽未明确提到，但文中所说的 scalar conservation equations 包括连续性方程（equation of continuity），因为后者描述质量守恒，而质量也是一种 scalar property。此外还应包括将温度、盐度和密度联系起来的状态方程（equation of state）。通常将动量方程和温盐方程整理成局地变化或倾向（tendency，$\partial/\partial t$）和平流项（advection，$\boldsymbol{u}\cdot\nabla$）在左边，驱动项或源汇项（forcing/source and sink）在右边的形式，只要给定右边各项的值，通过方程左边就能计算变量在空间和时间上的演化。因此，通常将方程左边的处理方法和方程右边的确定方法分开，前者叫作动力学核心（dynamical core，虽然实际上还包括热力学），后者叫作物理代码包（physical packages）。当然，一些易于确定的力，如压强梯度力和科氏力可纳入动力核心，而仅在物理部分涵盖需要依赖参数化来确定的力/源汇，例如地形阻力（bottom drag）、辐射吸/放热、内波混合等。

[3] ensemble 指在某规律控制下能出现的所有情况的“总体”，例如，统计学上符合某概率分布的某变量能出现的所有情况，以及此处提到的湍流运动能出现的所有情况。数值模拟中的一个案例则是相同强迫条件［forcing，或边界条件（boundary condition）］下由不同初始条件（initial condition）开始运行的多次模式模拟，也组成一个总体。相对的，某总体的某具体情况就是一个“样本（sample）”。由于现实中不可能涵盖所有情况，往往一个较大数量的样本集也可以叫作一个总体，因此样本统计量就是总体统计量的近似估计。例如，数学期望（expectation）是总体统计量，而平均值（average）是样本统计量，后者是前者的估计。文中所讲的“总体平均的雷诺型方程组”即放弃对各个单独湍流涡旋的计算，而将其平均起来［雷诺平均（Reynolds mean）］研究其总体规律的方程组。这样的方程组依据大于湍流尺度的背景场的状态来估算湍流运动的整体特征，以及此种特征的湍流对背景场的总体反馈作

用,因此,可不必采用能分辨湍流的精细网格。

[4] 由于湍流黏性系数(viscosity)和扩散系数(diffusivity)是未知的,导致未知数个数多于方程个数,使得方程不可解。确定湍流系数以使方程"封闭"可解的问题就是湍流封闭问题。最简单的方法是将湍流黏性系数和扩散系数视为常数,或者由背景场的某些特征量(如剪切等)决定,或者用更加复杂的公式来描述。每种方法都可以叫作一个湍流封闭方案(scheme),或湍流参数化方案。

[5] 为了能得到最佳的效果(optimum solution),计算资源(computer resources)越多越好。但计算资源毕竟是有限的,应该尽量节省。二者之间的矛盾使得我们不得不做出一些简化,使之既能保留最重要的特征,又是现实可行的(practical)。sought 是 seek 的过去分词,the solutions sought 即寻求中(想要得到)的解。

[6] aerodynamics(空气动力学)更多关注的是飞机、汽车、高楼、风力发电机叶片等结构附近的空气运动情况,往往速度快、尺度小,因此,不考虑地转效应和科氏力,此时的空气(而不是大气)不被看作地球流体。astrofluid dynamics 这一说法不甚规范,更标准的说法应该是 astrophysical fluid dynamics,是研究宇宙天体上的流体运动的学科。与地球物理流体力学(geo-)具有相似性,可互相借鉴。

[7] 声音的传播就是由于介质的周期性压缩/扩张(即密度变化)导致的,因此,声速可看作空气压缩性起重要作用的尺度。相同时间内,流体运动的距离远远小于声音传播的距离,可看作在一个声波波长上占据非常短的一段,空气的密度可看作不变。这导致连续性方程可简化为散度为 0,即将水平方向的辐聚、辐散与垂直方向的上升、下沉联系起来。more than counterbalanced 表示这一简化被层化海洋抵消,而且还有余,即层化海洋使问题复杂化的程度超过了不可压缩海洋使问题简化的程度。好在层化海洋也是薄层海洋,垂向速度通常很小,可通过静力平衡(hydrostatic balance)进一步简化。

[8] 这里描述的主要就是科氏力,即虚假的体力(fictitious body force)。体力(body force)是作用在流体微团整体上的力,与面力(surface force)或接触力(contact force),即作用在某个平面上的力相对。常见的体力有重力、磁力(magnetic force)、科氏力和向心力(centripetal force)等,面力有风应力和摩擦力等。nevertheless 表达的意思是虽然这些力是虚假的,但影响很大。

[9] 三维无结构网格(3D unstructured grid)在海洋中的确很少见到,但二维无结构网格已有不少应用,如在海洋环流模式 FVCOM 和风暴潮模式 ADCIRC 中。无结构网格即没有固定形状的网格,虽通常都是三角形,但各个三角形网格的形状可由模拟人(modeler)自由调整以适应区域加密(regional refinement)或更贴合岸线。此种网格仍是模拟开始之前事先确定的,而自适应网格(adaptive grid)则是每一步计算之后根据当时的流场情况自动确定在何处加密等问题,并在下一步计算时采用调整后的网格。文中说这些先进技术在海洋中未见广泛采用的原因仅仅是(simply)没有

观测数据来进行验证,这一说法未免失于片面。其他原因至少还包括海洋的层化特点使得采用三维无结构网格必要性不大[垂向上尝试用等压面(isobaric)坐标、等密面(isopycnic)坐标等意义更大]、自适应网格大幅增加计算成本导致性价比不高等。

[10] the former 后面的动词 result 用了复数形态,而没有加 s 的原因是 the former 指代的 round-off errors 是复数。计算机表示浮点数的方法是仅保留一定位数的有效数字(significant digits)和相应的指数,如 0.003 72 表示为 3.72×10^{-3},其中 3.72 为尾数(mantissa),-3 为指数(exponent),计算机分配一定的位数分别存储两个部分。这样一来,并非数轴上所有的实数都能表示,而是仅能表示一些离散的数值,相邻两个能表示的实数之间的数字则被向其两端能表达的数字中的一个舍入(round off,此舍入并非四舍五入)。相邻两个能表达的实数之间的距离就是计算机的精度(precision),数值越接近 0 精度越高。算术运算(arithmetic operation)与逻辑运算(logical operation)相对,表示数字的加减乘除和幂运算。

[11] 不仅空间上无法做到无限连续而不得不离散化,时间上也不可能无限连续,也需要离散化,即仅能给出预定的那些离散的空间位置和时间点上的值,这些时空点之间则只能认为不变,或者认为这些时空点上的值代表了其临近区域、前后一段时间内的平均值。预定的离散空间点构成了网格(grid),其两两之间的距离就是网格距(grid spacing)或分辨率(resolution)。离散的时间点之间的距离就是时间步长(time step 或 time interval),时间步长总是固定的。更大的问题出在微商(derivative)上。微商即两个微分的商,即导数,或空间上的梯度(gradient)或时间上的倾向(tendency)。离散化使得无法得到数学意义上的局地"微"分,因为此时的"微元"不够微小,只能是差分(difference)。差分是微分的近似,是假设局地的微分在有限大小的网格空间内满足某种分布(如定常)的情况下进行的估算。差分是用于从一个时刻推知下一个时刻、从边界推知海洋内部的工具。因此,差分的误差会在时间上和空间上传播、累积。

[12] 数字的阶乘英文为 factorial,如 2 的阶乘(2!)读作 factorial two。

[13] 此为最常见的方法——有限差分法(finite difference)的基本思想。有限体积法(finite volume)和有限元法(finite element)的思想与此不同,但当然也无法避免截断误差的存在。of order h 意为在 h 的量级上,也可用 on the order of h 。而 2nd order derivative 为二阶导数,first order accurate scheme 为一阶精确方案。$O(h^n)$ 表示量级,这叫作大 O 表示法(big O notation)。$\varepsilon = O(h)$ 读作 epsilon is big O of h ;注意用的是 is 而不是 equals。$O(h^n)$ 读作 big O of h to the power of n 。此处的差分是用 $j+1$ 处的值来计算的,当然也可以用 $j-1$ 处的值来计算,还可以用这两个差分结果的平均值。这三种格式分别叫作前差(forward difference)、后差(backward difference)、中央差(central difference)。

[14] 此处 convergence 表示数学或计算机中的收敛性,即找到一个分辨率使得舍入误差与截断误差的和最小。理论上存在这样一个最佳分辨率,但实际上不可能去寻找它,因为每一次数值模拟都很耗费资源。往往我们选定一个分辨率就开展模拟,并基于观测资料对模拟结果进行验证(model validation)。只要模拟结果的某些重要方面与观测结果相比不是很差,就用来进行科学分析。

[15] core 可以指 CPU 里的一个单元,也可以指内存里的一个单元,用于存储一个二进制位。core memory 是 20 世纪 50—70 年代的一种旧的内存技术,已经过时,但此说法仍保留下来,用于称呼计算机的内存(memory)。

[16] 显式(explicit)和隐式(implicit)方法或模式是针对时间微分进行求解的数值近似法。显式方法用系统目前的状态来计算下一个时刻的状态,在时间上逐步推进,因此又叫作时间行进法(time-marching)。隐式方法则无法通过目前的状态局地求解下一时刻的状态,需要在空间维度上连立有关下一时刻状态的方程组,通过迭代方式求解(iterative solution)。CFL 条件全称是 Courant-Friedrichs-Lewy convergence condition,即要求库朗数(Courant Number)小于某临界值 C_{max},数值计算才能收敛,否则模式就会不稳定,出现极大或极小的异常值直至溢出(overflow)。显式模式的 $C_{max}=1$,而隐式模式通常更稳定,C_{max} 较大。

[17] 根据 CFL 条件,显式模式的 $\Delta t \leqslant \Delta x/u$,其中 u 是最快的波动的波速,也就是信号传播速度。因此,对于固定的网格距 Δx,波动速度越快,时间步长必须越短。$\Delta x/u$ 实际上就是波动穿过一个长度为 Δx 的网格所需要的时间,而时间步长必须小于此时间,也就是说,波动穿过网格的过程应该跨越不止一个时间步,使得模式得以将此过程表示出来。否则,波动在一个时间步长内已经从此网格过去了,造成下一个时刻模式无法表达此运动,也无法对其速度进行耗散,使得模式不稳定。由于开尔文波(Kelvin waves)等表面波的速度很快,要能明确模拟它们就需要很小的时间步长。这种对步长要求很高的模式或方程可用一个专门的形容词 stiff(刚性的)来形容,对应的名词是 stiffness(刚性)。free surface model 是自由表面模型,表示模式里海面是可以起伏的,因此允许短而快的重力波存在。为此需要将海面高度也作为一个模式预测量(prognostic quantity),而不是诊断量(diagnostic quantity)。与之相对的是早先的刚盖近似(rigid lid approximation)模型,这种模式的海面总是平的,不能起伏,因此,对时间步长要求低(即刚性低),但缺失了很多重要的物理过程,现在已经很少使用。

[18] 为了提高时间步长(即降低刚性),可以用模态分解(mode splitting)的方法。此方法将流体运动分解为快速变化[fast,或外部(external),或正压(barotropic)]模态和缓慢变化[slow,或内部(internal),或斜压(baroclinic)]两个组分,其中缓慢变化部分用隐式格式和较长的时间步长求解,而快速变化部分则对时间步长进行加密,用较短的“子步长(sub-time step)”进行求解。这样可减少一部分计算量。

[19] 隐式模式的求解负担（computational load）主要来自每个时间步都要用到多次迭代，但这种额外的负担相比于显式模式的很小的时间步长造成的计算量倍增来讲有时可能是划算的。

[20] 此处涉及并行计算（parallel computing）的重要概念。并行性（parallelism）可分为数据并行性（data parallelism）和任务并行性（task parallelism）两种，前者又叫向量计算（vector computing），后者又叫并发计算（concurrent computing）。向量化（vectorization）就是将代码设计为能实现数据并行计算的操作过程，最常见的实现类型是“single instruction，multiple data（SIMD）”。数据并行计算时，每个 CPU 核心执行的是一样的操作，只是各自处理的数据不同。也就是说，针对一个巨大的数组进行计算时，将它拆分成多个部分，每个部分交由一个核心进行计算，不同核心的计算操作是一样的，最后将计算结果归并（merge）起来。并发化（concurrentization）则是将代码调整为可进行任务并行计算的过程。任务并行计算时，每个核心的计算任务不同，针对的数据也不同，相互之间没有必然联系，只是同时进行。这两种并行方式各有适合的应用场景，相互结合可大幅提高计算效率。实际上 vectorization 是 parallelization 的一种，但本文将二者相提并论，表明此处 parallelization 一词可能特指 concurrentization。文中提到的 serial 意为串行的，即没有并行能力，只能逐次顺序计算。有些专为串行计算机设计的隐式求解方案无法并行化。

[21] 逐次超松驰迭代法（successive overrelaxation）是一种经典的迭代算法。使用超松弛迭代法的关键在于选取合适的松弛因子，如果松弛因子选取合适，则会大大缩短计算时间。此方法公式简单，编制程序容易，很多工程学、计算数学中都有应用。共轭梯度法（conjugate gradient）是另一种解大型线性方程组的有效方法，其优点是所需存储量小，收敛快，稳定性高，而且不需要任何额外参数。

[22] 数据同化（data assimilation）是在模式运行过程中引入观测数据来矫正模式偏差（bias）的操作。由于相当于给模式强加了一个限制，与之相对的不进行数据同化的模拟操作就叫作 free-running model/simulation。数据同化的方法有多种，本文提到的就有卡曼滤波（Kalman filter）、伴随方法（adjoint method），以及广义上也可算作同化方法的最优插值（optimal interpolation）。显然，同化操作的引入大幅提高计算负担，但也能给出更符合观测事实的模拟结果。

17. Langmuir Circulation

Text

The surface of a wind-driven ocean often is marked by streaks roughly aligned with the wind direction. These streaks, or windrows, are visible manifestations of coherent subsurface motions extending throughout the bulk of the ocean surface mixed layer, from the surface down to the seasonal thermocline[1]. Windrows and their subsurface origins were first systematically studied and described by Irving Langmuir in 1938, and the phenomenon since has become known as Langmuir circulation[2]. The existence of a simple deterministic description making these large scales theoretically accessible distinguishes this problem from coherent structures in other turbulent flows[3]. The theory traces these patterns to a convective instability mechanically driven by the wind waves and currents[4]. Recent advances in instrumentation and computational data analysis have led to field observations of Langmuir circulation of unprecedented detail. Although the body of observational data obtained since Langmuir's own work is mainly qualitative, ocean experiments now can yield quantitative measurements of velocity fields in the near-surface region[5]. New measurement methods are capable of producing data comprehensive enough to characterize the phenomenon, and its effect on the stirring and maintenance of the mixed layer, although the labor and difficulties involved and the sheer complexity of the processes occurring in the surface layer leave much work to be done before this can be said to be accomplished[6]. Nevertheless, the combination of new experimental techniques and a simple and testable theoretical mechanism has stimulated rapid progress in the exploration of the stirring of the ocean surface mixed layer.

Langmuir circulation takes the ideal form of vortices with axes aligned the wind[7], as in the schematic drawing in Figure Ⅰ-10. The appearance resembles convective rolls driven by thermal convection[8], but all evidence indicates that the motions are due to mechanical processes through the action of the wind, as Langmuir originally indicated. At the surface, rolls act to sweep surface water from regions of surface divergence overlying upwelling water into convergence zones overlying downwelling water. Floating material is collected into lines of surface convergence visible as windrows.

In confined bodies of water, such as lakes and ponds, windrows are very nearly parallel to the wind, and can have a nearly uniform spacing. In the open ocean, evidence indicates that windrows tend to be oriented at small angles to the wind (typically to the right in the Northern Hemisphere), spacing is more variable, and individual windrows can be traced only for a

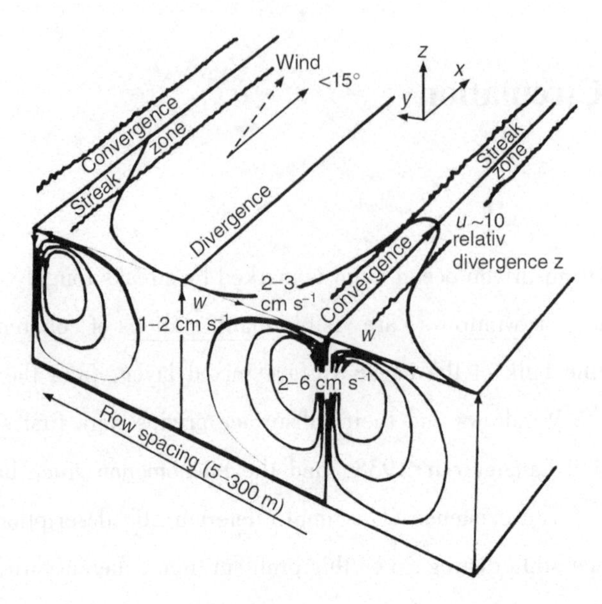

Figure Ⅰ-10: Sketch of Langmuir circulation. [Source: Pollard, 1977]

modest multiple of the mean spacing[9]. A windrow may either terminate, perhaps due to local absence of surface tracers, coalesce with an adjacent windrow, or split into two daughter windrows. Thus in the ocean, the general surface appearance is of a network of lines, occasionally intersecting, yet roughly aligned with the wind.

Windrows are visible in nature only when both Langmuir circulation and surface tracers are present. In the ocean, bubbles from breaking waves are the most readily available tracers, and Langmuir circulation and bubbles both appear to exist when speeds exceed some threshold. Threshold wind levels are not absolute, since swell, wind duration, fetch, and currents existing before the onset of wind forcing play a role, but windrows are commonly reported in winds of 3 m/s or more[10]. Tracers other than bubbles may produce windrows revealing underlying Langmuir circulation—all forms of flotsam serve. Langmuir first noticed windrows from the deck of a ship in the Sargasso Sea, which contained windrows of sargassum. Organic films on the water surface are compressed in windrows, causing capillary waves to be preferentially damped; in light winds, windrows thereby are made visible as slicks of smoother water[11]. Observations in the infrared reveal windrows due to variation of surface temperature created by Langmuir circulation.

A hierarchy of horizontal scales is observed, with windrow spacing ranging in the ocean from a few meters up to approximately 3 times the depth of the surface mixed layer. The largest scales are the most energetic and the most persistent, and extend to depths comparable to that of the mixed layer, so a cell extending from surface divergence to surface convergence, and from the surface to maximum depth of penetration, is approximately square. The maximum penetration depth in the open ocean is comparable to the depth of the seasonal thermocline. Langmuir reported the penetration depth in Lake George to be comparable to the epilimnion when the lake was stratified, and believed the epilimnion to be created by the mixing caused by the wind-driven convective motion. Whether the seasonal thermocline location is fixed by the scale of Langmuir circulation, or whether the penetration depth of the circulation is limited by the strong buoyancy at the thermocline acting like a bottom is not yet clear, although the latter seems more likely[12]. In shallow water, the seabed or lake floor of course fixes the maximum penetration depth. While horizontal scales of up to 3 times the mixed layer depth are reported, it is not clear that even larger scales exist that have not been detected in experiments[13].

The smaller scales are advected and presumably eventually swept up by the largest scales[14]. If a fixed number of permanent surface markers were used, as in some experiments where computer cards were released to serve as markers, the larger scales ultimately would be more prevalent. Most observations do depend on Lagrangian tracers, in particular bubbles with definite lifetimes that are continually but episodically created. The regeneration of bubbles on the surface between the large-scale windrows permits the smaller scales to be seen[15].

The largest observed windrow scales consolidate in about 20 min after a shift in wind direction, or in cases of sudden wind onset. This appears to establish the formation time, at least that required to sweep surface material into windrows. The largest windrow scales in the Pacific have been observed to persist for hours.

Downwelling speeds below windrows are substantially higher than upwelling speeds occurring below surface divergences. Furthermore, the downwind surface speed is larger in windrows than between them. Although the observed speed increase is commonly reported, it has not often been quantified. It appears, however, that the speed increases in surface jets seem to be comparable to the maximum downwelling speeds[16].

Theory promulgated in the 1970s has influenced experiments addressing Langmuir circulation. The theory most commonly utilized, and now commonly referred to as the Craik-Leibovich theory after its originators, begins with Langmuir's conclusion that the cellular motion bearing his name derives from the wind. Wind blowing over a water surface has two simultaneous consequences: currents are generated as horizontal momentum is transmitted from wind to water; and waves are generated on the water surface due to an instability of the air-sea interface under wind shear.

A detailed treatment of the shear flow in the presence of wind waves is not feasible. The eddy turnover timescale[17] (time required for a fluid particle to traverse the convective cell) in Langmuir circulation is on the order of tens of minutes. Surface waves have a substantially shorter timescale. Wind-driven water waves can be thought of as comprised of the superposition of wavelets with a continuous range of wavelengths and frequencies, and the amplitudes of waves in a given band of wavelength or frequency can be characterized by an energy spectrum[18]. For the Pierson-Moskowitz (PM) empirical wind wave spectrum, the waves at the peak of the spectrum of a wind-driven ocean under wind speed U have period approximately $7g/U$. This is a typical value for the energetic part of the wind wave spectrum[19]. For wind speeds leading to observable Langmuir circulation, a characteristic peak wave period is of the order of 10 s. Averaging over the waves is therefore useful. Orbital speeds of surface gravity waves near the surface are an order of magnitude larger than the mean speeds in the current system[20]. Although a wind-driven wave field is complex, and generally must be treated probabilistically, the theory depends only on averaged effects due to the net mass drift caused by the waves (the Stokes drift), and this often can be computed[21].

The driving force of the Langmuir cells is an interaction of the mean flow with wave averaged flows of the surface waves. Stokes drift velocity of the waves stretches and tilts the vorticity of the flow near the surface. The production of vorticity in the upper ocean is balanced by downward (often turbulent) diffusion υ_T. For a flow driven by a wind stress τ characterized by friction velocity u_* [22], the ratio of vorticity diffusion and production defines the Langmuir number:

$$L_a = \sqrt{\frac{\upsilon_T^3 k^6}{\omega a^2 u_*^2 k^4}}$$

or

$$L_a = \sqrt{\frac{\upsilon_T^3 \beta^6}{u_*^2 S_0 \beta^3}}$$

where the first definition is for a monochromatic wave field of amplitude a, frequency ω, and wavenumber k, and the second uses a generic inverse length scale β, and Stokes velocity scale S_0 [23]. This is exemplified by the Craik-Leibovich equations which are an approximation of the Lagrangian mean. In the open ocean conditions where there may not be a dominant length scale controlling the scale of the Langmuir cells the concept of Langmuir turbulence is advanced[24].

Vocabulary

windrow: (n) 风积丘

manifestation: (n) 证明,证据,例证

vt: manifest 表明,显示,证明

adj: manifest 显然的,清楚的

coherent: (adj) 协调的,一体的,相干的

seasonal thermocline: (n) 季节性温跃层

systematically: (adv) 系统性地

adj: systematic

deterministic: (adj) 确定性的

mechanically: (adv) 机械力的,机械学的,

力学的
adj：mechanical
field observation：（n）现场观测，实地观测，野外观测
unprecedented：（adj）前所未有的，史无前例的
comprehensive：（adj）全面的，详尽的，完整的
sheer：（adj）纯粹的，完全的，十足的（用于强调）
accomplish：（vt）达到，实现，完成
n：accomplishment 成就，成绩，实现，完成
ideal：（adj）理想的，完美的，最佳的
resemble /rɪˈzembl/：（vt）与……相似
n：resemblance 相似性，相似度
adj：resemblant 相似的
convective roll：（n）对流卷
orient：（vt/vi）（使）朝向
variable：（adj）可变的，变化的
terminate：（vt/vi）（使）终结、结束
n：termination
coalesce：（vt/vi）（使）结合、汇合
intersect：（vt）贯穿，横穿，横断
threshold：（n）门槛，临界点，阈值
wind duration：（n）风时
fetch：（n）风区
onset：（n）开始，开端
flotsam：（n）漂浮碎屑，漂浮残骸
sargassum：（n）马尾藻
film：（n）膜，薄膜，薄片
capillary wave：（n）毛细波
capillary：（adj）与毛细现象有关的
preferentially：（adv）优先地，倾向性地
adj：preferential
n：preference 偏爱，倾向，优先选择
damp：（vt）压制，使衰减，减幅，阻尼
n：damping
hierarchy：（n）（有层次分级的）体系
epilimnion：（n）湖上层（分层湖泊的最上水层）
pl：epilimnia
seabed：（n）海底，海床
lifetime：（n）生命期，存在期，有效期
syn：lifespan（n）寿命，生命期
episodically：（adv）断断续续地，偶尔
adj：episodic
syn：intermittent（adj）间歇的，断断续续的
n：episode（一系列事件中的）一个事件
permit：（vt）允许，许可，使成为可能；（n）许可，允许
n：permission
consolidate：（vt）巩固，强化，加强，整合
promulgate：（vt）公布，颁布，散布，传播
utilize/utilise：（vt）利用，使用
Craik-Leibovich theory：克雷克-列博维奇理论
cellular：（adj）由多个小单元组成的，分格式的
simultaneous：（adj）同时发生的，同时出现的，同时进行的，同步的
syn：synchronous
syn：concurrent
syn：coinstantaneous
syn：contemporary/contemporaneous 同时代的，同时期的
turnover：（n）更替，周转
traverse：（vt/n）横越，穿过
superposition：（n）重叠，叠加

=superimposition

vt：superpose

=superimpose

wavelet：(n) 小波

~wavelet transform 小波变换

~wavelet analysis 小波分析

~wavelet spectrum 小波谱

probabilistically：(adv) 可能性地，概率性地，以概率论的方式进行地

adj：probabilistic

Stokes drift：(n) 斯托克斯漂流

mean flow：(n) 平均流

或作 mean-flow

friction velocity：(n) 摩擦风速

Langmuir number：(n) 朗缪尔数

monochromatic：(adj) 单一的，单调的

syn：monotonic (adj)(数学函数或变量)单调的

Lagrangian mean：拉格朗日平均

Langmuir turbulence：朗缪尔湍流

Notes

[1] windrow(注意不是 window)一词本来指陆地上由于风的作用而堆积起来的干草或谷物等其他物体，译为“风积丘”或“干草列”。虽然用“丘”字，但其形状是长条形，而不是圆丘。海洋学上用在朗缪尔环流里面，指代被风的作用堆积起来的气泡或藻类等形成的条带，建议可采用“风积带”的译法。coherent 意为协调的、一体的、有序的，表明某结构清晰可见、完整有序、并不杂乱。物理学上也指两个波动的相干性。湍流理论里有 coherent eddy 的概念，译为“拟序涡”，因为它们是在杂乱无章的湍流运动中形成的似乎有序的涡旋结构，尺度比一般湍流涡旋大一些。近期开始有研究揭示海洋里也有类似的结构，目前多将其译为“相干涡”(因为 coherent 还可指波动的相干性)，但从概念上讲译为“拟序涡”似乎更贴切。

[2] 此处 since 不像通常情况一样作为介词使用，即后面不跟时间或事件，而是用作副词，表示“自从”“在……以后”，它所修饰的时间或事件出现在句子前面。比如此句，“自从”的对象是上半句里的“studied and described”这个事件。本句若将 since 放到句子的最后可能更容易理解，也是更常见的用法，即“…, and the phenomenon has been known as … since.”。另举一例，“I fell down this morning and have not been able to stand up since.”可以在 since 后面加 then，但没有必要(若加 then，则 since 还是介词)。

[3] 注意此句的结构。主语是 the existence，谓语是 distinguishes，而 making … accessible 是现在分词引导的定语从句，修饰 description。

[4] 风对海洋的作用是双重的，即动力的 (dynamic) 或机械的 (mechanical) 以及热力的 (thermodynamic)。用 mechanical 一词强调风向海洋的机械能 (mechanical energy) 或动量 (momentum) 输入，而不是通过风对热量和水汽的平流作用引起的海气热交换。

[5] 朗缪尔环流的观测涉及海洋近表层流速的观测，目前这种观测是较为困难的。ADCP（声学多普勒流速剖面仪，Acoustic Doppler Current Profiler）通常对几米深度以浅的近表层无能为力。新式的观测手段包括利用声呐（sonar）探测海洋中广泛存在的微气泡（microbubbles，微米量级）造成的声场传播异常，从而推测气泡的位置和移动速度，借此代表流速。此种手段可弥补上层几米范围内的流速测量缺口。

[6] stirring（搅拌）是海洋上混合层存在和维持的重要原因，朗缪尔环流就是产生此种搅拌的一种可能机制。注意区分此句中的 sheer 和另一个常见名词 "shear"（剪切），二者是同源词。sheer 是形容词，用于强调程度之大。sheer complexity 表示 "极高的复杂性"，而 shear complexity 则表示 "剪切的复杂性"。

[7] vortex 和 eddy 都表示涡旋，二者意思基本相等，没有严格区分，只是在不同的术语中有不同的习惯用法。因为中尺度涡的普遍性和重要性，eddy 一词往往在一些情况下隐含地专指中尺度涡或类似现象。此处虽然用了 vortex 来称呼朗缪尔涡旋，但下文也用了 eddy，因此，本文中这两个词汇是混用的。朗缪尔环流中的涡旋是在垂直切面上旋转的，旋转轴是水平方向的，且与风向基本平行（见 Figure Ⅰ-10）。但是，水质点的运动并非仅沿垂直于风向的切面（y-z 平面），而是存在沿风向的分量（x 方向），即从表面辐散带处斜向（曲线）运动至辐聚带处然后下沉。

[8] convective rolls（对流卷）又叫作 horizontal convective rolls，或水平翻滚涡旋（horizontal roll vortices）、云街（cloud street），是大气中出现的有组织的条状云带，与朗缪尔环流一样也与风向一致。

[9] modest multiple 意为少数几个、不多的倍数，即连续很多条并列的 windrow 不多见。

[10] 风较弱时朗缪尔环流也较弱，有观测指出，风速为 2 ~ 12 m/s 时垂向流速为 2 ~ 10 cm/s，垂向流速与风速的比例为 0.0025 ~ 0.0085。较弱的朗缪尔环流使漂浮物堆积的能力较低，因此更难观测。波浪的破碎和白冠（white cap）、气泡的出现都与风速有关，低风速波浪不破碎，没有气泡产生。

[11] 海洋有机膜（organic film）是漂浮在海洋表面的有机物，有来自石油污染的成分，也有天然成分。有机膜形成的条带（slicks）可以指示朗缪尔环流的存在，反过来这些条带也是观察有机膜本身存在性的常用标志物。

[12] 朗缪尔环流可延伸至混合层底/温跃层（温度主导密度），但朗缪尔环流在多大程度上控制混合层底的混合仍然未知，这也就导致了朗缪尔环流的深度和混合层底的深度二者哪个是控制方（determinant）、哪个是被控方（determinand）的问题。很可能二者是相互作用最终达到平衡的。

[13] 本句初读较难理解，主要是因为两个 that 的含义。第一个 that 较为清晰，引导后面的从句来代表前面的 it，即 "not clear" 的主体。第二个 that 的含义受到了 exist 的干扰，实际上应该将 exist 放到句子最后，这样就可看出第二个 that 引导的从句 "have not been detected in experiments" 是 even larger scales 的定语，而 exist 是 even

larger scales 的谓语动词。作者这样安排的原因应是考虑到 have not been … 这个从句较长,在最后突然出现 exist 可能让人以为它是 experiments 的谓语。

[14] 严格意义上 scale 只是“尺度”,不是某种现象或物质,不存在被输运等可能。这里(以及很多其他文章里)通常隐含地用 scale 表示某种特定尺度的过程、现象、结构等。比如,此处 the smaller scales 和 the largest scales 实际上应该是 the smaller-scale cells 和 the largest-scale cells。较小的 cell 在自身旋转的同时也被较大的 cell 带着移动,直至移动到大尺度 cell 的辐聚带处,被压缩、破坏,里面的水进入下沉流,完成大尺度 cell 的闭合循环。

[15] 这里讲的是气泡与其他示踪物的不同,主要在于它可以间断(episodically)但持续地(continually)再生。其他示踪物,如有人曾用过的硫黄,在下降流处将沉到海底,不能返回。或者如海藻等示踪物仅能在辐聚带处富集,因自身密度较小,不能跟随下降流一同下沉。

[16] 这些特征都可在 Figure Ⅰ-10 上看到。downwind surface speed 即沿(顺)风向的流速,前已阐明此分量的存在性。in windrows 即辐聚带处,between windrows 即辐散带处。所谓“speed increase”即这两处的流速差,在沿风向上等于辐聚带处的值,因为辐散带处为 0;而在与风向垂直的方向上也是辐聚处大于辐散处。此处提到的 surface jet 就是辐聚带处的沿风流速极大值形成的急流,或译射流。若没有朗缪尔环流,而仅有埃克曼流,则水平面上流速应该处处相等。

[17] 此即用 eddy 而不是 vortex 来称呼朗缪尔涡旋的例子。

[18] 此段描述海浪谱的概念。这里用的 wavelet 字面意思是小波,实际上表示的是海浪谱里面每个频率或波数上的波浪组分,例如,用傅里叶展开(Fourier expansion)的方法将海面分解成无数多个不同频率和波数的正弦波[即谐波(harmonic)]的叠加。风产生的不是某个特定频率的波而是存在一定的频散,中心频率附近的一定频段范围内的波动叠加形成实际的波面。用词方面之所以选择 wavelet 而不是 harmonic 可能是作者考虑到斯托克斯漂流(Stokes drift)对朗缪尔环流的重要性(见下文),而形成斯托克斯漂流的是非线性的斯托克斯波(Stokes wave),不是线性的谐波。wavelet 一词还有更普遍的用法,即小波变换(wavelet transform)和小波分析(wavelet analysis)、小波谱(wavelet spectrum),即用特定的“小波”代替正弦函数为正交基进行频域展开的方法。

[19] PM 谱是用经验的方式(即用观测资料拟合)得到的充分成长状态(fully developed)的海浪谱,即能量密度(power density)随频率的解析表达式。谱峰频率(peak frequency)即能量(振幅)最大的组成波的频率,频率高于和低于谱峰频率的波能量逐渐减小。PM 谱的谱峰频率与风速的关系是 $f_p = \dfrac{1.8\pi g}{U_{10}}$,其中,$g$ 是重力加速度(gravity acceleration),U_{10} 是 10 m 高度风速,f_p 的单位是赫兹(Hz)。文中给的表

达式未指定风速的测量高度。

[20] orbital speed 即沿轨道旋转速度,即水质点作近似圆周运动的速度,此速度的水平分量在海面上最大。

[21] 实际海浪的情况很复杂,很多时候必须以概率论的角度来研究（probabilistically）,也就是研究随机海浪（stochastic waves）的统计特征。斯托克斯漂流即非线性海浪的运动轨迹不闭合导致的水质点的净位移,可通过理论推知。对于深水波, $u_S \approx \omega k a^2 e^{2kz}$,其中,$\omega$ 是角频率, k 是波数, a 是振幅, z 是深度。海面上（$z = 0$）,对于固定频率和波数的组成波,斯托克斯漂流的速度与振幅的平方成正比。根据 PM 谱和频散关系（dispersion relation）,可以得到谱峰处和平均的漂流速度。

[22] 摩擦风速 u_* 与风应力 τ 的关系是 $u_* = \sqrt{\tau/\rho_a}$,ρ_a 为空气密度。摩擦风速本身没有意义,它只是风应力的另一种表达形式,用来更好地与湍应力和黏性应力保持一致性。

[23] 第一种定义只针对一个固定频率和波长的波,因此说 monochromatic（单一的）wave field。第二种则与频率和波数无关,是将风驱动下产生的波浪的统计特征进行了带入。

[24] 克雷克-列博维奇方程或理论将总流速分解为平均流速 $\boldsymbol{u}$ 和斯托克斯漂流速度 $\boldsymbol{u}_s$,由此在方程中引入了“克雷克-列博维奇涡旋力（Craik-Leibovich vortex force）”,即 $\boldsymbol{u}_s$ 与平均流涡度的叉积 $\boldsymbol{u}_s \times (\nabla \times \boldsymbol{u})$。同时,漂流的存在也产生一个额外的力——科里奥利-斯托克斯力（Coriolis-Stokes force）,可简称“科斯力”（注意不是科氏力）,表达形式是 $f\boldsymbol{k} \times \boldsymbol{u}_s$,与科氏力 $f\boldsymbol{k} \times \boldsymbol{u}$ 相似。压强梯度力也经过调整,多出 $\nabla(\boldsymbol{u} \cdot \boldsymbol{u}_s)$ 一项。拉格朗日平均（Lagrangian mean）,或一般化（generalized）拉格朗日平均,是一种将流场分解为平均流和波致振荡流（oscillatory flow）两部分,并表达振荡流的统计平均特征与平均流相互作用的方法。朗缪尔湍流（Langmuir turbulence）,表征许多不同尺度、杂乱无章的朗缪尔流环叠加在一起的湍流状态。

18. Estuarine Dynamics

Text

An estuary, in the strictest definition, is formed at the mouth of a river, where the river meets the sea. Cameron and Pritchard (1963) defined an estuary as "a semi-enclosed coastal body of water having a free connection to the open sea and within which the seawater is measurably diluted with fresh water deriving from land drainage."[1] They restrict the definition to coastal features and exclude large bodies of water such as the Baltic Sea. The river freshwater, which enters the estuary, mixes to some extent with the salt water therein to form brackish water and eventually flows out to the open sea in the upper layer. The mixing processes are mainly due to tides and the wind[2]. A corresponding inflow of seawater takes place below the upper layer. The inflow and outflow are dynamically associated so that while an increase in river flow tends to reduce the salinity of the estuary water, it also causes an increased inflow of seawater, which tends to increase the salinity[3]. Thus an approximate steady state prevails.

There are many types of estuaries and many types of flow in estuaries. Estuaries are classified in terms of both their shape and their stratification. They can also be classified in terms of tidal and wind forcing. The inland end of an estuary is called the head and the seaward end the mouth[4]. "Positive" estuaries have a river or rivers emptying into them, usually at the head[5]. In terms of geology, three specific types of estuary are recognized: the coastal plain (drowned river valleys, or rias), the deep basin (e.g., fjords), and the bar-built estuary (lagoon-type, or mouth bar-type)[6]; there are also types that do not fit in these categories. In terms of stratification and salinity structure, estuaries have been classified based on the distribution of water properties as (a) vertically mixed, (b) slightly stratified, (c) highly stratified, and (d) salt wedge estuaries (Figure Ⅰ-11)[7]. The stratification is due to salinity, because density in estuaries is determined mainly by salinity rather than by temperature. The classification system is not rigid.

In an estuary, the flow is out to the ocean in the upper layer and into the estuary in the bottom layer. In stratified estuaries, the depth of the halocline (thickness of the upper, low salinity layer) remains substantially constant from head to mouth of an estuary for a given river runoff. If the estuary width does not change much, then the depth remains constant, which means that the cross-sectional area of the upper layer outflow remains the same while its volume transport increases because of the entrainment of salt water from below. Consequently, the speed of the outflowing surface layer markedly increases along the estuary from head to mouth[8]. The increase in

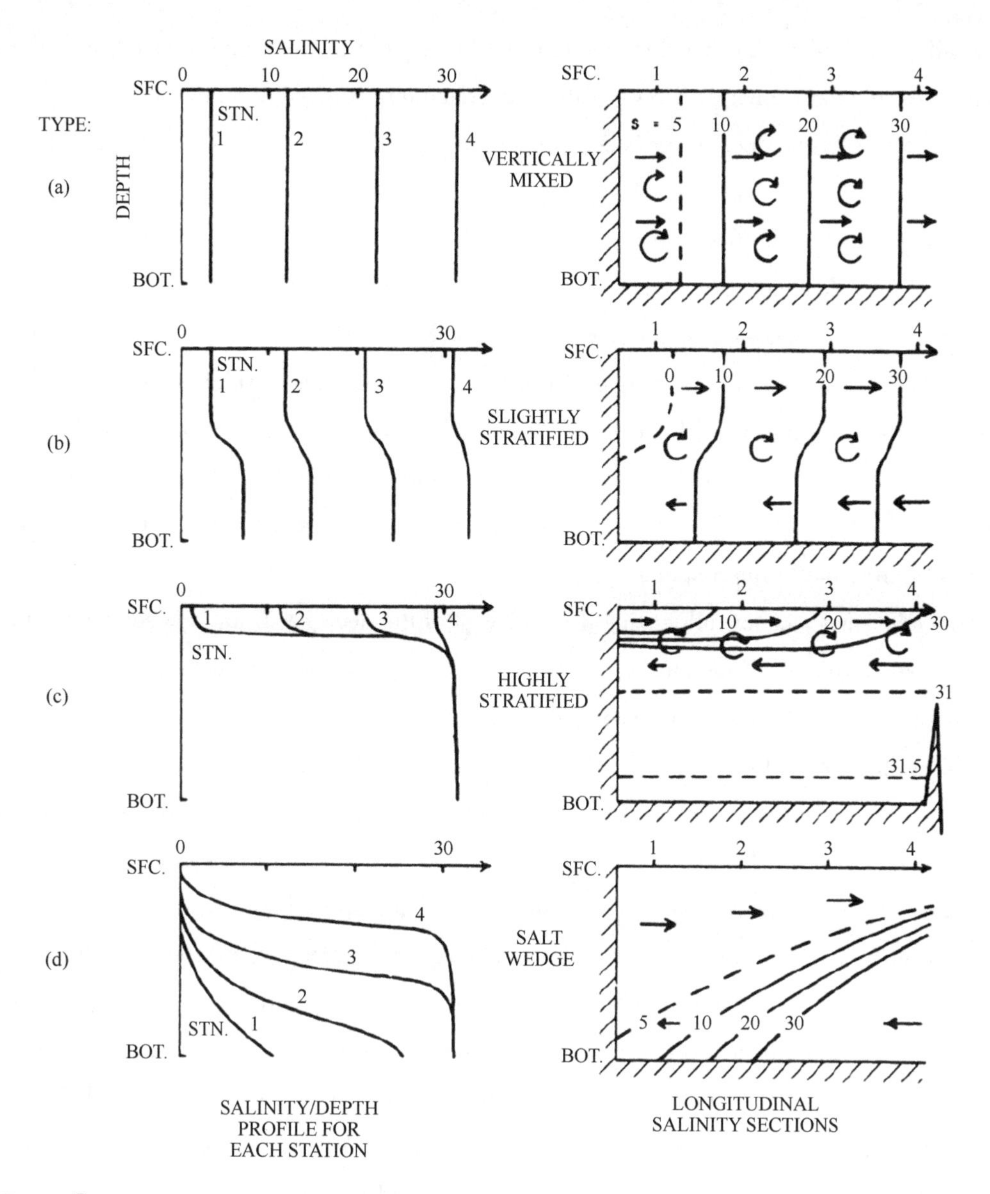

Figure Ⅰ-11: Typical salinity/depth profiles (left) and longitudinal salinity sections (right) in different types of estuaries: (a) vertically mixed, (b) slightly stratified, (c) highly stratified, and (d) salt wedge. [Source: Talley et al., 2011]

volume and speed can be considerable, with the outflow at the mouth as much as 10 to 30 times the volume flow of the river. In his classical study of Alberni Inlet—a typical, highly stratified, fjord-type estuary in British Columbia—Tully (1949) demonstrated the above features. He also showed that the depth of the upper layer decreased as the river runoff increased up to a critical value and thereafter increased as runoff increased.

Estuarine circulation depends on several factors: the sill depth, river runoff rate, and the character of the outside water density distribution. Tides and mixing also impact the circulation.

If the sill is so shallow that it penetrates into the low-salinity, out-flowing upper layer, the full estuarine circulation cannot develop and the subsurface inflow of saline water does not occur regularly. As a result, the deep water is not exchanged regularly and tends to become stagnant. This situation occurs in some of the smaller Norwegian fjords, but is by no means typical of deep basin estuaries. Most of the fjords in Norway, as well as on the west coasts of North and South America and New Zealand, have sills that are deeper than the upper layer. Therefore the estuarine circulation is developed sufficiently to affect continual renewal of the deep water and stagnation does not occur. The rate of renewal is proportional to the circulation, which is proportional to the river runoff. Fjord estuaries with small river runoff show more evidence of limited circulation in the form of low oxygen values than those with large runoff[9]. Hypoxic, and even anoxic waters, therefore, may form. The depth of the sill has little effect as long as it is greater than the depth of the low-salinity, outflowing upper layer.

The other major factor influencing the exchange of the deep basin water is seasonal variation in the density structure of the outside seawater. Although the downward mixing of fresh water in an estuary is small, it does occur to some extent. Therefore the salinity, and hence the density of the basin water, tends to decrease slowly. If a change then occurs in the outside water such that the density outside becomes greater than that inside at similar levels above the sill depth, then there will be an inflow of water from the sea. The inflowing water is likely to sink, although not necessarily to the bottom, in the estuary basin and displace upward and outward some of the previously resident water. In this way the basin water becomes refreshed. In deep-sill estuaries this refreshment may occur annually, but in shallow-sill estuaries it may occur only at intervals of many years; the disturbance to the biological regime may be cataclysmic on these occasions (by displacing upward into the biotic zone the low-oxygen water from the bottom)[10]. This type of basin-water replacement has been well documented for some Norwegian fjords (with very shallow sills), but it should not be considered characteristic of all fjord estuaries.

The time that it takes to replace the freshwater within an estuary through river discharge is called the flushing time. This is important for water quality within estuaries. The flushing time has significant temporal variation, especially since river flows have strong variability. Following Dyer (1997), the flushing time t_F is the time needed to replace its freshwater volume (V_F in units of m^3) at the rate of the river discharge (R, in units of m^3/s). Therefore, t_F equals V_F divided by R. Both the freshwater volume and the river discharge can be time dependent. Using observations of the average salinity $\langle S \rangle$ within the estuary compared with the seawater salinity S_0 outside the estuary, the freshwater fraction can be estimated as $F = (S_0 - \langle S \rangle)/S_0$ [11]. The freshwater volume is the total volume, V, multiplied by the freshwater fraction. The flushing time is then $t_F = V_F/R = FV/R$. Flushing times range from several days to a year. Observing the salinity at all locations in the estuary at all times is unrealistic, so various approximate methods are used to determine the flushing time. Dyer (1997) is a good source for these different methods.

The previous remarks only briefly describe some of the salient characteristics of stratified estuaries. Real distributions show fine and mesoscale structure and detailed features, some general and some local. In particular, because the density structure is determined largely by the salinity distribution, temperature maxima and minima are quite common in the water column[12]. Mixing between fresh and salt water is largely governed by tidal movements and the effects of internal waves. The circulation that was just reviewed for stratified estuaries is greatly modulated by the strong tidal currents in the estuaries. This brief description also neglects the horizontal variability and horizontal circulation in estuaries.

Vocabulary

estuary:(n) 河口(湾),三角湾
  adj: estuarine
  ~delta:(n) 三角洲
semi-enclosed:(adj) 半封闭的
  enclosed:(adj) 封闭的,隔绝的
    vt: enclose
drainage:(n) 排水,排泄,放水
  vt: drain
therein:(adv) 在其中,在那里
brackish:(adj)(水)略含盐分的,微咸的
classify:(vt) 把……分类
  syn: categorize
empty:(vi)(河流)注入,流入;(vt) 倒出,倒空
coastal plain:(n) 沿海平原
  plain:(n) 平原
drowned river valley:(n) 溺谷,溺湾,水淹河谷
  drown:(vt/vi) 淹没,浸没
ria:(n) 溺谷,溺湾,水淹河谷
fjord/fiord:(n) 峡湾
bar-built estuary:(n) 障壁坝型河口湾
lagoon:(n) 潟(xì) 湖
mouth bar:(n) 拦门沙,河口坝,障壁坝
salt wedge:(n) 盐楔
  wedge:(n) 楔子,楔形物
rigid:(adj) 坚硬的,刚性的,固定不变的
entrainment:(n) 卷挟,卷走,卷入
  vt: entrain
classical:(adj) 经典的,传统的
inlet:(n) 小海湾,小河湾,进口,入口
critical value:(n) 临界值
  critical:(adj) 临界的
thereafter:(adv) 其后,此后,之后
sill:(n) 海槛
stagnant:(adj)(水)不流动的,停滞的
  n: stagnation
by no means: 绝不,一点也不,肯定不
renewal:(n) 更新,更换
  vt: renew
hypoxic:(adj) 低氧的
  n: hypoxia

anoxic：(adj) 缺氧的

n：anoxia

refresh：(n) 更新，刷新

n：refreshment

cataclysmic：(adj) 天灾的，自然剧变的

syn：catastrophic (adj) 灾难性的

biotic：(adj) 生命的，生物的，生态的

ant：abiotic

~biotic zone：生物区

~biotic factor：生物因子

~abiotic factor：非生物因子

document：(vt)(书面)记录，记载，证明

n：documentation

flushing time：冲刷时间，净化时间

divide：(vt)(数学)除

n：divisor 因子，除数

n：dividend 被除数

~denominator：(n)(分数的)分母

~numerator：(n)(分数的)分子

multiply：(vt)(数学)乘

remark：(n/vt) 评论，谈论，论证

salient：(adj) 突出的，显著的，重要的

Notes

[1] 河口湾 (estuary) 是一种重要的近海海洋环境类型，是河口的一种形态。如果河口区没有大量的沉积物堆积，而是被海水淹没，则称之为河口湾，中文又叫三角湾，典型的例子如钱塘江口。另一种常见的河口类型是三角洲 (delta)，是大量泥沙沉积形成的露出水面的地形，典型例子如尼罗河三角洲。长江口几千年前是喇叭口状的河口湾，但泥沙淤积使得现在已形成三角洲地貌。semi-enclosed 意为半封闭的，semi-前缀表示“半”的意思。河口湾是河流环境 (river environment) 和海洋环境 (maritime environment) 的过渡带，也是一种生态过渡带 (ecotone)。

[2] 陆源淡水由于密度低于海洋咸水，因此浮在上层，淡水和咸水的边界形成垂直方向上的盐跃层 (halocline)，以及水平方向上的锋面 (front)，也是混合发生的地方。

[3] 盐度从河口向外海逐渐增大，因此密度也逐渐增大，形成水平方向的压强梯度力，迫使下层海水向岸运动 (inflow)。上层海水则保持河流入海的惯性而向外运动 (outflow)。出流增大，则盐度梯度增大，因此入流也增大。河流的汛期 (flood season) 和枯水期 (dry season) 导致出流流量的变化。

[4] 河口湾(以及其他河口海洋环境)的上界(湾顶，head)通常用感潮界 (tidal limit 或 head of tide 或 tidehead) 来定义，即河流能感受到潮汐作用的最远距离，可用潮流(潮流界，tidal current limit)或潮差(潮区界，tidal range limit)来定义。感潮界虽然随着大、小潮 (spring tide 和 neap tide)、汛期和枯水期 (flood season 和 dry season) 等变化，但可定义一个平均位置，即正常感潮界 (normal tidal limit)。长江在枯水期的潮流界约在江苏镇江，潮区界约在安徽大通；汛期的潮流界约在江苏江阴，潮区界约在安徽芜湖。另外，还可用海水入侵界 (seawater intrusion limit，或称盐分入侵界，salt intrusion limit) 来定义，即盐度与淡水盐度之差接近 0 的地方。

[5] 与正河口湾（positive estuary）相对的负（negative）或反（inverse）河口湾通常在干燥气候区（arid/dry climate）形成，此处由于蒸发大大超过降水，形成一个高盐度区，高盐水沉于底层并向河与海两侧扩展，而海水和河水都在上层流向此区域，例如，澳大利亚南部的斯潘塞湾（Spencer Gulf）。地中海（Mediterranean Sea）也可看作一个大的负/反河口湾（negative/inverse estuary），直布罗陀海峡（Strait of Gibraltar）处地中海水从底层溢出，而大西洋水从表层流入。

[6] 溺谷（drowned river valley 或 ria）型河口湾又叫沿海平原，是海水入侵河谷（river valley）形成的，也是温带气候条件（temperate climate）下最常见的河口湾类型。此种类型的宽度往往远大于深度（宽度－深度比很大），深度很少超过 30 m，形成喇叭口形状，或称楔形（wedge-shaped，即楔子或尖刺一样的形状）。峡湾型（fjord-type）河口湾是在冰川期（ice age）由冰川在事先存在的河谷的基础上加深和加宽形成的，此即冰川的侵蚀（erosion）作用。此种类型的特点是水平方向呈狭长的"U"形（U-shaped），湾侧陡峭，湾口处往往有冰川残留物质［即冰碛（qì），moraine］形成的海槛（sill），因此，湾口处水深反而最浅，内部水深可超过 300 m，宽度/深度比较小。此种类型往往在高纬度地区（如阿拉斯加、加拿大、格陵兰、冰岛、新西兰、挪威等）出现。潟湖型（lagoon-type）或障壁坝型（bar-built）河口湾在泥沙沉积较多、海底地形平坦处形成，被沙嘴（spit 或 sand spit 或 barrier spit）或障壁岛（barrier island）与外海阻隔开［沙嘴、障壁岛统称障壁沙滩（barrier beach）或障壁坝（barrier bar）或拦门坝（mouth bar）］，仅通过狭窄的水道（inlet）与外海连接，是很浅的半独立河口湾，多见于热带和亚热带。

[7] 有些分类方法将高度层结型（highly stratified）和盐楔型（salt wedge）合并，因为盐楔型也有很强的层结，而高度层结型在倾斜海底（从河口向外逐渐加深）时底层高盐水也呈楔形（上平下尖）。Figure Ⅰ－11 没有反映倾斜海底，仅用了平底地形来示意，不要被其误导。

[8] 强的层结是峡湾型河口湾的典型特征，因此内外宽度变化不大。如果径流流量固定，那么单位时间进入的淡水质量和体积不变，淡水进入后在湾内表层流动过程中在垂向截面上的分布面积也守恒。但是随着越来越多的下层海水被卷挟（entrain）进入上层，上层面积不变的情况下只能加速将其向下游输运，因此流量和流速增大。河口湾内的强层结处常有内波发生，可导致混合。

[9] 沉积物中有大量的细菌（bacteria），它们消耗大量氧气，因此底层海水氧含量低，若与外海交换不畅则逐渐形成低氧区［low-oxygen zone，或称死亡区（dead zone）］。开阔外海也可因为物理和生物双重作用形成低氧区［oxygen minimum zone，OMZ，或称阴影区（shadow zone）］。根据海水低氧现象的程度不同，可分为 hypoxic/hypoxia 和 anoxic/anoxia（斜杠前后分别为形容词和名词），分别表示氧气含量低于 2 mg/L 和 0.5 mg/L的情况，可见 anoxia 比 hypoxia 更严重。hypoxia 和 anoxia 可分别译为低氧

和缺氧,但并不严格。dead zone 和 OMZ 都用低氧(hypoxia)指标来定义。与人类活动相关的气候尺度海洋低氧化(ocean deoxygenation)则是指 dead zone 和 OMZ 都有增多的趋势。

[10] 由于海槛的阻挡,河口湾内海盆的底层海水相对独立,随着上层淡水的混合逐渐变淡变轻。海槛以外的温盐季节变化导致(通常)冬季密度更高,因此可越过海槛溢流(overflow)进入湾内海盆。若海槛较高,则外部温度、盐度、密度的季节变化通常不能影响内部,因为上层海水密度总是较轻。但当季节变化与年际甚至年代际变化叠加,可能出现多年一次的极端事件(extreme event),使得海槛以上部分的外海海水密度也足以溢入湾内,由此形成极端的低氧甚至缺氧事件(hypoxic/anoxic event),导致上层海洋生物大量死亡。biotic zone 即上层有生命的区域,粗略指代透光带[photic zone,或真光层(euphotic zone),或光照层(sunlight zone),或光合带(epipelagic zone)]。

[11] 此种估计方法的思路:假设湾内海水本来与外海一样盐度是 S_0,由于盐度基本等于盐分与淡水的质量比例,盐度 $S_0 = s/M_0$,其中,s 是盐分质量,M_0 是淡水质量。后来河流开始向湾内注入一定质量的淡水 M,在此过程中盐分含量 s 不变,淡水质量增加至 $M_0 + M$,盐度降低至 $\langle S \rangle$。因此,$\langle S \rangle = s/(M_0 + M)$。现在要求湾内新增淡水质量 M 与总质量 $M_0 + M$ 之比。经过简单推导可知此比例就等于文中给出的 $F = (S_0 - \langle S \rangle)/S_0$。这个质量比例也等于体积比例,因为淡水密度不变。

[12] 若密度由温度主导,则温度变化导致密度变化,低温水下沉,垂向上形成温度分层、逐渐变化。而河口湾由盐度主导分层,因此,在海面上被加热或冷却的水可能依据盐度被沉至不同水层,造成温度在垂向上不再单调降低,极大值和极小值不再出现在水面和水底,而是出现在中间水层。

19. Empirical Orthogonal Functions

Text

In studies of oceanic variability, we may be presented with a large data set from a grid of time-series stations, which we wish to compress into a smaller number of independent pieces of information[1]. For example, in studies of climate change, it is necessary to deal with time series of spatial maps, such as surface temperature. A useful obvious choice would involve a linear combination of orthogonal spatial "predictors," or modes, whose net response as a function of time would explain the combined variance in all of the observations[2]. The signals we wish to examine may all consist of the same variable, such as temperature, or they may be a mixture of variables such as temperature and wind velocity or current and sea level. The data may be in the form of concurrent time-series records from a grid (regular or irregular) of stations on a horizontal plane or time-series records at a selection of depths on a vertical cross-section. Examples of time series from cross-sectional data include those from a string of current meters on a single mooring or from moorings of upward-looking bottom-mounted Acoustic Doppler Current Profilers (ACDPs) strung across-channel[3].

A useful technique for compressing the variability in this type of time-series data is principal component analysis (PCA). In oceanography, the method is commonly known as empirical orthogonal function (EOF) analysis. The EOF procedure is one of a larger class of inverse techniques and is equivalent to a data reduction method widely used in the social sciences known as factor analysis (FA)[4]. The first reference we could find to the application of EOF analysis to geophysical fluid dynamics is a report by Edward Lorenz (1956) in which he develops the technique for statistical weather prediction and coins the term "EOF."

As discussed by Preisendorfer (1988), one of the essential aspects of the PCA method was developed by an Italian geometer, Beltrami, in 1873. He formulated a modern form of the resolution of a general square matrix into its singular value decomposition (SVD), which stands at the core of PCA[5]. This same discovery was made independently by the French algebraist, Jordan, in 1874. PCA appears to have made its first appearance in the United States as an exercise in abstract algebra when Sylvester (1889) considered the problem of the reduction of a square matrix into its SVD. A decade later, Pearson (1901) recast linear regression analysis into a new form to avoid the common asymmetrical relationship between "dependent" and "independent" variables. In his paper, Pearson introduced a clear geometric visualization of PCA in Euclidean space[6]. The first application of PCA to meteorology appears to have been made at the Massachusetts Institute of Technology (MIT) by G. P. Wadsworth and his colleagues in 1948. The goal of their study was to develop a short-term prediction method for sea level atmospheric pressure over the northern hemisphere. In test calculations over the North Atlantic, Wadsworth was faced, in 1944, with the daunting task of hand-calculating the 91 eigenvalues of a 91 * 91 matrix. Confronted with this unmanageable numerical task, Wadsworth dropped the PCA approach and went on to use theoretical orthogonal functions (Chebyshev polynomials) to complete the project[7]. It is interesting to note that, about the same time, a completely independent use of PCA in meteorology was being carried out by Fukuoka (1951).

When the Whirlwind general-purpose computer became available at MIT in the 1950s, E. N. Lorenz, starting with the work of Wadsworth and colleagues, undertook prediction studies of the 500 mbar height anomaly for January (1947–1952) for a grid of 64 points covering the mainland United States, Southern Canada, and portions of the surrounding oceans[8]. Lorenz (1956) is now a classic in the field of statistical-dynamical approaches to weather prediction. The Statistical Forecasting Project at MIT under Lorenz's direction produced some outstanding early applications of PCA to short-range forecasting[9]. Applications of PCA to oceanographic data sets began to appear about a decade after Lorenz's work. Trenberth (1975) related southern hemisphere atmospheric oscillations to SST observations. PCA studies based on SSTs in the Pacific by Barnett and Davis also appeared in the 1970s along with similar work by Weare et al. (1976). An interesting idea involving the use of extended EOFs (EEOFs) for moving pattern detection in tropical Pacific Ocean temperatures is explored in Weare and Nasstrom (1982)[10].

The advantage of EOF analysis is that it provides a compact description of the spatial and temporal variability of data series in terms of orthogonal functions, or statistical "modes." Usually, most of the variance of a spatially distributed series is in the first few orthogonal functions whose patterns may then be linked to possible dynamical mechanisms[11]. It should be emphasized that no direct physical or mathematical relationship necessarily exists between the statistical EOFs and any related dynamical modes. Dynamical modes conform to physical constraints through the governing equations and associated boundary conditions (LeBlond and Mysak, 1979); EOFs are simply a method for partitioning the variance of a spatially distributed group of concurrent time series[12]. They are called "empirical" to reflect the fact that they are defined by the covariance structure of the specific data set being analyzed.

In interpreting the meaning of EOFs, it is worth keeping in mind that, while EOFs offer the most efficient statistical compression of the data field, empirical modes do not necessarily correspond to true dynamical modes or modes of physical behavior. Often, a single physical process may be spread over more than one EOF. In other cases, more than one physical process may be contributing to the variance contained in a single EOF[13]. The statistical construct derived from this procedure must be considered in light of accepted physical mechanisms rather than as physical modes themselves. It often is likely that the strong variability associated with the dominant modes is attributable to several identifiable physical mechanisms. Another possible clue to the physical mechanisms associated with the EOF patterns can be found in the timeseries coefficients. The temporal variability of certain processes might resemble the time series of the EOF coefficients, which would then suggest a causal relationship not readily apparent in the spatial structure of the EOF[14].

In oceanography and meteorology, EOF analysis has found wide application in both the time and frequency domains. Conventional EOF analysis can be used to detect standing oscillations only. To study propagating wave phenomena, we need to use lagged covariance matrix (Weare and Nasstrom, 1982), or complex PCA in the frequency domain (Wallace and Dickinson, 1972; Horel, 1984). Our discussion focus on space/time domain applications. Readers seeking more detailed descriptions of both the procedural aspects and their applications are referred to Lorenz (1956), Davis (1976), and Preisendorfer (1988)[15].

The best analogy to describe the advantages of EOF analysis is the classical vibrating drum problem. Using mathematical concepts presented in most undergraduate texts, we know that we can describe the eigenmodes of drumhead oscillations through a series of two-dimensional orthogonal patterns. These modes are defined by the eigenvectors and eigenfunctions of the drumhead. Generally, the lowest modes have the largest spatial scales and represent the most dominant (most prevalent) modes of variability[16]. Typically, the drumhead has, as its largest mode, an oscillation in which the whole drumhead moves up and down, with the greatest amplitude in the center and zero motion at the rim where the drum is clamped, i.e., a monopole pattern. The next highest mode has the drumhead separated in the center with one side 180° out of phase with the other side (one side is up when the other is down, i.e., a dipole or see-saw). Higher modes have more complex patterns with additional maxima and minima[17]. Now, suppose we had no mathematical theory, and were required to describe the drumhead oscillations in terms of a set of observations; we would look for the kinds of eigenvalues in our data that we obtain from our mathematical analysis. Instead of the analytical or dynamical solutions that can be derived for the drum, we wish to examine "empirical" solutions based strictly on a measured data set. Since we are ignorant of the actual dynamical analysis, we call the resulting modes of oscillation EOFs.

Vocabulary

linear combination: (n) 线性叠加

n: orthogonality

orthogonal: (adj) 正交的,直角的,独立的

predictor: (n) 预报因子

~predictand：预报对象
variance：(n) 方差
string：(n) 序列,一串；(vt) 使排成一行/列/串
current meter：(n) 海流计
mooring：(n) 锚系(设备/浮标)
mount：(vt) 把……固定/安装到位
Acoustic Doppler Current Profiler：声学多普勒流速剖面仪
profiler：(n) 剖面仪
principal component analysis：主成分分析
empirical orthogonal function：经验正交函数
procedure：(n) 程序,步骤,手续
adj：procedural
data reduction：(n) 数据规约
social science：(n) 社会科学
~natural science：(n) 自然科学
factor analysis：因子分析
reference：(n) 资料,文献,引文
geometer：(n) 几何学家
n：geometry 几何,几何学,几何形状
formulate：(vt) 制订,阐述,(简练、系统、数学地)表述
n：formulation
singular value decomposition：奇异值分解
singular：(adj) 非凡的,奇异的,单个的,独特的
decomposition：(n) 分解,腐败
vt/vi：decompose(使)分解/腐败
[de-]：向下,去除,相反
algebraist：(n) 代数学家
n：algebra 代数,代数学
~linear algebra 线性代数
abstract algebra：抽象代数
recast：(vt) 重新塑造,重订,改写
linear regression：(n) 线性回归
regression：(n)(统计)回归,倒退,消退
vt：regress 回归
dependent variable：(n) 因变量
=response variable
independent variable：(n) 自变量
visualization：(n) 可视化,形象化
vt：visualize/visualise
Euclidean space：(n) 欧几里得空间
Euclidean：(adj) 欧几里得的
~Euclidean geometry
eigenvalue：(n) 特征值,本征值
[eigen-]：本证,特征,固有
Chebyshev polynomial：(n) 切比雪夫多项式
polynomial：(n/adj) 多项式(的)
[poly-] 多
whirlwind：(n) 旋风
outstanding：(adj) 杰出的,出众的
short-range (weather) forecast/forecasting：短程天气预报
~medium-range forecast：中程天气预报
compact：(adj) 紧凑的,密集的,简练的,简洁的；(vt) 使紧凑、密集、简练或简洁
conform：(vt) 遵从,符合(规律)
partition：(vt) 分割,隔开,分开；(n) 部分,分区
covariance：(n) 协方差
[co-]：相互,联合,共有
in light of：考虑到,鉴于,按照,基于
clue：(n) 线索
causal relationship：(n) 因果关系
causal：(adj) 与原因有关的,构成原因的
n：causality 因果关系,因果律,因果性

conventional：（adj）传统的，习惯的

n：convention 惯例，准则，约定，公约，协定，会议，大会

syn：canonical（adj）公认的，正统的，公理的

standing：（adj）持久的，停滞的，不动的

~standing wave：驻波

=stationary wave

syn：stationary（adj）不动的，静止的，稳定的

lagged covariance matrix：（n）滞后协方差矩阵

complex：（adj）复数的

vibrate：（vi/vt）（使）震动，颤动，抖动，摆动

n：vibration

undergraduate：（adj/n）大学生（的）

~postgraduate：（adj/n）研究生（的）

~diploma：（n）毕业/学位/结业证书，文凭

~degree：（n）学位

eigenmode：（n）本征模，特征模态

eigenvector：（n）特征向量

eigenfunction：（n）特征函数

rim：（n）边缘，边界

clamp：（vt）夹住，夹紧

monopole：（n）单极子

adj：-polar

[mono-/mon-]：单，一，单一

dipole：（n）偶极子

adj：dipolar

see-saw：（n）跷跷板形态

可去掉连词符成为一个单词

ignorant：（adj）无知的，不知道的

Notes

[1] 从海量数据中分析和挖掘出其中的最主要信息的技术在各学科中都是非常重要的，在不同的学科和应用领域里有不同的名称，例如，数据挖掘（data mining）、数据降阶（data reduction，或译为规约、缩减）、数据压缩（data compression）、模态识别（pattern recognition）等，这些领域很大程度上有重合。本文要讲的 empirical orthogonal functions 或 principal component analysis 就可归为上面领域中的任何一个。不管是挖掘、规约、压缩还是识别，都表达了发现海量数据中的少数主要特征的意图。

[2] 预报因子（predictor）用来预测、推算预报对象（predictand）的变化。此处并不涉及"预"报，仅表示将一个量表达为其他几个量的线性叠加，因此预报因子又叫基底（basis）或基底函数（basis function）。理想的基底相互之间应该是独立的，即正交（orthogonal）的。用这些基底构成的线性叠加应该能完全表达预报量的变化，即构成完备正交基（complete orthogonal basis）。基底的选择有多种方式，例如，本文介绍的 EOF 方法就是用经验的方法从观测事实中求得基函数，如此基函数的具体形式不定，且没有解析表达式，但个数是确定的，等于空间点的个数。如果事先指定基函数为某些特定频率上的正弦函数并以经验的方式获取这些正弦函数的振幅和相位，即为调和分析（harmonic analysis）。球面上的球谐函数（spherical harmonics）分解也

是此类。net response 即这些基函数叠加的总和。explain the combined variance 即解释总的方差,每个基函数解释的方差（explained variance,或译方差贡献）之和等于总方差。observation 此处指每次观测,也就是每个时刻的值。每个空间点(二维或三维)视为一个观测站,每个时刻的值就是一次观测。

[3] mooring 作为 moor 的动名词,不仅表示“锚定”这个动作,而更成为单独的名词,表示锚系(或称系泊式)观测设施。一个 mooring 通常由浮子（float/buoy)、绳索(cable/rope/wire)、重物(即锚,anchor),以及挂在绳子上的各种仪器组成。安装在海底的锚系 ADCP 也需要有重物和框架使之固定,只是可省去浮子和绳索;也有将 ADCP 安装在浮子上的。ADCP 可观测垂向流速廓线,原理是通过多普勒频移确定流速,通过声波反射回来所需的时间确定距离。

[4] 大气科学中多用 PCA 这一术语,PCA 与 EOF 和 FA 是一回事。此种方法在不同领域内的其他变种还包括:discrete Karhunen-Loève transform(离散 K-L 变换)、Hotelling transform(霍特林变换)、proper orthogonal decomposition(本征正交分解)、empirical eigenfunction decomposition(经验特征函数分解)、quasi-harmonic modes(准调和模态)、empirical modal analysis(经验模态分析)等。逆向技术（inverse technique）是指从结果发现原因、从总体发现因子的过程,例如,拆解一架飞机以得到其零部件和核心技术的做法叫作逆向工程（inverse engineering)。EOF 方法从混杂的结果中拆解出其构成因子,因此也是逆向工程。

[5] 此处 resolution 不是分辨率的意思,而是“解析”,即分解、拆解,将一个一般的方阵分解为几个矩阵的乘积。SVD 属于矩阵分解（matrix decomposition 或 factorization）的一种,是 EOF 的核心。下文相似的场景下用了 reduction 一词,还可用 factorization。

[6] 一般认为 PCA 是 Pearson 在主轴定理（principal axis theorem)的基础上发明的。主轴定理表明多维椭球（ellipsoid）的多个轴［即主轴（principal axis),类比于椭圆的长短轴］相互之间是垂直的,并给出了求得主轴的过程。所谓的 clear geometric visualization 即类比于主轴定理,将信号(预报对象)看作由多个主轴(预报因子)构成的多维椭球,并用数据来拟合（fit）椭球的各个主轴的长度。这些主轴构成完备正交基,每个主轴对应一个主成分（principal component)。若仅考虑前几个主要主成分,则拟合的过程就是使误差(剩余被忽略的主成分)最小的过程,类似于线性回归(linear regression)。

[7] 切比雪夫多项式（Chebyshev polynomials）是以递归定义的一系列正交多项式序列,在逼近理论（approximation theory）里很有用处。切比雪夫这一俄语姓氏也有其他拉丁化转写如 Tchebycheff、Tchebyshev、Tshebyschev、Tschebyschow 等。

[8] 旋风计算机（Whirlwind）是麻省理工学院（MIT）于 1951 年研制成功的早期计算机,是当时最先进的计算机。用了 5000 根真空管（vacuum tube),内存 2 Kb。500 mbar height anomaly,即 500 毫巴(百帕)等压面的高度异常。

[9] 按照提前时间的长短,天气预报分为临近预报(nowcast,<3 h)、极短期(very short-range,<6 h)、短期[short-range,<48 h,又称日预报(daily forecast)]、中期(medium-range,2~7 d)、长期(long-range,>7 d)。其中,长期又可分为次季节[sub-seasonal 或 intra-seasonal,<3 个月,又称延伸期(extended-range)]和季节(seasonal,3~9 个月)。这些划分标准并不严格,可能因人而异,互有重叠。

[10] EOF 仅依赖各空间点的协方差,即不考虑各点关系在时间上的变化。而 EEOF 考虑了这一点,因此,可以表征随时间变化的特征,即移动的或传播的信号。EEOF 在其他领域也有其他名称,即奇异系统分析(singular system analysis,一维问题)或多元/多通道奇异系统分析(multivariate/multichannel singular system analysis)。

[11] 将特征向量(即特征函数或模态)按照对应的特征值的大小排列,特征值最大的模态即为第一模态(EOF-1),或称主导模态(leading mode),特征值的大小反映了方差贡献的大小。将各模态的特征值绘制成曲线,这种模态编号与特征值的关系图叫作 scree plot,中文可译为碎石图、陡坡图或特征值图。scree 意为山体下方由崩裂的碎石组成的斜坡。因为特征值常随模态编号的增加先迅速降低,而后平缓下降,曲线形式看起来与这种碎石坡类似。常用特征值由陡变缓的拐点(形象地称为 elbow)来决定保留哪些最主要模态用于后续分析,拐点以后的模态被舍弃。

[12] EOF 方法寻找的是一组正交模态,而且每个模态都尽量解释更多的统计方差,而使后续的模态可解释的方差最小。而物理模态不受解释最大方差的条件限制,仅受制于运动方程,各模态实际上是不同尺度上不同外力作用的结果。因此统计模态不一定是物理模态。这是 EOF 方法被挑战的主要方面。necessarily 表示必要性,no … relationship necessarily exists 意为"没有任何(物理/数学)关系是一定存在的",或"不保证存在……关系",即存在的必要性不成立。

[13] 若一个物理模态对应多个 EOF 模态,则如果可能,可将多个 EOF 模态组合起来得到一个模态,而这仅通过模态的线性组合就可以实现。现实困难是选择哪些模态进行合并,一个常见做法是将方差贡献很接近的模态进行组合。这样合并得到的新模态与其余模态仍是正交的。另一种情况,即一个 EOF 模态对应多个物理模态,则表明这些物理模态之间不是正交的。一般地,物理模态往往并不正交,即相互之间有关系。例如,海盆尺度的 SST 变化可能影响区域尺度的 SST 变化(例如,通过风场),或者也可能存在区域尺度对海盆尺度的影响。EOF 对正交性的执着导致有着相互关系的物理模态被拆分、合并、混合、杂糅,使其难以从物理上进行解释。当然,若物理模态之间恰好独立或近似独立,则 EOF 模态仍是真实的。为了更加符合物理特征,20 世纪 80 年代以来旋转 EOF(rotated EOF)方法得到了广泛应用。通过对 EOF 模态进行旋转,放弃了严格正交性,但使获得的模态更简单更易于解释。

[14] 总的信号(各观测点上的时间序列)等于各 EOF 模态与其时间序列乘积的叠加。通常将时间序列叫作主成分(principal component,如与 EOF-1 对应的 PC1 或

leading PC),空间模态叫作系数(coefficient)。为了更好地反映各模态的方差贡献,通常将 PC 进行标准化(normalization,即除以标准差),而将标准差乘在空间模态上。这样,模态可同时反映空间分布(pattern)和振幅大小,而时间序列可看作此模态的伸缩比例。

[15] complex PCA 与 EEOF 都试图解决 EOF 仅反映驻波(standing/stationary)信号、而对传播或移动信号无能为力的问题,但使用了不同的技巧。要表达"请参考……"或"见……",可用被动态:the readers are referred to …,或主动态:please refer to …。

[16] 鼓/膜振动问题(vibration of drum/membrane)研究二维圆形鼓面(drumhead)或薄膜(membrane)振动时的形状问题,是数学和物理领域的经典问题。鼓膜的振动满足波动方程(wave equation),配合边界条件(boundary condition)求解可知膜的振动可分解为一系列由贝塞尔函数(Bessel function)构成的特征模态(normal mode,或译为正则模态、简正模),其中最低频模态叫作基本模态(fundamental mode,或译为基谐模),最低频模态的波数也最低,即空间尺度最大。大尺度振动的振幅也大,这是自然界的基本规律,原因是物质在时间长、空间大的运动中才可能积累足够的能量,在极小的时空尺度集中较大的能量不太现实。这里用膜振动问题来类比 EOF 实际上不是特别合适,因为在已知膜的模态的解析形式的情况下,将膜的振动分解为不同模态这一操作更接近于频域上的调和分析、一维的弦振动(vibration of string)问题、球面上的球谐函数展开(spherical harmonics expansion)问题等。当然,如果假设不知道这些模态,确实可以与 EOF 同样以经验的方式进行分解,如下文所述。

[17] 此规律对 EOF 得出的海洋和大气变量经验模态常也成立,但并不一定。

20. Laboratory Experiments

Text

Laboratory experiments have provided considerable insight and quantitative information about many of the physical processes which affect the fluid ocean. Although often made with the purpose of investigating some fundamental process in fluid dynamics, motivation for making laboratory experiments frequently comes directly from a need to improve understanding of processes in the oceans, or in some other geophysical fluid such as the fluid interior or atmosphere of the Earth and other planets[1], and such studies consequently belong to the broad field of geophysical fluid dynamics. Laboratory experiments are particularly valuable in testing theory and in providing quantitative, if empirical, estimates of, for example, constants of proportionality which cannot presently be determined by theory or numerical computations[2]. They are therefore an essential component of geophysical fluid dynamics in relating theory to reliable application.

Laboratory experiments as a means of illuminating oceanographic processes have a long history which can be traced back at least as far as the experiment by Marsigli, reported in 1681, which demonstrated the way in which density differences drive exchange flows in the Bosphorus between the Mediterranean and the less dense Black Sea[3]. The purpose of making laboratory experiments is rarely, however, to reproduce some aspect of ocean circulation. More often it is to study a particular process in isolation from others which occur in the natural environment. In addition to density differences or stratification, laboratory studies have been made of processes which result, for example, as a consequence of the Earth's rotation (including the β effect; e.g., Ekman spiral and Rossby waves) and from the effect of free or fixed boundaries (e.g., promotion of turbulence or waves)[4].

Several general objectives in making laboratory experiments may be identified and some are briefly described in the following. The particular experiments mentioned as examples are perhaps not always the best which might be chosen, but they are ones (among many) which demonstrate some particular value of making laboratory studies.

Testing Predictions

A beautiful example is the study made by Mowbray and Rarity (1967) of internal gravity wave propagation in a tank filled with salt-stratified water. Waves were generated by the slow oscillation of a horizontal cylinder and made visible using an optical Schlieren system[5]. The experiments demonstrated beautifully the theoretical prediction that internal waves propagate away from the cylinder in a vertical plane at an angle to the horizontal given by $\sin\theta = \sigma/N$, where σ is the frequency of the cylinder and N the buoyancy frequency in the stratified fluid[6].

Providing Measures of Unknown Coefficients or Values

An example is provided by the studies of Britter and Linden (1980) on the flow of a dense gravity current down a sloping boundary. They released dense salty fluid at a constant rate from the top of a uniform slope in a tank filled with fresh water, and measured the speed of advance of the "head" of the gravity current formed by the salty water as it descended the slope[7]. It may be deduced on dimensional grounds that the speed U will be proportional to the buoyancy flux, B , to the one-third power, and to some function of the angle of inclination of the slope to the horizontal, α . The flux, B , is a product of the measured flow rate of salty water into the tank and its reduced gravity (the acceleration due to gravity times the fractional density difference between the ambient water and the dense water)[8]. The experiment shows that the speed is uniform and equal to $(0.5 \pm 0.2) B^{1/3}$ independent of α provided it is greater than about 5°. The experiment nicely supplements the dimensional arguments and provides a value of the coefficient of proportionality (and its uncertainty) which are of use in predicting the descent of dense water down slopes in the ocean provided that the conditions of the laboratory experiment are met, one being that the stratification of the ambient water is negligible[9].

Provision of Information Beyond the Range of Existing Theory or Numerical Calculation

The classic example in this category is the experiment by Osborne Reynolds (1883) on the transition from laminar to turbulent flow in a pipe. It was not made with oceanographic applications in mind but is fundamental, not only demonstrating the nature of flow that will occur in the ocean but providing a measure of the conditions under which a laminar flow becomes turbulent. This "measure" is what is now called the Reynolds number. As a direct consequence of these studies, turbulent transfer and what are now known as Reynolds stresses, are now understood to be active in the processes of flux of momentum, heat, solutes, and suspensions, particularly (but not only) in the boundary layers at the sea surface and the ocean floor[10]. Reynolds was also the first to investigate the onset of mixing in stratified shear flows, now generally called Kelvin-Helmholtz instability (KHI), well before others had advanced the theory sufficiently to identify a critical parameter, the Richardson number, below which instability may occur[11].

An experiment which has had profound influence on the development of analytical and numerical models is that of Morton, Taylor and Turner (1956) on the rise of buoyant thermals, mainly because of their successful demonstration of the validity of a hypothesis attributed to Sir

G. I. Taylor, that the rate of flux of fluid into a turbulent plume (the entrainment rate) is proportional to the mean excess speed of the turbulent fluid through the ambient, and their determination of the proportionality factor. The small-scale laboratory experiment proves useful in providing estimates valid at far greater scale in the atmosphere and the ocean because it is able to identify and quantify the parameters of importance in the process of convection and entrainment[12].

In the experiments of Rapp and Melville (1990) a train of waves is produced by an oscillating wavemaker with its frequency controlled so that the lower frequency, faster-moving waves catch up with the higher-frequency waves produced early in the experiment, all combining to create a breaking wave at a predetermined location in the wave channel. The energy lost from the wave in breaking and the rate of dissipation of energy within the water are carefully measured[13]. These experiments typify an important aspect of laboratory studies which is unattainable in the natural environment: Conditions are controlled and reproducible, and given processes may be carefully measured in the absence of extraneous factors (e.g., wind forcing of waves). The experiment goes beyond what is presently possible in numerical studies in that it provides measures of the evolution and decay of turbulence and the generation of circulation within the water column[14].

Unexpected Advances

Discoveries are sometimes made by chance through making laboratory experiments. The experiments now associated with the discovery of Benjamin-Feir instability (Benjamin and Feir, 1967) were originally performed with the intension of examining wave reflection phenomena at a sloping beach. Very regular waves were required, and considerable care was therefore taken over the design of an oscillating-paddle wavemaker. Quite unexpectedly the regular wave train it produced broke up into short groups. This prompted theoretical advances proving that the classical Stokes waves can be unstable and initiated substantial revision of the understanding of oceanographic wave phenomena[15].

Influences on Ocean Observation

Double-diffusive convection has been intensively studied in the laboratory and, perhaps more so than for any other process, the experiments have been very influential in establishing the importance of double-diffusive convection in the ocean, in interpreting the observations of the associated "steps and layers" in the temperature and salinity structure of the ocean, and in providing guidance for further measurement[16]. Although the dynamical transfers in the relatively high vertical gradient "steps" containing salt fingers can be described by theory and numerical simulation, insufficient can be yet predicted of the formation and properties of the intermediate thick turbulent layers[17].

Laboratory studies often go far beyond a (perhaps) preconceived and limited objective of testing theory, for example by revealing information about the stages involved in the transition from a laminar to a turbulent flow in stratified or rotating flows. Transitional phenomena are of particular relevance in the ocean where, except near boundaries, turbulence is rarely well developed or continuously sustained and where mixing processes are consequently transient and less-than-perfectly efficient or effective in homogenizing the water column[18]. Studies of processes in turbulent boundary layers, such as "bursting" and "ejection" events with relatively rapid transfer of slow-moving fluid from near a rigid boundary, have had significant impact on measurement strategies and observational requirements needed to provide accurate estimates of quantities like turbulent dissipation rates, which are log-normally distributed and whose mean magnitude is determined by highly intermittent events[19].

Conclusion

Laboratory experiments provide cost-effective means of quantifying processes and of examining the bounds of validity of theory, especially when nonlinearity is important and approximations are required to make progress in developing theoretical analysis. They often reveal unexpected or previously unknown features, processes or phenomena which, without the insight provided by the experiments, might not have been recognized or discovered by other means. They are consequently an important component of geophysical fluid dynamics. Laboratory experiments may suggest measurements which should be made in the ocean to test theories and ways of designing sea-going experiments so that the data collected are the most appropriate and useful.

Vocabulary

insight: (n) 洞察力,洞悉力,深刻见解

investigate: (vt) 调查,审查,调查,研究

n: investigation

illuminate: (vt) 照明,照亮,阐明,澄清,启发

n: illumination

reproduce: (vt) 再生,再现,复制,繁殖

adj: reproducible 可再生的,可复现的

isolation: (n) 隔离,孤立,脱离,分离

vt: isolate

optical: (adj) 光学的

n: optics 光学

schlieren: (n) 纹影(技术)

纹影即透明介质中因密度或成分不同而出现的可视的条纹。德语词,原意为"条纹"。

descend: (vt/vi) 下降,落下

ant: ascend 上升

inclination：(n) 倾斜,倾角,斜坡,倾度
  vt/vi：incline 倾斜,偏离,倾向
provided：(conj) 假如,若……,以……为条件
provision：(n/vt) 供应,提供
transition：(n) 过渡,转变,变革,跃迁
  vt：transit 经过,穿过(某地区)
  n：transit 运送,运输,运载,经过,通过
solute：(n) 溶质,溶解物
  ant：solvent (n) 溶剂;(adj) 有溶解力的
Kelvin-Helmholtz instability：开尔文-亥姆霍兹不稳定
wavemaker：(n) 造波机
typify：(vt) 作为……的典型,具有……的特点,代表,象征
unattainable：(adj) 达不到的,不可获得的,无法实现的
  ant：attainable
extraneous：(adj) 外来的,外源的,无关的
Benjamin-Feir instability：本杰明-费尔不稳定
paddle：(n) 短桨,搅拌桨,叶片
initiate：(vt) 开始,发起,创始
  n：initiation
  n：initiative 首创精神,创造力,主动性,积极性,先机,对策,新举措,提议
revision：(n) 修改,修正,修订
  vt：revise
double-diffusive convection：(n) 双扩散对流
influential：(adj) 有影响力的
guidance：(n) 指导,引导,指引
salt finger：(n) 盐指
preconceive：(vt) 预想
  conceive：(vt) 认为,想象,想出,构思,设想
relevance：(n) 相关性,关联性
  adj：relevant 紧密相关的,切题的,适宜的
homogenize/homogenise：(vt) 使均匀化
  adj：homogeneous 同质的,均质的,均匀的
  n：homogeneity 均匀性,均质性
  n：homogenization 均质化(操作、处理)
burst：(vt) 爆炸,炸裂,崩裂,破裂
ejection：(n) 喷射,弹射,发射
  vt：eject
strategy：(n) 策略
log-normally：(adv) 对数正态地
  adj：log-normal
cost-effective：(adj) 合算的,划算的
sea-going：(adj) 航海的,海上的

Notes

[1] 此处 fluid interior of the Earth 具体是指地核的外核 (outer core) 部分,温度极高,呈液体状态。

[2] 此句用了多个插入语,请注意分辨句子主体结构,可将 if empirical 和 for example 两部分暂时去掉。注意 testing theory 和 providing quantitative estimates of …两个动名词短语都是 valuable 这一形容词修饰的对象,因此二者前面的 in 都不能省略。引入 if

empirical 这个插入语的含义:实验室实验可得出定量估计,这很好,只是有时这种估计仅是经验性的,因此可能使一些人认为不够完美。这里 if 的意思是“尽管”“虽然”“即使”,而不是“如果”。constants of proportionality 即比例系数,或称 proportional coefficient。

[3] density differences drive exchange flows 即密度差异驱动交换流。in which 可以省略。Bosphorus 即沟通地中海和黑海的博斯普鲁斯海峡。

[4] 实验室实验往往无法复现复杂的实际外海环境,这既是其缺点也是其优势来源,因为能方便地将不同因子分离,仅关注某一个或某几个量的影响。result as a consequence of 等于 result from 或 be a consequence of 等。实验室设备通常用 apparatus 一词来称呼。如果要在实验室里模拟地球的旋转效应,必须借助旋转的水槽或水缸(rotating tank)来将地转效应放大,因为空间尺度太小,自然的地转效应难以观察到。水槽通常是圆柱形,将高速相机固定在水槽上面或侧面,与水槽一起随旋转平台(turntable)旋转,则可模拟在地球这一旋转参照系中观察到的海洋现象。地转流、热成风、热盐环流、埃克曼流、罗斯贝波等现象都可观察到。free boundary(自由边界)即上边界、与大气之间的界面;fixed boundary 则指侧边界和底边界,即与固体之间的边界。

[5] optical schlieren system 即光学纹影系统,也可叫作 schlieren imaging 或 schlieren photography,是一种重要的流动可视化(flow visualization)技术。schlieren 是从德语借入的词,原意是“条纹”,可以大写也可以小写。纹影系统的基本原理(Figure Ⅰ-12):用一束平行光照射实验对象(即水槽里的流体),由于流体运动导致其密度变得不均匀,从而影响光线的折射,使得光线在实验对象后面的屏幕或相机上留下条纹。实际操作时首先使用一个凹面镜(concave mirror)将点光源变为平行光(collimated light),穿过对象后再用另一个凹面镜进行汇聚。如果流体密度没有变化,则光线仍是平行光,经过第二个凹面镜之后将汇聚于一个焦点(focus),在此处放置一个刀片(knife edge)可阻挡全部光线,因此后方屏幕或相机上得到的是黑暗影像。如果流体密度发生不均匀,则有一部分光线经过折射可以穿过刀片照亮屏幕,因此留下明暗条纹(streak),可反映流体内部的密度梯度[gradient,即一阶导数(first derivative)]。如果不用刀片,则成为所谓的“阴影照相”(shadowgraph)系统,此时的图像反映的是流体密度的二阶导数(second derivative)。

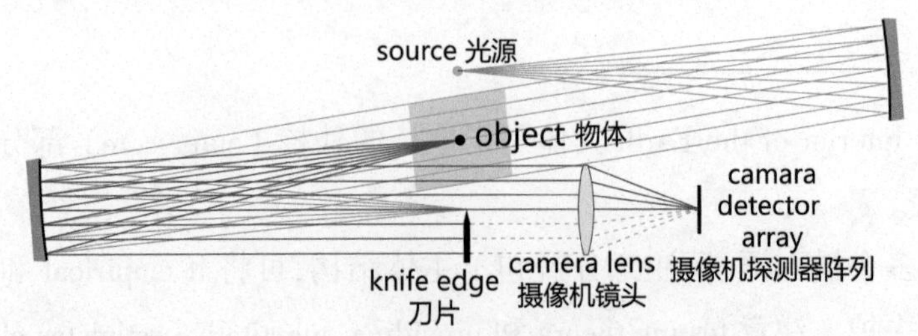

Figure Ⅰ-12: Sketch of schlieren system.

[6] 此实验验证了内波的传播方向与其频率的对应关系。固定频率的内波沿某固定方向传播,这是内波的独特特性。严格来讲,频率决定的是传播方向角的正弦,因此,实际上内波从波源开始向四个象限都有传播,形成“X”形的传播特征。通过对密度的影响,在纹影系统中形成“X”形的条纹。由于“X”形状类似西方熟知的圣安德鲁十字(St. Andrew's cross),因此该实验被叫作圣安德鲁十字实验。当然,此实验的条件是浮力频率为常数,即连续线性层化(continuous linear stratification)。

[7] 此实验用了阴影照相(shadowgraph)法来识别盐水的前锋(即与淡水的分界处),并追踪和测量其速度。此“锋”(front)即本文所说的“头”(head)。盐水在重力作用下流下斜坡,因此叫作 gravity current。注意此句用了多个 of 来表达多层的所属关系:speed→advance→head→gravity current,即重力流头部的前进速度。

[8] on dimensional grounds 即用量纲的方法,ground 此处意为“理由”“根据”。B to the one-third power 即 B 的 1/3 次方。angle of inclination of the slope to the horizontal 即斜坡与水平方向的夹角。to 表示了角度的参考面,如果去掉 to the horizontal 部分,则不知道倾角是以水平面还是垂直面为参考的。这里用语言描述的关系如果用公式表达就是 $U = B^{1/3} f(\alpha)$ 。而 $B = Qg' = 2Qg(\rho_2 - \rho_1)/(\rho_2 + \rho_1)$ 。flow rate(Q)表示单位时间、单位宽度内流下斜坡的盐水体积,reduced gravity(g')即约化重力,the acceleration due to gravity 就是 g ,而 $2(\rho_2 - \rho_1)/(\rho_2 + \rho_1)$ 是 fractional density difference [等于密度差与平均密度之比,即阿特伍德数(Atwood number)的两倍],ρ_1 是 density of the ambient water,ρ_2 是 density of the dense water。

[9] 当 $\alpha \geqslant 5°$ 时,浮力(buoyancy force,实际上此处指负的净浮力,即重力与浮力的差)足以克服摩擦力(frictional force),因此流速稳定。因此,上面公式中的 $f(\alpha)$ 实际上是与 α 无关的常数。而当 $\alpha \leqslant 0.5°$ 时由于摩擦力大于浮力,重力流的速度越向前越减小。注意此段两次使用了 provided 一词,其意思不是“提供”,而是表示“只要”,类似于 if,但比 if 更强。这两次使用 provided 都是用后面的从句来说明条件,即“it is greater than 5°”和“the conditions of the …”,只是第一次省略了引导从句的 that,而第二次没有省略。句子里也出现了 provides a value of …,这里的 provide 确为“提供”的意思。be of use 意为“有用”,of 此处表达“有”,而不是“属于”之意。类似的用法还有 of importance(有重要性)、of interest(有兴趣)、of relevance(有关系)等,分别等于 useful、important、interesting、relevant 等对应的形容词。当然还可以插入修饰程度的副词,如 of great use、of sheer importance、of prime interest、of particular relevance 等。

[10] “This measure is what is now called the Reynolds number.”一句中的 what is 去掉也不影响理解,但加在这里构成主语从句的目的是表达类似于“这就是现在称为雷诺数的那个东西”之意。下面出现的 what are now known as Reynolds stresses are now understood to be active 也是类似的结构和含义,表示“现在称为雷诺应力的那个东西”。solutes 和 suspensions 分别表示溶解和悬浮在水里的物质,它们的通量与动量

和热量一样,也受到雷诺应力的作用而发生扩散。

[11] 开尔文-亥姆霍兹不稳定(Kelvin-Helmholtz instability, KH 不稳定)是发生在有速度剪切的连续流体内部或有速度差的两个不同流体界面上的不稳定现象,不稳定的发生可以很快产生混合。理查森数(Richardson number)详见"Submesoscale Processes"(Text 12)注释[10]。理查森数小于 0.25 是 KH 不稳定发生的判据。KH 不稳定在稳定层结的情况下仍可发生,只要剪切足够强。稳定层结即轻的流体在上、重的在下,若反过来则发生不稳定,即瑞利-泰勒不稳定(Rayleigh-Taylor instability, RT 不稳定)。RT 不稳定可用阿特伍德数来表征。

[12] 此段中的 thermal 是名词,意为热空气。the rise of buoyant thermals 即具有浮性的热空气(如蒸汽)的上升运动。此实验旨在确定卷挟(entrainment)系数,以及从管道释放的热空气柱的高度。实验是通过在稳定连续线性层化的盐水底部释放工业酒精(methylated spirits,即甲基化酒精,酒精与 10%的甲醇的混合物)来实现的。此段第一句特别长,前半部分讲述热气上升实验的重要,然后从 mainly because of 开始讲述重要性的原因,即 their successful demonstration of …和 their determination of …。其中前者的对象是 the validity of a hypothesis,并用一个 that 从句来解释此 hypothesis 的内涵:卷挟率与速度差(excess speed)成正比。后者的对象则是 the proportionality factor,即比例系数。因此,总结起来此句的主要思想:此实验很重要,因为验证了一个正比关系并确定了系数。热气或低密度流体上升过程中与周围流体发生卷挟混合,其上升速度在不同高度上并不均匀,但平均速度决定了卷挟速率,上升越快卷挟混合越剧烈,这与直观认知相符。

[13] 此实验针对深水波(即水深大于 1/2 波长)。因为深水波的频散性(dispersion),低频的长波速度快,高频的短波速度慢。故意先激发高频波,一段时间后再激发低频波,使低频波追上高频波,二者叠加后发生破碎。此种制造破碎波的原理是使波浪能量在空间上汇聚,即使能量密度(energy density)在某空间位置处增强。除了利用频散原理,还可以使水槽的宽度逐渐变小。这两种方法激发的是不稳定的卷波(unsteady plunging breaker)。利用水翼(hydrofoil)在水下相对于波浪运动可产生稳定的溢波(steady spilling breaker)。

[14] 与海洋的自然情况相比,进行实验的优点是可以控制变量,即固定其他因素不变,单独观测某一个因素的影响,即此处描述的 given processes may be carefully measured in the absence of extraneous factors。这里 given 意为"指定的"。为了分离指定因素的影响,必须对环境进行控制(control)并能重现(reproduce)。仅能实现一次而不能重现的实验结果不被认为具有科学意义。同样是实验,实验室里进行的物理实验(physical experiment)和数值模式里进行的数值实验(numerical experiment)都具有以上优点。而物理实验和数值实验相比,还具有可完全真实地测量湍流生消的优点,而数值模式需要依赖参数化来考虑湍流的效应。

[15] 本杰明-费尔不稳定（Benjamin-Feir instability，BF 不稳定）又称调制不稳定（modulation instability）或边带不稳定（sideband instability），是指波形的微小的不规则性（如斯托克斯波）会被非线性相互作用放大，导致频谱上主频侧面的边带被加强，最终使波形破裂成为一系列波包（wave packets 或 pulses）。BF 不稳定可能是畸形波（rogue wave）的成因。

[16] double-diffusive convection 即双扩散对流，是指由于热传导远快于盐扩散，使得原本暖而咸（hot saline）在上、冷而淡（cold fresh）在下的两团水发生彼此穿插形成“盐指”（salt fingers），或使冷而淡在上、暖而咸在下的两团水在各自内部发生对流。这两种情况原本都是稳定层结，前者叫作盐指型（finger/fingering type），后者叫作扩散型（diffusive type）。more so than …是一个常用句型，其中的 so（这样、如此）是副词，其所表示的“样子”或“情况”可以在前面出现，也可以在后面出现。例如，此处的意思是双扩散的实验比其他任何物理现象的实验都更大程度上影响了……，可以理解为对于双扩散实验而言后面描述的情况更加真实、程度更高（so）。

[17] 此句后半部分的语法存在问题，导致难以理解。主干部分为 insufficient can be yet predicted，其中，insufficient 是形容词，却作了主语，严格来讲这是错误的。insufficient 后面实际上省略了名词，即“不足”的主体、“预报”的对象，可以是 information 或 process 等。of 后面的部分实际上应该属于这个被省略的主语，即 insufficient information of the … can be yet predicted。另外，the relatively high vertical gradient steps 是指那些垂向梯度相对较强的“阶跃”处，而 the intermediate thick turbulent layers 指两个阶跃处之间的湍流较强、混合得相对均匀的区域。

[18] efficient 与 effective 两个词都可翻译为“有效的”，二者意思相近但不同。efficient 的“效”是指效率，强调速度快；而 effective 的“效”是指效果，强调实现结果。若 efficient 却不 effective，表明虽然效率高但白忙一场；若 effective 却不 efficient，表明虽能最终达成效果但过程很慢。很多时候二者是统一的。远离上下边界和侧边界，海洋受到的外来搅拌很弱，mixing processes 很大程度上依赖内波的破碎，而这些过程往往是由间断式的短暂事件组成的。homogenize the water column 即使水柱均匀，就是混合的意思。

[19] 这里讲的湍流边界层内的 bursting 和 ejection 事件，实际上是指气泡（bubble）或液滴（droplet）的破碎和喷射，是发生在海洋上层和大气底层这两个主要的 turbulent boundary layer 内的常见事件，历来研究难度大但需求高。所谓 relatively rapid transfer of slow-moving fluid 是指由气泡或液滴导致的物质交换比起空气或海水自身的移动速度快得多。由于气泡或液滴的存在对一些观测仪器而言是误差来源（如超声波风速计，ultrasonic anemometer），因此需要设计一些策略来规避它；但在另一些情况下又是观测的重点对象（如研究湍流混合和气体交换），因此需要想办法观测它。例如，本书“Langmuir Circulation”（Text 17）曾提及用声学手段测量微

气泡以反映流速，以及这里阐述的用来推算湍耗散率（turbulent dissipation rate）。对数正态分布（logarithmic/log normal distribution）即变量的对数符合正态分布，则变量本身符合对数正态分布，是一个明显的偏态（skewed）分布。由于其偏态性（skewness），其概率密度函数（probability density function）峰值右侧的尾部（tail）较长，也就是说出现很大值的可能性很高，适合描述间断性较强、大部分情况都取低值，但平均值被偶尔出现的大值主导的信号。

Text Sources

Caldwell, Moum, 1995. Turbulence and Mixing in the Ocean. Review of Geophysics (supp), 1385-1394.

Carlson, Bates, Hansell, Steinberg, 2011. Carbon Cycle// Steele, Thorpe, Turekian (eds). Encyclopedia of Ocean Sciences. Cambridge: Academic Press.

Kantha, Clayson, 2000. Numerical Models of Oceans and Oceanic Processes. Cambridge: Academic Press.

Knauss, Garfield, 2017. Introduction to Physical Oceanography. Long Grove: Waveland Press.

Leibovich, 2011. Langmuir Circulation and Instability// Steele, Thorpe, Turekian (eds). Encyclopedia of Ocean Sciences. Cambridge: Academic Press.

Marshall, Plumb, 2008. Atmosphere, Ocean, and Climate Dynamics: An Introductory Text. Cambridge: Academic Press.

NOAA Ocean Exploration lessons 7-10, 13. https://oceanexplorer.noaa.gov/edu/learning/.

Pope, 2000. Turbulent Flows. London: Cambridge University Press.

Rhines, 2011. Mesoscale Eddies// Steele, Thorpe, Turekian (eds). Encyclopedia of Ocean Sciences. Cambridge: Academic Press.

Talley, Pickard, Emery, Swift, 2011. Descriptive Physical Oceanography. Cambridge: Academic Press.

Thomson, Emery, 2014. Data Analysis Methods in Physical Oceanography. Amsterdam: Elsevier Science.

Thorpe, 2011. Fluid Dynamics, Introduction, and Laboratory Experiments// Steele, Thorpe, Turekian (eds). Encyclopedia of Ocean Sciences. Cambridge: Academic Press.

Wadhams, 2011. Sea Ice// Steele, Thorpe, Turekian (eds). Encyclopedia of Ocean Sciences. Cambridge: Academic Press.

PART II

MARINE BIOLOGY

1.Marine Biology

Text

Marine biology, the science that focuses on animals and plants that live in the sea. Many more organisms of different kinds live in the ocean than on land. The life in the ocean glow, swim, swarm, squish, spout, wave, hide, drift, or pounce. They range from gigantic whales to microscopic phytoplankton, and includes organisms such as corals that affect the shape of the seafloor[1]. In the broadest sense it attempts to describe all vital phenomena pertaining to the myriads of living things that dwell in the vast oceans of the world[2]. For example, many medical advances have been underpinned by research on marine organisms, such as studies of the animal immune system in sea anemones and sea star larvae, the fertilization of sea urchin eggs, nerve conduction in squids, and barnacle muscles[3]. Marine life is also a vast source of human wealth. It provides food, medicines, and supports tourism all over the world.

At a much more fundamental level, marine life helps determine the very nature of the planet. Marine organisms produce around half the oxygen we breathe and help regulate Earth's climate. The shorelines are shaped and protected by marine life, and some marine organisms even can create new land. In economic terms, it's been estimated that the ocean's living systems are worth more than $20 trillion a year. To make full and wise use of the sea, to solve the problems that marine organisms create, and to predict the effects of human activities on the ocean, it's important to learn about marine life. In addition, marine organisms provide valuable clues to Earth's past, the history of life, and even our own bodies. This is the challenge, the adventure, of marine biology.

Marine biology is the more general science of biology applied to the sea rather than a separate science. Nearly all the disciplines of biology are represented in marine biology. It is most closely related to biological oceanography that the two are difficult to separate. A principal aim of marine biology is to discover how ocean phenomena control the distribution of organisms. Marine biologists study the way in which particular organisms are adapted to the various chemical and physical properties of the seawater, to the movements and currents of the ocean, to the availability of light at various depths, and to the solid surfaces that make up the seafloor[4]. Special attention is given to determining the dynamics of marine ecosystems, particularly to the understanding of food chains and predator-prey relationships[5]. Marine biological information on the distribution of fish and crustacean populations is of great importance to fisheries. Marine biology is also concerned with the effects of certain forms of pollution on the fish and plant life of the oceans, particularly the effects of pesticide and fertilizer runoff from land sources, accidental spills from oil tankers, and silting from coastline construction activities[6].

Morphological and taxonomic studies of marine organisms are generally performed on preserved materials in connection with the work in universities. Physiological and embryological investigations requiring the use of living material are generally pursued at biological stations. These are situated on the seacoast, thus facilitating the rapid transfer of specimens to the laboratory where they may be maintained in seawater provided by special circulating systems. A marine biologist's interests may also overlap broadly with those of biologists who study terrestrial organisms. Many of the basic ways in which living things make use of energy, for example, are similar whether an organism lives on land or in the sea. Nevertheless, marine biology does have a flavor all its own, partly because of its history.

Vocabulary

swarm:(vi) 成群浮游
squish:(vt) 挤压
spout:(vt) 喷射,喷出
wave:(vi) 摇摆,起伏
hide:(vt) 隐藏,躲避
drift:(vi) 漂流,漂移
pounce:(vi) 突袭,猛扑

gigantic:(adj) 巨大的,庞大的
microscopic:(adj) 微观的,极小的
[micro-]:微小的,显微的
phytoplankton:(n) 浮游植物
[phyt-]:植物
[-plankton]:浮游生物
dwell:(vi) 居住,栖身

anemone：(n) 海葵

larva：(n) 幼体，幼虫

pl：larvae

fertilization：(n) 受孕/受精(现象)

squid：(n) 鱿鱼

barnacle：(n) 藤壶(小甲壳动物，附着于水下岩石或船底等)

shoreline：(n) 海(或湖)岸线

separate：(adj) 独立的，分开的

predator：(n) 猎食者

prey：(n) 猎物

predator-prey relationship：猎食者-猎物关系

crustacean：(adj/n) 甲壳动物(的)

morphological：(adj) 形态学(上)的

taxonomic：(adj) 分类学的

adv：taxonomically

physiological：(adj) 生理学(上)的

embryological：(adj) 胚胎学的

specimen：(n) 标本，样品

Notes

[1] 海洋生物的范围较广，从巨大的鲸鱼到微小的浮游植物，还包括影响海底地形的珊瑚等。在下文中会讲到浮游植物、珊瑚等相关概念。

[2] 从最广义上讲，海洋生物学旨在描述与生活在广阔海洋中的无数生物相关的所有重要现象。此句用 that 引导定语从句，在全句中做插入语。

[3] 海洋生物学的研究对医学的发展具有重要支撑作用。例如，对海葵和海星幼虫免疫系统的研究、海胆卵的受精过程、鱿鱼的神经传导和藤壶肌肉的研究都为医学提供了宝贵的见解。藤壶（barnacle），又称为“马牙”或“蚵沕仔”，属节肢动物门甲壳纲围胸目。藤壶是营固着生活的特殊甲壳动物，体外围有坚硬的壳板，中间留有一小口，形似小山，依靠过滤海水中的有机物为生。由于其独特的形态结构、生活史和种群生态，藤壶已成为主要的海洋污损生物之一。成体藤壶既不能游泳，也不能爬行，主要生活在潮间带，附着在海水中固定或漂浮的硬质物体上，如船体、浮标、桥墩、码头、网箱及渔具等。附着在网箱上的藤壶会导致网衣堵塞，减少海水的流动交换，恶化养殖环境，引发养殖动物的疾病。此外，还会增大网衣阻力，导致网箱漂移，并在大潮汛和台风期间造成鱼体擦伤。藤壶的附着对牡蛎、珍珠贝等贝类养殖也有极大的危害。

[4] 此句解释了海洋生物学家主要研究特定海洋生物如何适应海洋环境，包括海水的各种物理和化学特性、海洋的运动和洋流、不同深度的光照条件以及海底地形特征。

[5] 需特别关注的是明确海洋生态系统的动态变化特征，理解海洋(生态系统)的食物链和捕食者-猎物关系。Lotka-Volterra 模型，通常称为捕食者-猎物（predator-prey）模型，其核心是两个微分方程，即 Lotka-Volterra 方程，可描述捕食者和猎物的种群规模（population size）如何取决于双方的数量和互动。

[6] 渔业也是海洋生物学的研究对象，包括研究海洋污染对海洋鱼类和浮游植物的影响。句中“certain forms of pollution”指的是来自陆地的农药和化肥径流、油轮意外泄漏以及海岸线建设活动导致的海洋污染，这些都会影响海洋渔业。

2. Cells and Organelles

Text

Cells contain all the molecules needed for life packaged in a living wrapper called the cell membrane, or plasma membrane and of which all living things are composed[1]. A single cell is often a complete organism in itself, such as a bacterium or yeast. Other cells acquire specialized functions as they mature. These cells cooperate with other specialized cells and become the building blocks of large multicellular organisms. Although cells are much larger than atoms, they are still very small. The smallest known cells are a group of tiny bacteria called mycoplasmas; some of these single-celled organisms are spheres as small as 0.2 μm in diameter, with a total mass of 10^{-14} gram—equal to that of 8,000,000,000 hydrogen atoms. Cells of humans typically have a mass 400,000 times larger than the mass of a single mycoplasma bacterium, but even human cells are only about 20 μm across. It would require a sheet of about 10,000 human cells to cover the head of a pin, and each human organism is composed of more than 30,000,000,000,000 cells.

The interior of the cell is organized into many specialized compartments, or organelles, each surrounded by a separate membrane (Figure Ⅱ-1). Known as the cell's "command center," the nucleus is a large organelle that stores the cell's DNA (deoxyribonucleic acid)[2]. The nucleus controls all of the cell's activities, such as growth and metabolism, using the DNA's genetic information. Within the nucleus is a smaller structure called the nucleolus, which houses the RNA (ribonucleic acid)[3]. RNA helps convey the DNA's orders to the rest of the cell and serves as a template for protein synthesis. Each cell contains only one nucleus.

Other types of organelles are present in multiple copies in the cellular contents, or cytoplasm. Organelles include mitochondria, which are responsible for the energy transactions necessary for cell survival; lysosomes, which digest unwanted materials within the cell; and the endoplasmic reticulum and the Golgi apparatus, which play important roles in the internal organization of the cell by synthesizing selected molecules and then processing, sorting, and directing them to their proper locations[4]. Ribosomes sit in some of the endoplasmic reticulum. Organic molecules, water, and other materials are often stored in vacuoles.

In addition, plant and algal cells have two important structures that animals do not. First, they contain chloroplasts, which are responsible for photosynthesis, whereby the energy of sunlight is used to convert molecules of carbon dioxide (CO_2) and water (H_2O) into carbohydrates. Second, except for some single-celled algae, they have a cell wall. Between all these organelles is the space in the cytoplasm called the cytosol. The cytosol contains an organized framework of

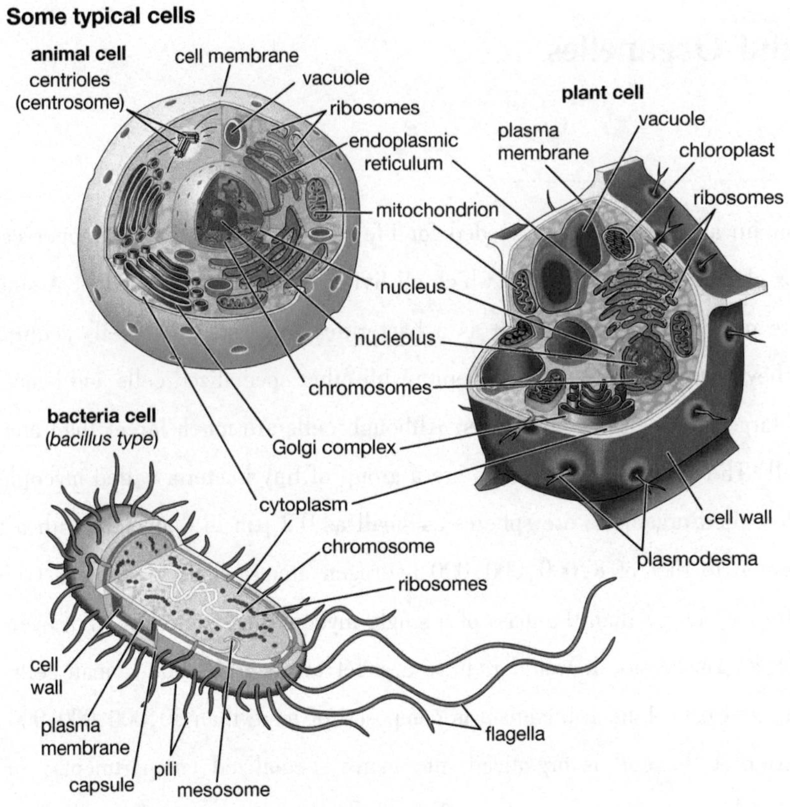

Figure Ⅱ-1: Animal cells and plant cells contain membrane-bound organelles, including a distinct nucleus. In contrast, bacterial cells do not contain organelles. [Source: Encyclopedia Britannica]

fibrous molecules that constitute the cytoskeleton, which gives a cell its shape, enables organelles to move within the cell, and provides a mechanism by which the cell itself can move[5]. The cytosol also contains more than 10,000 different kinds of molecules that are involved in cellular biosynthesis, the process of making large biological molecules from small ones.

Specialized organelles are a characteristic of cells of organisms known as eukaryotes. In contrast, cells of organisms known as prokaryotes do not contain organelles and are generally smaller than eukaryotic cells. However, all cells share strong similarities in biochemical function.

Vocabulary

cell membrane: (n) 细胞膜

plasma: (n) 血浆,原生质

bacterium: (n) 细菌,细菌类

pl: bacteria

yeast: (n) 酵母,酵母菌

multicellular: (adj) 多细胞的

[multi]: (n) 多种,多数,多元

n: multicellularity 多细胞体

atom：(n) 原子，微粒，微量
mycoplasma：(n) 霉形体，支原体
organelle：(n) 细胞器
nucleus：(n) 核，核心，细胞核
  [nucle-]：核
  pl：nucleuses 或 nuclei
deoxyribonucleic acid：(n) 脱氧核糖核酸
nucleolus：(n) 核仁
  pl：nucleoli
ribonucleic acid：(n) 核糖核酸
cytoplasm：(n) 细胞质、胞浆
  [cyto-]：细胞
mitochondrion：(n) 线粒体
  pl：mitochondria
lysosome：(n) 溶(酶)体
endoplasmic reticulum：(n) 内质网
Golgi apparatus：(n) 高尔基体
ribosome：(n) 核糖体
vacuole：(n) 液泡
chloroplast：(n) 叶绿体
photosynthesis：(n) 光合作用
carbohydrates：(n) 碳水化合物
cytosol：(n) 细胞液，胞浆基质
fibrous：(adj) 纤维构成的，纤维状的
cytoskeleton：(n) 细胞骨架
biosynthesis：(n) 生物合成
eukaryote：(n) 真核生物
  adj：eukaryotic
prokaryote：(n) 原核生物
  [-karyote]：有核细胞
  adj：prokaryotic

Notes

[1] 细胞包含了生命所需的各种分子，这些分子被包裹在被称为细胞膜或质膜的“生命外壳”中，且所有生物都是由细胞构成。在 Figure Ⅱ-1 的图题中，membrane-bound 是一个形容词，描述被膜包裹或依附在膜上的结构或物质。图题可理解为，动物细胞和植物细胞中含有被膜包裹的细胞器，其中包括一个明确的被核膜包裹的细胞核。与之相反，细菌细胞没有细胞器。

[2] 细胞核（nucleus）被称为细胞的“指挥中心”，是储存遗传物质 DNA（脱氧核糖核酸）的大型细胞器。DNA 包含遗传信息，下面将详细介绍。

[3] 细胞核内有一个更小的结构，称为核仁（nucleolus），其中含有 RNA（核糖核酸）。核仁由蛋白质、DNA 和 RNA 组成，是核糖体组装的场所。除此之外，核仁还参与信号识别颗粒的形成，并在细胞的应激反应中发挥作用。

[4] 这句话采用了排比的修辞手法，介绍了细胞中其他细胞器。细胞器包括线粒体，负责为细胞提供能量；溶酶体，负责消化细胞内不需要的物质；内质网和高尔基体，合成特定功能的蛋白质，进行加工、分类，并将其输送到适当的位置，对细胞的内部组织起重要作用。

[5] 细胞质含有由纤维状分子组成的结构框架，称为细胞骨架，赋予细胞形状，使细胞器能够在细胞内移动，并为细胞自身的运动提供机制。

3. Genetic Information

Text

DNA is the genetic material. During the early 19th century, it became widely accepted that all living organisms are composed of cells arising only from the growth and division of other cells. The improvement of the microscope then led to an era during which many biologists made intensive observations of the microscopic structure of cells. By 1885 a substantial amount of indirect evidence indicated that chromosomes—dark-staining threads in the cell nucleus—carried the information for cell heredity. It was later shown that chromosomes are about half DNA and half protein by weight. The revolutionary discovery suggesting that DNA molecules could provide the information for their own replication came in 1953, when American geneticist and biophysicist James Watson and British biophysicist Francis Crick proposed a model for the structure of the double-stranded DNA molecule (called the DNA double helix, see Figure Ⅱ-2). In this model, each strand serves as a template in the synthesis of a complementary strand[1]. Subsequent research confirmed the Watson and Crick model of DNA replication and showed that DNA carries the genetic information for reproduction of the entire cell.

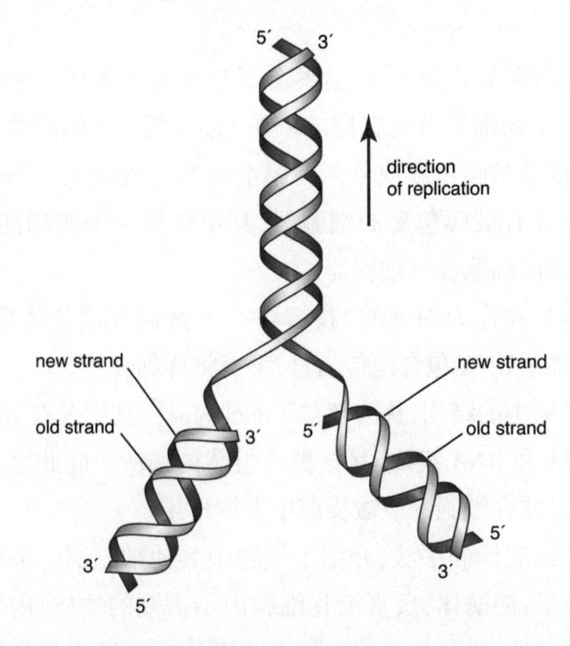

Figure Ⅱ-2: The initial proposal of the structure of DNA by James Watson and Francis Crick was accompanied by a suggestion on the means of replication. [Source: Encyclopedia Britannica]

The genetic information of an organism is stored in DNA molecules. How can one kind of molecule contain all the instructions for making complicated living beings like ourselves? What component or feature of DNA can contain this information? It has to come from the nitrogen bases, because the backbone of all DNA molecules is the same. There are only four bases found in DNA: adenine (A), thymine (T), guanine (G), and cytosine (C). DNA, located in the cell nucleus, is made up of nucleotides that contain the bases. The sequence of these four bases can provide all the instructions needed to build any living organism[2]. The DNA alphabet can encode very complex instructions using just four letters, though the messages end up being really long. For example, the *Escherichia coli* (*E. coli*) bacterium carries its genetic instructions in a DNA molecule that contains more than five million nucleotides. The human genome (all the DNA of an organism) consists of around three billion nucleotides divided up between 23 paired DNA molecules, or chromosomes[3].

Structure of DNA double helix (Figure Ⅱ -3). Sugar-phosphate backbone is shown in yellow, specific base pairings via hydrogen bonds (red lines) are colored in green and purple (A–T pair) and red and blue (C–G). The sugar deoxyribose with the phosphate group forms the scaffold or backbone of the molecule (highlighted in yellow in Figure Ⅱ -3). Bases point inward. Complementary bases form hydrogen bonds with each other within the double helix. See how the bigger bases (purines) pair with the smaller ones (pyrimidines). This keeps the width of the double helix constant. More specifically, A pairs with T and C pairs with G. As we discuss the function of DNA in subsequent sections, keep in mind that there is a chemical reason for specific pairing of bases.

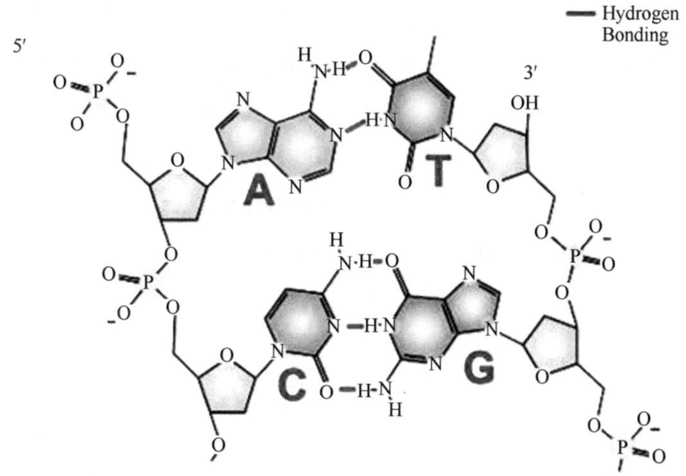

Figure Ⅱ -3: DNA's double helix structure. [Source: Madeleine Price Ball]

In order for DNA to function effectively at storing information, two key processes are required. First, information stored in the DNA molecule must be copied, with minimal errors, every time a cell divide. This ensures that both daughter cells inherit the complete set of genetic information from the parent cell. Second, the information stored in the DNA molecule must be translated, or expressed. In order for the stored information to be useful, cells must be able to access the instructions for making specific proteins, so the correct proteins are made in the right place at the right time[4].

It is possible for RNA to replicate itself by mechanisms related to those used by DNA, even though it has a single-stranded instead of a double-stranded structure. In early cells RNA is thought to have replicated itself in this way. However, all of the RNA in present-day cells is synthesized by special enzymes that construct a single-stranded RNA chain by using one strand of the DNA helix as a template[5]. Although RNA molecules are synthesized in the cell nucleus, where the DNA is located, most of them are transported to the cytoplasm before they carry out their functions[6].

The RNA molecules in cells have two main roles. Some, the ribozymes, fold up in ways that allow them to serve as catalysts for specific chemical reactions. Others serve as "messenger RNA," which provides templates specifying the synthesis of proteins. Ribosomes, tiny protein-synthesizing machines located in the cytoplasm, "read" the messenger RNA molecules and "translate" them into proteins by using the genetic code[7]. In this translation, the sequence of nucleotides in the messenger RNA chain is decoded three nucleotides at a time, and each nucleotide triplet (called a codon) specifies a particular amino acid. Thus, a nucleotide sequence in the DNA specifies a protein provided that a messenger RNA molecule is produced from that DNA sequence. Each region of the DNA sequence specifying a protein in this way is called a gene[8].

By the above mechanisms, DNA molecules catalyze not only their own duplication but also dictate the structures of all protein molecules. A single human cell contains about 10,000 different proteins produced by the expression of 10,000 different genes. Actually, a set of human chromosomes is thought to contain DNA with enough information to express between 30,000 and 100,000 proteins, but most of these proteins seem to be made only in specialized types of cells and are therefore not present throughout the body.

Vocabulary

division:(n) 分配,分开
  vt: divide 分隔,分配
chromosome:(n) 染色体
cell heredity:(n) 细胞遗传
heredity:(n) 遗传,遗传特征
  adj: hereditary 遗传的
replication:(n) 复制
template:(n) 模板
complementary strand:(n) 互补链
base:(n) 碱基

adenine:(n) 腺嘌呤
thymine:(n) 胸腺嘧啶
guanine:(n) 鸟嘌呤
cytosine:(n) 胞嘧啶
nucleotide:(n) 核苷酸
instruction:(n) 命令,指令
genome:(n) 基因组
double helix:(n) 双螺旋
sugar-phosphate backbone:(n) 磷酸骨架
deoxyribose:(n) 脱氧核糖
scaffold:(n) 染色体骨架
purine:(n) 嘌呤
pyrimidine:(n) 嘧啶
translate:(v) 翻译
express:(v) 表达
enzyme:(n) 酶
ribozyme:(n) 核糖酶
nucleotide triplet:核苷酸三连体
codon:(n) 密码子
amino acid:氨基酸
catalyze:(vt/vi) 催化

Notes

[1] 在 DNA 双螺旋模型中,每条单链都作为互补链合成的模板。

[2] 该句翻译为,DNA 只含有四种碱基:腺嘌呤(A)、胸腺嘧啶(T)、鸟嘌呤(G)和胞嘧啶(C)。这四种碱基的序列提供了构建任何生物体所需的全部指令。脱氧核糖核酸(Deoxyribonucleic acid,DNA),由含氮碱基、脱氧核糖和磷酸组成。由于脱氧核糖和磷酸在 DNA 中是恒定的,而碱基有四种(A、G、T、C),因此 DNA 的多样性源于碱基序列的不同。在书写 DNA 序列时,可以用碱基的缩写字母代替。

[3] 人类基因组(即人类所有 DNA 的总和)由大约 30 亿个核苷酸对组成,分布在 23 对 DNA 分子或染色体上。人类基因组计划(human genome project, HGP)是一项规模宏大,跨国跨学科的科学探索工程,旨在测定人类单倍体染色体中约 30 亿个碱基对的核苷酸序列,绘制人类基因组图谱,识别其中的基因及其序列,最终破译人类遗传信息。HGP 于 1990 年启动,2003 年宣布完成,中国是参与该计划的六个国家之一。人类基因组计划所建立的策略、思想和技术,形成了生命科学领域的新学科——基因组学,可应用于微生物、植物及其他动物的研究。

[4] 为了使存储的信息发挥作用,细胞必须能够读取这些合成特定蛋白质的指令,以便在正确的时间和地点合成所需的蛋白质。

[5] 细胞中的所有 RNA 都是由专门的酶合成的,这些酶以 DNA 双螺旋中的一条链为模板,合成对应的单链 RNA。

[6] 虽然 RNA 分子在细胞核中合成,但大多数在发挥功能前会被转运到细胞质中。RNA 在 DNA 和核糖体之间充当核酸信使。一些 RNA 分子通过催化生物反应、控制基因表达或感知和传递细胞信号,在细胞内发挥积极作用。RNA 有以下三种主要类型。① 信使 RNA(mRNA):将 DNA 上的遗传信息传递到细胞质中的核糖体(蛋白

质合成部位)。mRNA 在细胞核中由 DNA 转录而来,携带编码蛋白质的序列。② 核糖体 RNA (rRNA):与蛋白质一起构成核糖体的主要成分,参与 mRNA 的读取和解码。③ 转运 RNA (tRNA):在翻译过程中,将特定的氨基酸运输到核糖体上正在合成的多肽链。tRNA 根据 mRNA 的编码序列,将氨基酸添加到新生蛋白质中。

[7] 核糖体是位于细胞质中的微小蛋白质合成机器,它“读取”信使 RNA 分子,并使用遗传密码将其“翻译”成蛋白质。

[8] 在翻译过程中,信使 RNA 链中的核苷酸序列以每三个核苷酸为一组被解码,每个三联体(称为密码子)指定一种特定的氨基酸。因此,DNA 中的核苷酸序列,通过信使 RNA 的转录和翻译,指定了一种蛋白质,以这种方式编码蛋白质的 DNA 序列的每个区域称为基因。

4. Reproduction

Text

Reproduction is one of the most important biological process by which an organism reproduces an offspring that is biologically similar to the organism[1]. Reproduction enables and ensures the continuity of species, generation after generation. It is only by reproducing that a species ensures its survival. The production of organisms must do two things. First, they must produce new individuals to perpetuate the species. Second, they must pass on to this new generation the characteristics of the species in the form of genetic information. The transfer of genetic information from one generation to the next is called heredity. Organisms achieve the first objective, the production of offspring, in many different ways. But they use relatively few ways to pass on genetic information. This similarity of hereditary mechanisms is evidence that all living things are fundamentally related.

Individual cells reproduce mainly by dividing to form new daughter cells (Figure Ⅱ-4). In prokaryotes, where the DNA that carries the genetic information is not enclosed in a nucleus, the cell divides into two new ones by a simple process called cell fission[2]. Before dividing, the cell copies, or replicates, its DNA. Each daughter cell gets one of the copies, as well as the rest of the cellular machinery it needs to survive. In eukaryotes the DNA lies on several different chromosomes. Before a cell divides, each chromosome is replicated. The most common form of eukaryotic cell division, mitosis, is a complex process that takes place in the nucleus. Mitosis ensures that each daughter cell gets a copy of every chromosome. As in bacterial cell division, the end result is two daughter cells that are exact duplicates of the original, with the same genetic information.

Asexual Reproduction

Asexual reproduction refers to the type of reproduction in which only a single organism gives rise to a new individual[3]. Cell division is the primary way that single-celled organisms reproduce. Because a single individual can reproduce itself without the involvement of a partner, this form of reproduction is called asexual reproduction. All forms of asexual reproduction are similar in that the offspring inherit all the genetic characteristics of the parent. They are, in fact, exact copies, or clones.

Many multicellular organisms also reproduce asexually. Some sea anemones for example, split in half to create two smaller anemones. This process is known as fission. Another common

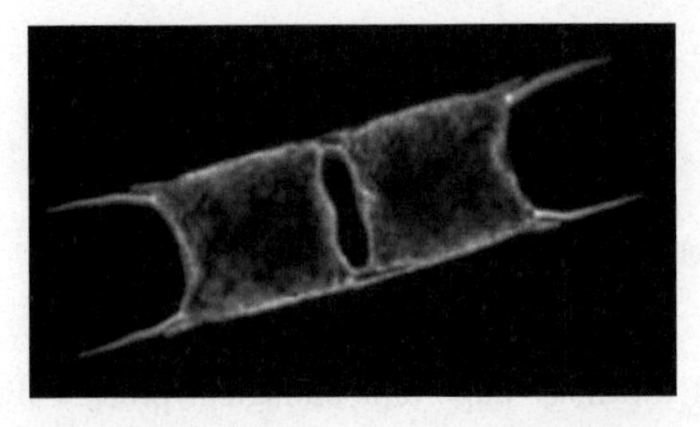

Figure Ⅱ-4: Two cells of a microscopic organism called a diatom (Odontella sinensis). The cells were formed by the division of a single "mother" cell. The two cells may separate into individual cells, or continue to divide to form a chain. [Source: D. P. Wilson/FLPA/Science Source]

form of asexual reproduction is budding. Instead of dividing into two new individuals, the parent develops small growths, or buds, that break away and become separate individuals [Figure Ⅱ-5(a)]. Many plants reproduce asexually by sending out various kinds of runners that take root and then sever their connection to the parent [Figure Ⅱ-5(b)]. Because it is so common in plants, asexual reproduction is sometimes called vegetative reproduction, even when it occurs in animals.

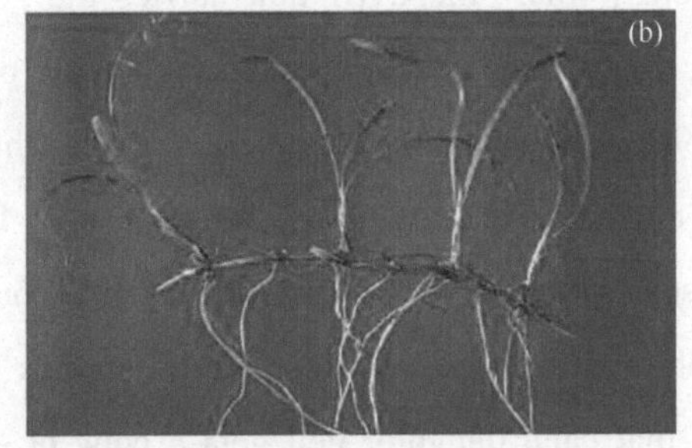

Figure Ⅱ-5: (a) Budding is a common form of reproduction in marine animals, including this mushroom coral (*Fungia fungites*). The parent coral is nearly dead, but the many buds on its surface will soon break off and become free-living adults. (b) This seagrass (*Halodule pinifolia*) reproduces asexually by sending out rhizomes, or "runners," that take root and form new plants. [Source: Australian Institute of Marine Science]

Sexual Reproduction

Most multicellular and some unicellular organisms reproduce part or all of the time by sexual reproduction, in which new offspring arise from the union of two separate cells called gametes[4]. Usually each of the gametes comes from a different parent. Organisms that reproduce sexually have a special kind of tissue called germ tissue. Whereas all the other cells in the body divide only by mitosis, germ cells are capable of a second type of cell division, meiosis[5].

In most cells of eukaryotic organisms the chromosomes occur in pairs, with each chromosome of a pair storing similar genetic information. Such cells are diploid, designated 2n. Meiosis produces daughter cells that have copies of only half the parents' chromosomes—one of each pair. Cells with half the normal number of chromosomes are haploid, designated n or 1n. The haploid daughter cells produced in meiosis are the gametes. In some seaweeds and microbes, all the gametes are the same. Sexual reproduction usually, however, involves two types of gametes: female gametes called eggs and male gametes called sperm. In animals the germ tissue is usually contained in gonads, the organs that produce the gametes. Ovaries are the female gonads, which produce eggs. In animals the male gonads, or testes, produce sperm. Seaweeds also have gonads that produce sperm, while in flowering plants sperm develop within pollen, which is produced by organs in the flower[6].

Vocabulary

reproduction: (n) 繁殖,生殖
offspring: (n) 后代,子代
perpetuate: (vt/vi) 使永久化,使持续
fission: (n) 分裂生殖
mitosis: (n) 有丝分裂
asexual: (adj) 无性的
  adv: asexually
involvement: (n) 参与
vegetative reproduction: 营养繁殖
sexual: (adj) 性的,有性繁殖的
gamete: (n) 配子(形成受精卵的精子或卵子)
meiosis: (n)(细胞的)减数分裂
diploid: (n) 二倍体;(adj) 二倍体的(含有两套染色体)
egg: (n) 卵子
sperm: (n) 精子
germ: (n) 胚芽、胚胎
ovary: (n) 卵巢
gonad: (n) 性腺、生殖腺
pollen: (n) 花粉

Notes

[1] 在本文中,繁殖或生殖(reproduction)是最重要的生物过程之一,生物体通过繁殖产生与自身生物学上相似的后代。繁殖是所有生命体的基本现象,每个现存的个体都是上一代繁殖的结果。繁殖方式主要分为两大类:有性生殖(sexual reproduction)和无性生殖(asexual reproduction)。在后文中对有性生殖和无性生殖均有介绍。

[2] 在原核生物中,携带遗传信息的 DNA 并未被封闭在细胞核中,其细胞是通过一种简单的过程,称为二元裂变,分裂成两个新细胞。细菌等原核生物通过二元裂变进行繁殖。对于单细胞生物来说,细胞分裂是产生新个体的唯一方法。

[3] 无性生殖是指仅由单一生物体产生新个体的繁殖方式。无性生殖主要包括分裂生殖、出芽生殖、孢子生殖、营养生殖和克隆等。其中,分裂生殖和出芽生殖是海洋生物最主要的两种无性繁殖方式。例如,海葵通过分裂生殖繁殖,水螅和海绵通过出芽生殖繁殖。

[4] 大多数多细胞生物和一些单细胞生物通过有性生殖进行繁殖,这种方式由两个称为配子的独立细胞结合产生新的后代。与无性生殖相比,有性生殖需要由雌性和雄性个体产生生殖细胞。当二者的生殖细胞结合后,最终形成受精卵,进而产生后代。以海洋鱼类为例,它们的繁殖方式属于有性繁殖,根据不同的特征,可以细分为卵生、胎生和卵胎生。

[5] 虽然生物体内的所有其他细胞仅通过有丝分裂进行分裂,但生殖细胞能够进行另一种类型的细胞分裂,称为减数分裂。减数分裂是一种特殊的细胞分裂方式,使染色体数目减半,产生单倍体细胞,每条染色体来自亲代细胞。需要注意的是,减数分裂不同于有丝分裂,是专门用于产生生殖细胞的分裂方式。

[6] 海藻具有产生精子和卵细胞的性器官,而在被子植物(开花植物)中,精子在花粉粒内发育,花粉由花的雄蕊(花药)产生。

5.Natural Selection and Adaptation

Text

Evolution occurs because individual organisms have genetic differences in their ability to find food and avoid being eaten, in their success at producing offspring, in their metabolism, and in countless other attributes[1]. The best-adapted individuals, those most successful at meeting the challenges of the environment, produce more offspring on average than those that are not so well adapted[2]. English naturalist Charles Darwin firstly proposed the modern theory of evolution, called this process natural selection.

Natural selection is the process through which populations of living organisms adapt and change. Individuals in a population are naturally variable, meaning that they are all different in some ways. This variation means that some individuals have traits better suited to the environment than others. Individuals with adaptive traits—traits that give them some advantage—are more likely to survive and reproduce[3]. These individuals then pass the adaptive traits on to their offspring. Over time, these advantageous traits become more common in the population. Through this process of natural selection, favorable traits are transmitted through generations.

Natural selection can lead to speciation, where one species gives rise to a new and distinctly different species. Every population is constantly adapting to its environment. The world is an ever-changing place, however, and organisms are continually faced with new challenges. Populations either adapt to the changes in the environment or become extinct, making way for others. Evolution is an endless process.

Adaptation, the process by which a species becomes fitted to its environment; it is the result of natural selection's acting upon heritable variation over several generations (Figure Ⅱ-6)[4]. Organisms are adapted to their environments in a great variety of ways: in their structure, physiology, and genetics, in their locomotion or dispersal, in their means of defense and attack, in their reproduction and development, and in other respects[5].

Adaptation has three meanings. First, in a physiological sense, an animal or plant can adapt by adjusting to its immediate environment—for instance, by changing its temperature or metabolism with an increase in altitude. Second, and more commonly, the word adaptation refers either to the process of becoming adapted or to the features of organisms that promote reproductive success relative to other possible features[6]. Here the process of adaptation is driven by genetic variations among individuals that become adapted to—that is, have greater success in—a specific environmental context[7].

Figure Ⅱ-6：Adaptations：the habitat adaptations of walruses（thick skin to protect against cold conditions），hippopotamuses（nostrils on the top of the snout），and ducks（webbed feet）.［Source：Encyclopedia Britannica］

The third and more popular view of adaptation is in regard to the form of a feature that has evolved by natural selection for a specific function. Examples include the long necks of giraffes for feeding in the tops of trees，the streamlined bodies of aquatic fish and mammals，the light bones of flying birds and mammals，and the long daggerlike canine teeth of carnivores. Thus，before explaining that a trait is an adaptation，it is necessary to identify whether it is also shown in ancestors and therefore may have evolved historically for different functions from those that it now serves[8].

Vocabulary

individual：（n）个体，个人；（adj）单独的，个别的

evolution：（n）演变，进化

trait：（n）特点，特性

generation：（n）一代，世代

speciation：（n）物种形成

population：（n）群体，种群

extinct：（adj）（动植物）灭绝的

adaptation：（n）调适，适应性

heritable variation：遗传变异

physiology：（n）生理学

genetics：（n）遗传学，基因学

locomotion：（n）移动（力），运动

dispersal：（n）分散，扩散，传播

defense：（n）防御，防守

attack：（n）攻击，袭击

streamlined body：流线型体

mammal：（n）哺乳动物

carnivore：（n）食肉动物

Notes

[1] 进化的发生是因为个体生物在寻找食物、避免被捕食、繁殖成功率、新陈代谢以及其他众多特征上存在的基因差异。

[2] 那些最适应环境、能够成功应对环境挑战的个体,平均产生的后代比适应性较差的个体更多。

[3] 这种差异意味着某些个体比其他个体具有更适合环境的特征。具有适应性特征(赋予它们优势的特征)的个体更有可能生存和繁殖。

[4] 该句翻译为,适应(adaptation)是一个物种逐渐适应其环境的过程,是自然选择作用于多代遗传变异的结果。适应指生物体通过遗传变异和自然选择,逐渐发展出对环境的适应性特征。这些特征使生物体更好地生存于特定环境中,包括形态、生理、行为特征等,提高了获取食物、避免捕食、应对气候变化或其他环境压力的能力,从而提高生存和繁殖成功率。适应是生物进化的重要驱动力,使物种不断适应变化的环境。

[5] 生物以多种方式适应环境,包括结构、生理和遗传、运动或扩散、防御和攻击、繁殖和发育等方面。

[6] 适应有两个含义:一是指适应的过程;二是指促进繁殖成功的生物体特征,相对于其他可能的特征而言。

[7] 适应过程由个体间的遗传变异驱动,这些个体在特定环境背景下更成功。换句话说,适应是基因变异与环境交互,使一些个体在特定环境中更有效地生存和繁殖的过程。

[8] 在解释物种特性是否为一种适应之前,有必要确定该特性是否也在祖先中表现,因此可能在历史上进化出与现在不同的功能。本句可理解为,要理解某个特征是否是适应性特征,需要考察其在演化历史中的起源和功能变化。

6. Phytoplankton

Text

The plant-like community of plankton is called phytoplankton, and the animal-like community is known as zooplankton. This convenient division is not without fault, for, strictly speaking, many planktonic organisms are neither clearly plant nor animal but rather are better described as protists. When size is used as a criterion, plankton can be subdivided into macroplankton, microplankton, and nannoplankton, though no sharp lines can be drawn between these categories[1]. Macroplankton can be collected with a coarse net, and morphological details of individual organisms are easily discernible. These forms, 1 mm or more in length, ordinarily do not include phytoplankton. Microplankton (also called net plankton) is composed of organisms between 0.05 and 1 mm in size and is a mixture of phytoplankton and zooplankton. The lower limit of its size range is fixed by the aperture of the finest cloth used for plankton nets. Nannoplankton passes through all nets and consists of forms of a size less than 0.05 mm.

In detail, phytoplankton are buoyant and float in the upper part of the ocean. They drift about in the water, allowing tides, currents, and other factors determine where they go. Phytoplankton make their energy through photosynthesis, the process of using chlorophyll and sunlight to create energy. Like terrestrial plants, phytoplankton take in carbon dioxide and release oxygen. Phytoplankton account for about half of the photosynthesis on the planet, making them one of the world's most important producers of oxygen. Phytoplankton also require inorganic nutrients such as nitrates, phosphates, and sulfur which they convert into proteins, fats, and carbohydrates[2].

The chief components of marine phytoplankton are found within the algal groups and include diatoms, dinoflagellates and coccolithophorids. Silicoflagellates, cryptomonads, and green algae are found in most plankton samples[3]. Dinoflagellates and diatoms are the most important two main classes of phytoplankton in marine.

In the oceans, phytoplankton biomass rises and falls according to multiyear cycles and appears to be sensitive to changes in sea surface temperatures, climate change, and ocean acidification[4]. However, scientists examining records of phytoplankton kept from 1899 to 2008 noted that phytoplankton biomass fell by 1 percent per year in 8 of Earth's 10 ocean basins, resulting in a cumulative loss of roughly 40 percent. Rising sea surface temperatures over the same period are thought to be the primary cause of this decline.

In a balanced ecosystem, phytoplankton provide food for a wide range of sea creatures including shrimp, snails, and jellyfish. When too many nutrients are available, phytoplankton may grow out of control and form harmful algal blooms (HABs)[5]. These blooms can produce extremely toxic compounds that have harmful effects on fish, shellfish, mammals, birds, and even people. The National Centers for Coastal Ocean Science conduct extensive research on harmful algal blooms. Scientists use a range of technologies to predict where and when HABs are likely to form and how they will affect the areas where they occur. Scientists use this information to inform coastal authorities on how to best respond in order to minimize negative impacts.

Vocabulary

plankton: (n) 浮游生物
　adj: planktonic 浮游的
　~zooplankton: (n) 浮游动物
　~phytoplankton: (n) 浮游植物
　~macroplankton: (n) 大型浮游生物
　~microplankton: (n) 小型浮游生物
　~nannoplankton: (n) 超微浮游生物
protist: (n) 原生生物
discernible: (adj) 可辨别的
chlorophyll: (n) 叶绿素
terrestrial plant 陆生植物
inorganic nutrient 无机营养盐
diatom: (n) 硅藻
dinoflagellate: (n) 甲藻,腰鞭毛虫
coccolithophorid: (n) 球石藻类
silicoflagellate: (n) 硅鞭毛虫,硅鞭藻
cryptomonad: (n) 隐藻类

Notes

[1] 按照大小尺度,浮游生物可以细分为大型浮游生物(macroplankton)、微型浮游生物(microplankton)和微小浮游生物(nanoplankton),但这些类别之间并没有明确的界限。大型浮游生物通常指较大且易被肉眼观察到的浮游生物,如大型水母和浮游水草等。微型浮游生物指体型较小但仍可能被肉眼看到的浮游生物,如硅藻、甲藻和浮游动物的幼体等。微小浮游生物指体型微小,只能在显微镜下观察到的浮游生物,如细菌和微小的硅藻等。

[2] 浮游植物需要无机营养物质,例如硝酸盐、磷酸盐和硫酸盐,它们利用这些营养物质合成蛋白质、脂类和碳水化合物。

[3] 海洋浮游植物主要由藻类组成,包括硅藻、甲藻和颗石藻。大多数海洋浮游生物样本中还存在硅鞭毛虫、隐藻和绿藻。其中,硅藻(diatoms)和甲藻(dinoflagellates)将在后文中详细介绍。

[4] 在海洋生态系统中,浮游植物的生物量呈多年周期性波动,并对海表温度、气候变化和海洋酸化等因素高度敏感性。

[5] 当营养物质过盛时，浮游植物可能会迅速增殖，形成有害藻华（harmful algal blooms）。藻华暴发是藻类在局部区域内的大量繁殖，发生于最优的生长条件且捕食压力减轻的情况下。当特定的微藻类（特别是蓝藻细菌）在暴发中达到高浓度时，这种现象会呈现出有害特征。这些微藻物种在自然界中只占很小比例，但它们能够产生有害或有毒物质，造成各种不利影响，包括破坏食物链、导致动物死亡以及对人类健康造成严重危害。

7. Zooplankton

Text

Zooplankton are a type of heterotrophic plankton that range from microscopic organisms to large species, such as jellyfish (Figure Ⅱ-7b). Zooplankton are found within large bodies of water, including oceans and freshwater systems. Zooplankton, small floating or weakly swimming organisms that drift with water currents and, with phytoplankton, make up the planktonic food supply upon which almost all oceanic organisms are ultimately dependent. They are usually larger than phytoplankton, ranging from tiny copepods, less than a centimeter long, to jellyfishes and colonial salps that may be meters long. There are two major types of zooplankton: those that spend their entire lives as part of the plankton (called Holoplankton) and those that only spend a larval or reproductive stage as part of the plankton (called Meroplankton)[1].

Zooplankton are the favorite food of a great many marine animals so camouflaging themselves is a very important survival strategy. Zooplankton consume a variety of bacterioplankton, phytoplankton, and even other zooplankton species. Since such organisms reside at the surface of bodies of water, zooplankton are also typically found in the upper waters. Developing effective camouflage when you live in clear, blue water is no easy matter. The best solution and the one most often used by members of the zooplankton is to be as transparent as possible or, in the case of many surface floating jellyfishes, blue.

Many animals, from single-celled Radiolaria to the eggs or larvae of herrings, crabs, and lobsters, are found among the zooplankton. Permanent plankton, or holoplankton, such as protozoa and copepods (an important food for larger animals), spend their lives as plankton. Temporary plankton, or meroplankton, such as young starfish, clams, worms, and other bottom-dwelling animals, live and feed as plankton until they leave to become adults in their proper habitats[2]. The most important types of zooplankton include the radiolarians, foraminiferans, and dinoflagellates, cnidarians, crustaceans, chordates, and molluscs (Figure Ⅱ-7)[3].

Radiolarians are small protozoan species that are characterized by the production of mineral skeletons made of silica[4]. The remains of these organisms can be found at the bottom of oceans, comprising a large part of the sediment.

For aminiferans are a type of amoeboid protest that exhibit an external shell and ectoplasm used to obtain food[5]. While the shell is typically comprised of calcium carbonate, the shells of some species contain other minerals. These zooplankton can be found in the sediment or drifting about the upper surface waters.

Dinoflagellates are considered a mixotrophic species, meaning than they can be both photosynthetic or ingest other species. This type of zooplankton is extremely small and represent a significant portion of marine eukaryotes and are important for the health of coral reefs.

Cnidarians are marine species that are characterized by specialized cells called "cnidocytes", which are used to capture their prey. They have bodies consisting of a jelly-like substance called mesoglea, a mouth, and tentacles that contain the cnidocytes.

Crustaceans are a type of arthropod that consists of crabs, krill, shrimp, and barnacles [Figure Ⅱ-7(b)]. Crustaceans range in terms of size, and comprise a significant part of the food chain. Krill and copepods in particular, are important zooplankton species.

Chordates are animals that possess anotochord, norsal nerve chord, endostyle, post-anal tail, and pharyngeal slits. This is a highly diverse family that includes sea stars, scallops, and many other species[6]. Molluscs are a highly diverse group of organisms, which include squid species as well as sea slugs, and sea snails. Molluscs comprise a large component of all marine life.

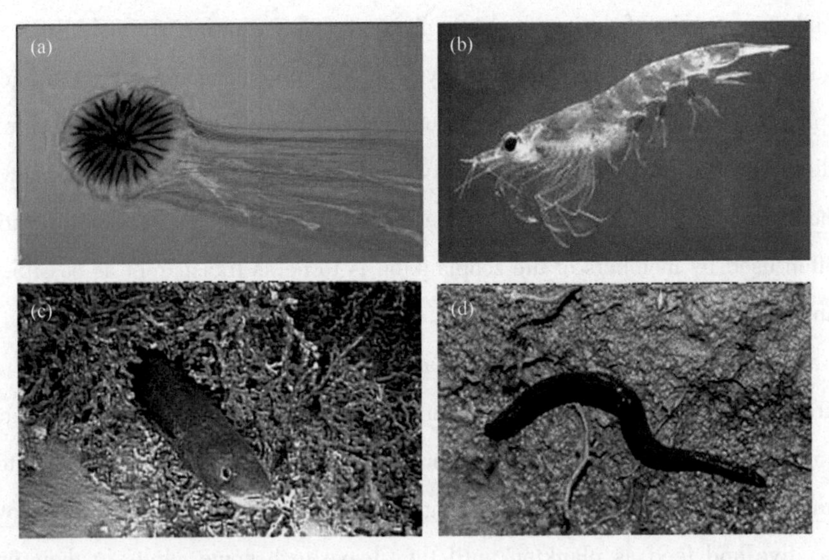

Figure Ⅱ-7: (a) Jellyfish. They are most commonly found near coastal regions throughout the world. (b) Krill. They are consumed by larger marine animals, thus making them a significant contributor to the lower food chain in marine environments. (Source: Weisse, 2017)

Vocabulary

heterotrophic: (adj) 非自养的

jellyfish: (n) 水母,海蜇

copepod: (n) 桡足动物

salp: (n) 樽海鞘,被囊动物,海樽类功能群

holoplankton: (n) 全浮游性生物

larval：(adj) 幼虫的，潜在的，隐蔽的

meroplankton：(n) 季节浮游生物

camouflage：(n) 伪装，隐藏；(vt/vi) 隐蔽，隐藏，掩盖

Radiolaria：(n) 放射虫类，放射虫门(纲)

herring：(n) 鲱鱼

crab：(n) 螃蟹，蟹肉；(vi) 捕蟹

lobster：(n) 龙虾

permanent plankton：(n) 永久性浮游生物

protozoan：(n/adj) 原生动物(的)

pl：protozoa 或 protozoans

temporary plankton：(n) 暂时性浮游生物

clam：(n) 蚌，蛤

worm：(n) 蠕虫

bottom-dwelling 底栖的

radiolarian：(n) 散线虫类的各类动物，放射虫门

foraminiferan：(n) 有孔虫目

cnidarian：(n) 刺胞动物

chordate：(n/adj) 脊索动物(的)，具有脊索的

mollusc/mollusk：(n) 软体动物(门)

silica：(n) 硅石，二氧化硅

amoeboid：(adj) 变形虫样的，似变形虫的

ectoplasm：(n) 外质(指细胞基质外部的胶化区)

mixotrophic：(n) 混合营养的，兼养的

coral：(n) 珊瑚

reef：(n) 暗礁

mesoglea：(n) 中胶层

cnidocyte：(n) 刺细胞

krill：(n) 磷虾

pharyngeal slit：(n) 咽裂，腮裂

Notes

[1] 浮游动物有两种主要类型：一种是整个生命周期都作为浮游生物的一部分称为全浮游动物(holoplankton)；另一种是只在幼虫或繁殖阶段作为浮游生物的一部分，称为季节性浮游动物(meroplankton)。具体而言，meroplankton 是指在生命周期的某些阶段以浮游生物的形式存在，而在其他阶段则转变为定居生活方式或适合其成体生活的环境。

[2] 临时浮游生物，或季节性浮游生物，如幼年的海星、蛤蜊、蠕虫和其他栖息在海底的动物，以浮游生物的形式生活和进食，直到它们发育成成体并返回适合它们生活的栖息地。

[3] 最重要的浮游动物类型包括放射虫、有孔虫、甲藻、刺胞动物、甲壳动物、脊索动物和软体动物。

[4] 放射虫是一类微小的原生动物，以其特有的放射状硅质骨骼而闻名。它们生活在海洋中，通过细胞吞噬微小的浮游生物来获取营养。放射虫在海洋中广泛分布，是重要的浮游生物之一。

[5] 有孔虫是一种变形虫状的原生动物，具有外壳和用于获取食物的伪足。它们的外壳由钙质或硅质构成，具有复杂的孔洞结构。它们生活在海洋中的底层沉积物中或浮

游水层，以浮游生物或有机碎屑为食。有孔虫在古生物学中具有重要地位，其化石可用于古环境和古气候研究。

[6] 脊索动物是具有脊索、背神经管、内胃沟、肛后尾和咽裂等特征的动物。这是一个高度多样化的动物类群，包括海星、扇贝以及其他物种。

8. Diatom

Text

Diatom, class Bacillariophyceae, any member of the algal class Bacillariophyceae (division Chromophyta), with about 16,000 species found in sediments or attached to solid substances in all the waters of Earth. Diatoms are among the most important and prolific microscopic sea organisms and serve directly or indirectly as food for many animals. Diatoms may be either unicellular or colonial (Figure Ⅱ－8). The silicified cell wall forms a pillbox-like shell (frustule) composed of overlapping halves (epitheca and hypotheca) perforated by intricate and delicate patterns[1]. Food is stored as oil droplets, and the golden-brown pigment fucoxanthin masks the chlorophyll and carotenoid pigments that are also present. Diatoms are commonly divided into two orders on the basis of symmetry and shape: the round nonmotile Centrales have radial markings; the elongated Pennales, which move with a gliding motion, have pinnate (featherlike) markings.

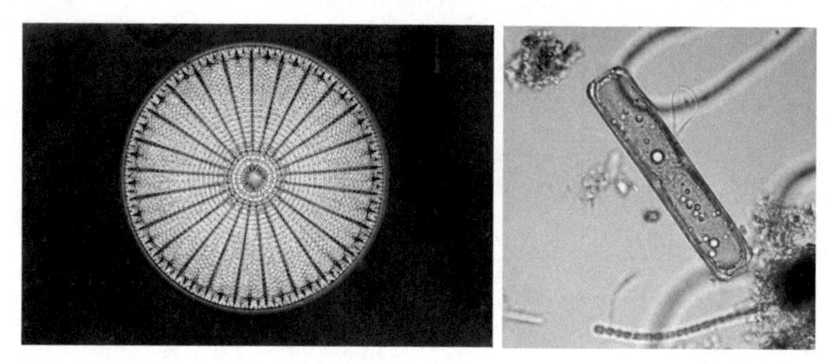

Figure Ⅱ－8: Diatom (highly magnified), Magnified diatom (Pinnularia species).
[Source: Encyclopedia Britannica]

Diatoms primarily reproduce asexually via binary fission. Each daughter cell receives one of the frustules from the parent cell, which forms the larger frustule, and provides the basis for the construction of the second, slightly smaller frustule[2]. Since, the two frustules are not the same size, with each replication, one daughter cell will be slightly smaller than the other. Thus, after several rounds of reproduction. The population of diatoms will be smaller than the original. To avoid a size-reduction or re-establish the original size of the diatom population, sexual reproduction must occur. Since the vegetative cells of diatoms are diploid, haploid gametes can be produced via meiosis. The fusion of the male and female gametes forms a zygote, which develops a membrane, termed the auxospore. Within the auxospore, a new diatom forms, which will then to produce new daughter diatoms. When environmental conditions change and resources become limited, diatoms can also produce resting spores, which germinate once the conditions become favorable. While diatoms themselves are not mobile, the male gametes of some diatom species are motile due to the presence of flagella[3]. The female gametes of all species are large, immobile cells.

Favorable environmental conditions, such as abundant nutrients and light, trigger periods of rapid reproduction called blooms. Such blooms may even occur under Arctic and Antarctic ice. This is a general phenomenon that also occurs in other algae. During blooms most diatoms get progressively smaller, not only because of asexual reproduction but also as a result of depletion of silicate from the water by the growing population[4].

The glassy frustules of dead diatoms eventually settle to the bottom of the sea floor, where they can form thick deposits of siliceous material that cover large portions of the ocean floor[5]. Such biogenous sediments are known as diatomaceous ooze, a type of siliceous ooze[6]. Huge fossil deposits of these sediments can be found inland in various parts of the world. The siliceous material, or diatomaceous earth, is mined and used in products such as filters for swimming pools, for clarifying beer, as temperature and sound insulators, and as mild abrasives in toothpaste.

Vocabulary

Bacillariophyceae:(n) 芽孢杆菌科
unicellular:(adj) 单细胞的
colonial:(adj) 殖民地的
silicified:(adj) 硅化的
frustule:(n)(硅)藻细胞,硅藻细胞膜
epitheca:(n) 上壳,外鞘(腔肠动物)
hypotheca:(n) 下壳
pigment:(n) 色素
fucoxanthin:(n) 岩藻黄质
carotenoid:(n) 类胡萝卜素
pennales:(n) 羽纹硅藻目
pinnate:(adj) 翼、鳍之类的,羽状的
binary:(adj) 二元的
vegetative:(n) 植物的
zygote:(n) 受精卵
auxospore:(n) 复大孢子,发育孢子
biogenous:(adj) 生命产生的
diatomaceous:(adj)(含)硅藻的
siliceous:(adj)(化)含硅的,硅质的
ooze:(n) 泥浆,淤泥;(vt/vi) 缓缓流出、渗出
~siliceous ooze: 硅质渗出物
insulator:(n) 绝缘、隔热或隔音等的物质

Notes

[1] 硅藻的硅化细胞壁形成了类似药盒状的壳体(称为硅壳),由相互重叠的两半(上壳

和下壳)组成,其表面具有复杂精细的穿孔图案(Figure Ⅱ-9)。

[2] 这两句话描述了硅藻的生殖方式:硅藻主要通过二分裂的方式进行无性繁殖。每个子细胞从母细胞继承一个硅壳,形成较大的壳,并为构建略小的第二个硅壳提供基础。关于硅藻的生殖方式,可参见"Reproduction"(Text 4)。

[3] 当环境条件发生变化并且资源变得有限时,硅藻可以产生休眠孢子,一旦条件转好,孢子便会萌发。虽然硅藻自身无法移动,但某些硅藻物种的雄性配子因具有鞭毛而能够运动。例如,一些海洋硅藻的雄性配子具有鞭毛,能够在水中游动。此外,属于辐射硅藻目(Centrales)和羽纹硅藻目(Pennales)的一些硅藻,其雄性配子通过鞭毛运动来寻找并受精卵细胞。

[4] 在藻华期间,大多数硅藻的体型逐渐变小,这不仅是由于无性繁殖造成的,还因为不断增长的种群消耗了水中的硅酸盐。硅酸盐对硅藻的生长至关重要,是其生长所必需的物质之一。硅藻是一类单细胞藻类,细胞壁主要由二氧化硅(硅酸盐)组成,形成坚硬的外壳或硅壳。硅藻利用水中的硅酸盐来合成和构建这些硅质细胞壁,因此硅酸盐是硅藻生长和繁殖的关键要素。

[5] 死亡硅藻的玻璃质硅壳最终沉降到海底,可形成厚厚的硅质沉积物,覆盖海洋底部的大片区域。

[6] 这种生物源沉积物被称为硅藻土,是硅质软泥的一种类型。除硅藻土(diatomaceous ooze)外,硅质软泥还包括其他类型,主要根据其来源和组成物质进行分类。常见的硅质泥浆类型包括,放射虫软泥(radiolarian ooze),由死去的放射虫(radiolarian)的硅质壳体沉积而成;硅藻-放射虫软泥(diatom-adiolarian ooze),由硅藻和放射虫残骸混合组成的沉积物,常见于特定的海洋区域;硅质海绵软泥(siliceous sponge ooze),由死去的硅质海绵(siliceous sponge)的残骸组成的沉积物。

9. Dinoflagellates

Text

Dinoflagellate, division Dinoflagellata, any of numerous one-celled aquatic organisms bearing two dissimilar flagella and having characteristics of both plants and animals[1]. Most are marine, though some live in freshwater habitats. The group is an important component of phytoplankton in all but the colder seas and is an important link in the food chain. Dinoflagellates also produce some of the bioluminescence sometimes seen in the sea. Under certain conditions, several species can reproduce rapidly to form water blooms or red tides that discolour the water and may poison fish and other animals [Figure Ⅱ-10(a)][2]. Some dinoflagellates produce toxins that are among the most poisonous known.

Dinoflagellates range in size from about 5 to 2,000 micrometers. Most are microscopic, but some form visible colonies. Nutrition among dinoflagellates is autotrophic, heterotrophic, or mixed; some species are parasitic or commensal. About one-half of the species are photosynthetic; even among those, however, many are also predatory. Although sexual processes have been demonstrated in a few genera, reproduction is largely by binary or multiple fission. Under favourable conditions, dinoflagellate populations may reach 60 million organisms per liter of water.

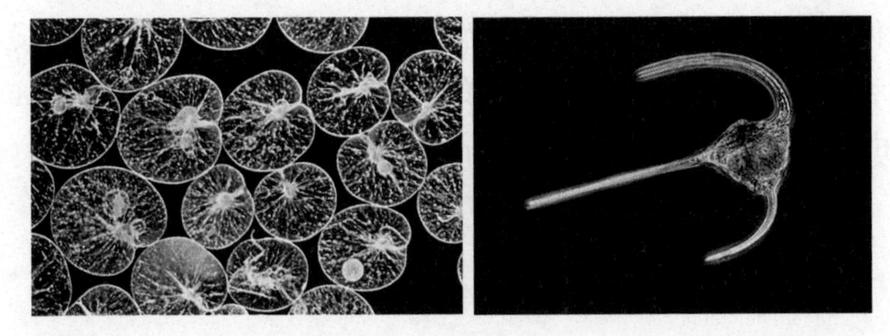

Figure Ⅱ-10: (a) Sea sparkle. (b) Dinoflagellate. [Source: Eric Grave/Photo Researchers; Blickwinkel/age footstock]

The dinoflagellate cell is banded by a median or coiled groove, the annulus, which contains a flagellum [Figure Ⅱ-10(b)]. A longitudinal groove, the sulcus, extends from the annulus posteriorly to the point at which a second flagellum is attached. The nuclei of dinoflagellates are larger than those of other eukaryotes. So-called armoured dinoflagellates are covered with cellulose plates, which may have long spiny extensions; some species lacking armour have a thin pellicle (protective layer)[3]. Photosynthetic dinoflagellates have yellowish or brownish plastids (pigment-containing bodies) and may store food in the form of starches, starchlike compounds, or oils[4].

Dinoflagellates reproduce almost exclusively by simple cell division. They sometimes form blooms that color the water red, reddish-brown, yellow, or other unusual shades. Some of these dinoflagellates release toxic substances, and seafood collected during red tide periods can be poisonous, as in the case of ciguatera, a tropical fish poisoning. Other dinoflagellates are noted for the production of light, or bioluminescence. Though bioluminescence has also been observed in some bacteria and many types of animals, dinoflagellates are generally responsible for the diffuse bioluminescence sometimes seen on the sea surface[5]. This effect is, of course, seen only at night. It is especially bright if the water is disturbed by a boat or when a wave crashes on the shore.

Vocabulary

aquatic:(adj) 水生的,水产的

bioluminescence:(n) 生物体发光

　luminescence:(n)(物)发光,冷发光,荧光

toxin:(n) 毒素,毒质(尤指生物体内细菌产生的毒物)

colony:(n) 群体

autotrophic:(adj) 自给营养的

　trophic:(adj) 营养的,与营养有关的

parasitic:(adj) 寄生的

commensal:(adj) 共生的;(n) 共食者,(生)共生体

predatory:(adj) 掠夺性的,捕食性的

longitudinal:(adj) 纵的,纵向的

groove:(n) 凹槽,沟

　~sulcus:(n) 沟

annulus:(n) 体(环)

flagellum:(n) 鞭毛

　pl: flagella

armour:(n) 盔甲

　adj: armoured 有装甲的

pellicle:(n) 薄膜

plastid:(n) 质体

ciguatera:(n) 鱼肉毒,热带海鱼毒

Notes

[1] 甲藻(dinoflagellata)是一类单细胞水生生物,具有两根不同类型的鞭毛,兼具植物和动物的特征。植物特征:许多甲藻含有叶绿素和其他色素,能够进行光合作用,利用光能合成有机物质。动物特征:部分甲藻能够摄取有机物,表现出异养的特征,有些甚至具有捕食性。甲藻在海洋和淡水中广泛分布,是浮游生物中的重要组成部分,对海洋生态系统和食物链起着关键作用。

[2] 藻华(water blooms)是由于水体中氮、磷等营养元素过高,导致藻类突然大量增殖的自然现象,水面可能呈现蓝色、红色、棕色或乳白色等颜色。参与藻华的藻类包括

蓝藻、绿藻、硅藻等。当水温度高于20℃、水体pH值偏高、光照强度强且持续时间长时,藻类会迅速繁殖,形成藻华。赤潮(red tides)是在特定环境条件下,海水中某些浮游植物、原生动物或细菌暴发性增殖或高度聚集,导致水体变色的一种有害生态现象。

[3] 甲藻细胞具有一个中间(median)或螺旋状(coiled)的环状凹槽(环带,annulus),其中包含一根鞭毛。另一个纵向凹槽(纵带,sulcus)从环带延伸,容纳第二根鞭毛。甲藻的细胞核比其他真核生物的细胞核更大。所谓的“有甲”甲藻,其细胞被由纤维素板组成的板片覆盖,有的还具有长而尖的突起;而“无甲”甲藻则具有薄的皮层(保护层)。

[4] 进行光合作用的甲藻具有黄色或棕色的质体(含有色素),并能以淀粉、淀粉类化合物或油的形式储存养分。

[5] 虽然生物发光现象也存在于一些细菌和许多类型的动物中,但海面上弥漫性的生物发光(the diffuse bioluminescence)通常是由甲藻引起的。当大量发光的甲藻使聚集时,海面会呈现蓝色的发光景象,称为蓝眼泪(blue tears),这是一种特殊的海洋现象。甲藻体内含有荧光素类化合物,在受到机械或化学刺激时会发光。当海水中这种甲藻大量聚集时,就会形成蓝眼泪现象,通常在夜晚或清晨出现,主要分布在热带和亚热带海域。大量甲藻的聚集可能对生态系统产生影响,它们可能释放毒素,对其他海洋生物如鱼类造成危害。

10. Protozoans

Text

Protozoans are usually unicellular and heterotrophic (using organic carbon as a source of energy) eukaryotic organisms, belonging to any of the major lineages of protists and, like most protists, typically microscopic. All protozoans are eukaryotes and therefore possess a "true," or membrane-bound, nucleus. They also are nonfilamentous (in contrast to organisms such as molds, a group of fungi, which have filaments called hyphae) and are confined to moist or aquatic habitats, being ubiquitous in such environments worldwide, from the South Pole to the North Pole[1]. Many are symbionts of other organisms, and some species are parasites.

Modern ultrastructural, biochemical, and genetic evidence has rendered the term protozoan highly problematic. For example, protozoan historically referred to a protist that has animal-like traits, such as the ability to move through water as though "swimming" like an animal. Protozoans traditionally were thought to be the progenitors of modern animals, but contemporary evidence has revealed that this is not the case for most protozoans. In fact, modern science has shown that the protozoans represent a very complicated grouping of organisms that do not necessarily share a common evolutionary history. This unrelated, or paraphyletic, nature of the protozoans has caused scientists to abandon the term protozoan in formal classification schemes. Hence, the subkingdom Protozoa is now considered obsolete. Today the term protozoan is used informally in reference to nonfilamentous heterotrophic protists.

Commonly known protozoans include representative dinoflagellates, amoebas, paramecia, and the malaria-causing *Plasmodium* (Figure Ⅱ-11)[2].

Although protozoans are no longer recognized as a formal group in current biological classification systems, protozoan can still be useful as a strictly descriptive term. The protozoans are unified by their heterotrophic mode of nutrition, meaning that these organisms acquire carbon in reduced form from their surrounding environment. However, this is not a unique feature of protozoans. Furthermore, this description is not as straightforward as it seems. For instance, many protists are mixotrophs, capable of both heterotrophy (secondary energy derivation through the consumption of other organisms) and autotrophy (primary energy derivation, such as through the capture of sunlight or metabolism of chemicals in the environment)[3]. Examples of protozoan mixotrophs include many chrysophytes. Some protozoans, such as Paramecium bursaria, have developed symbiotic relationships with eukaryotic algae, while the amoeba Paulinella chromatophora remarkably appears to have acquired autotrophy via relatively recent endosymbiosis of a

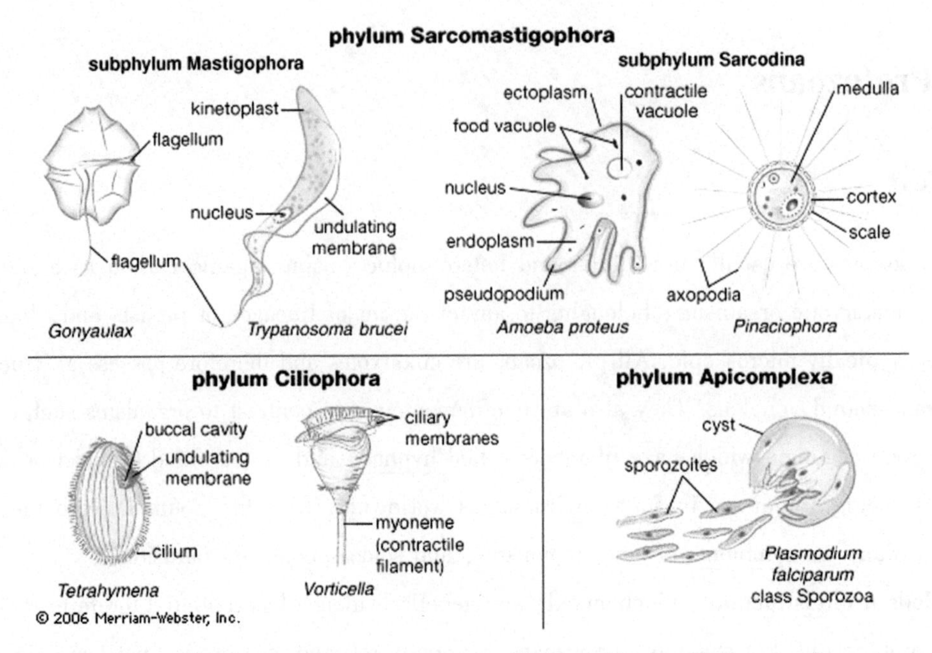

Figure Ⅱ-11: Representative protozoans. [Source: Merriam-Webster Inc.]

cyanobacterium (a blue-green alga). Hence, many protozoans either perform photosynthesis themselves or benefit from the photosynthetic capabilities of other organisms. Some algal species of protozoans, however, have lost the ability to photosynthesize (e.g., Polytomella species and many dinoflagellates), further complicating the concept of "protozoan."

Protozoans are motile; nearly all possess flagella, cilia, or pseudopodia that allow them to navigate their aqueous habitats. However, this commonality does not represent a unique trait among protozoans; for example, organisms that are clearly not protozoans also produce flagella at various stages in their life cycles (e.g., most brown algae). Protozoans are also strictly non-multicellular and exist as either solitary cells or cell colonies. Nevertheless, some colonial organisms (e. g., *Dictyostelium discoideum*, supergroup *Amoebozoa*) exhibit high levels of cell specialization that border on multicellularity.

The descriptive guidelines presented above exclude many organisms, such as flagellated photosynthetic taxa (formerly *Phytomastigophora*), that were considered protozoans by older classification schemes. Organisms that fit the contemporary definition of a protozoan are found in all major groups of protists that are recognized by protistologists, reflecting the paraphyletic nature of protozoans[4].

The most important groups of free-living protozoans are found within several major evolutionary clusters of protists, including the ciliates (supergroup Chromalveolata), the lobose amoebae (supergroup Amoebozoa), the filose amoebae (supergroup Rhizaria), the cryptomonads (supergroup Chromalveolata), the excavates (supergroup Excavata), the opisthokonts (supergroup

Opisthokonta), and the euglenids (Euglenozoa). These groups of organisms are important ecologically for their role in microbial nutrient cycles and are found in a wide variety of environments, from terrestrial soils to freshwater and marine habitats to aquatic sediments and sea ice. Significant protozoan parasites include representatives from Apicomplexa (supergroup Chromalveolata) and the trypanosomes (Euglenozoa). Organisms from these groups are the causative agents of human diseases such as malaria and African sleeping sickness. Owing to the prevalence of these human pathogens, and to the ecological importance of the free-living protozoan groups mentioned above, much is known about these groups. This article therefore concentrates on the biology of these comparatively well-characterized protozoans. At the end of this article is a summary of the contemporary protistan classification scheme.

Vocabulary

lineage:(n) 谱系,世系
nonfilamentous:(adj) 无细丝状的
  filamentous:(adj) 细丝状的,丝状体型的,细丝的
hyphae:(n) 菌丝,真菌菌丝
ubiquitous:(adj) 无处不在的
symbiont:(n) 共生体,共栖生物
ultrastructural:(adj) 超微结构的
  structural:(adj) 结构上的
paraphyletic:(adj) 分类群的
obsolete:(adj) 过时的
amoeba:(n) 变形虫
  pl: amoebae/amoebas
paramecium:(n) 草履虫
  pl: paramecia
malaria-causing:(adj) 引起疟疾的
plasmodium:(n) 疟原虫
  pl: plasmodia
metabolism:(n) 代谢
chrysophytes:(n) 金藻
Paramecium bursaria 绿草履虫
autotrophy:(n) 自养作用
endosymbiosis:(n) 内共生作用
motile:(adj) 能动的,游动的
cilia:(n) 纤毛
pseudopodia:(n) 伪足
solitary:(adj) 独立的
colonial organisms: 群体藻类
Dictyostelium discoideum: 盘基网柄菌
Phytomastigophora:(植)鞭毛虫类
cluster:(n) 分枝
ciliate:(n) 纤毛虫
lobose amoebae:(n) 叶状变形虫
filose amoebae:(n) 丝状变形虫
cryptomonad:(n) 隐滴虫
excavate:(n) 挖掘虫
opisthokont:(n) 后囊虫
euglenid:(n) 眼虫
Apicomplexa:(n) 端复分亚门
trypanosome:(n) 锥体虫

Notes

[1] 它们也是非丝状的［与具有菌丝体的霉菌（molds）等真菌（fungi）生物相反］,仅限于潮湿或水生栖息地,在全球从南极到北极的此类环境中无处不在。

[2] 众所周知的原生动物包括典型的甲藻（Dinoflagellates）、变形虫（Amoebas）、草履虫（Paramecia）和引起疟疾的疟原虫（Plasmodium）。甲藻是一类海洋或淡水中的单细胞生物,具有鞭毛,能够运动。有些甲藻能进行光合作用,有些则为异养生物。它们在海洋生态系统中扮演重要角色,也是赤潮现象的主要原因之一。变形虫是形状不断变化的单细胞原生动物,通过伪足运动,并通过吞噬作用摄食。某些变形虫如阿米巴原虫可寄生在人类体内,引起阿米巴痢疾。草履虫是具有草鞋形状的单细胞生物,表面覆盖纤毛,通过纤毛摆动来移动和摄食。草履虫是细胞生物学和基因表达研究的常用模型生物。疟原虫是一类寄生在红细胞中的单细胞生物,是疟疾的病原体。疟原虫通过蚊子的叮咬传播,引起人类疟疾,包括引起恶性疟疾的恶性疟原虫（Plasmodium falciparum）。

[3] 例如,许多原生生物是混合营养生物（mixotrophs）,既能通过消耗其他生物体获取能量(异养,heterotrophy）,也能通过光合作用或化学合成获取能量(自养,autotrophy）。

[4] 符合当代原生动物定义的生物体存在于原生动物学家认可的所有主要原生生物群体中,反映了原生动物的并系性质（paraphyletic nature）。并系性（paraphyletic）是一个分类学术语,描述某一类群包含了其共同祖先和部分但不是全部的后代。例如,原生动物（Protozoa）被认为是并系性的,因为尽管它们具有共同祖先,但该祖先的所有后代并未都被包括在原生动物中。一些后代可能被归入其他不同的群体。这样的分类方式与单系性（monophyletic）相对,单系群包括一个共同祖先及其所有后代。并系性也不同于多系性（polyphyletic）,后者指一组生物由多个不同的祖先独立演化而来。

11. Fungi

Text

Fungi (kingdom Fungi) are eukaryotic and mostly multicellular, though some, such as molds and yeasts, are unicellular. Multicellular fungi typically form long filaments called hyphae. Fungi are heterotrophs that lack chloroplasts and chlorophyll and cannot perform photosynthesis. As in plants however, fungi have cell walls. The cell wall of fungi is made of chitin, a highly resistant carbohydrate found in the skeleton of many animals, whereas it is made of cellulose in plants.

There are at least 1,500 known species of marine fungi, mostly microscopic, and mostly living in or on other organisms. Fungi absorb nutrients from their environment by secreting digestive enzymes, and many, as in bacteria, decompose detritus. They can decompose the cellulose in the cell walls of seaweeds and plants. Cellulose, which is abundant in driftwood and the dropped leaves of seagrasses and salt-marsh plants, is not digested by most bacteria. Fungi are the most important decomposers of dropped mangrove leaves and thus contribute to the recycling of nutrients in mangrove forests[1]. A similar process occurs with dropped seagrass leaves, drift seaweed, and in the driftwood that accumulates in deep water along tropical rain forests. Marine fungi are also important in the breakdown of detritus and other organic matter that accumulates as the result of plankton blooms and pollution. Some marine fungi are parasites of seagrasses or borers in mollusc shells. Others are parasites that cause diseases of economically important seaweeds, sponges, molluscs, lobsters, and fishes, as well as corals and marine mammals. Some marine fungi are being investigated as a source of antibiotics for use in medicine.

Many fungi live in symbiotic associations, mostly with green algae or cyanobacteria, to form unique entities, the lichens. In lichens the long hyphae of the fungi provide support, whereas the algae or cyanobacteria provide food from photosynthesis[2]. Marine lichens can be typically found as thick, dark brown or black patches in the wave-splashed zone of exposed rocky shores (Figure Ⅱ-12). They are tolerant to exposure to air for long periods of time.

By comparison with the multitude of lichens on land, there are very few types of marine lichens. Their role on the ecology of rocky shores is largely unknown. The cyanobacteria of some marine lichens are known to be nitrogen-fixers. Some lichens are able to loosen bits of the rocks where they live by growing into the rock. A few types live on the shells of rocky shore barnacles and molluscs.

Molecular techniques similar to those used in identifying viruses and marine microbes are

Figure Ⅱ-12: Encrusting marine lichens, cyanobacteria, and some microscopic algae are often visible as a black band on wave-splashed rocky shores, as in this exposed rocky point south of Monterey Bay, California. [Source: Castro and Huber, 2018]

being applied to the sampling of fungi in places such as deep-sea sediments and anoxic environments. These are fungi that are often difficult to culture in the lab. The number of species of marine fungi will undoubtedly increase with time.

Most fungi can reproduce through both sexual and asexual reproduction. Asexual reproduction occurs through the release of spores or through mycelial fragmentation, which is when the mycelium separates into multiple pieces that grow separately. In sexual reproduction, separate individuals fuse their hyphae together. The exact life cycle depends on the species, but generally multicellular fungi have a haploid stage (where they have one set of chromosomes), a diploid stage, and a dikaryotic stage where they have two sets of chromosomes but the sets remain separate.

All fungi reproduce using spores. Spores are microscopic cells or groups of cells that disperse from their parent fungus, usually through wind or water. Spores can become dormant for a long time until conditions are favorable for growth. This is an adaptation for opportunism; with a sometimes-unpredictable food source availability, spores can be dormant until they are able to colonize a new food source[3]. Fungi produce spores through sexual and asexual reproduction.

Vocabulary

eukaryotic:(adj) 真核生物的
heterotroph:(n) 异养生物
chitin:(n) 壳质,角素
carbohydrate:(n) 碳水化合物,糖类
skeleton:(n) 骨架,骨骼
cellulose:(n) 纤维素,细胞膜质
detritus:(n) 碎石,岩屑
driftwood:(n) 浮木,流木
seagrass:(n) 海草
salt-marsh:(n) 盐沼

mangrove:(n) 红树林
parasite:(n) 寄生虫
borers:(n) 蛀虫
antibiotic:(n) 抗生素
cyanobacteria:(n) 蓝藻
lichen:(n) 地衣,苔藓
mycelial:(adj) 菌丝的
fragmentation:(n) 破碎,(染色体)断裂
dikaryotic:(adj) 双核的
dormant:(adj) 休眠的,潜伏的

Notes

[1] 真菌是降解红树林落叶的最重要分解者,对红树林生态系统的养分循环起着关键用。通过分解落叶,真菌释放并循环养分,确保其他植物和生物获得所需资源,维持生态系统的平衡和生产力。

[2] 在地衣中,真菌的长菌丝提供结构支持,而绿藻或蓝藻则通过光合作用提供有机物。

[3] 这是对机会主义环境的一种适应。由于食物来源有时不可预测,真菌的孢子可以进入休眠状态,直到能够定殖于新的食物来源。机会主义是指生物在资源不稳定或不确定的环境中,利用任何可利用资源的策略。适应是指生物为生存和繁殖,逐渐演化出某些特性或行为以应对环境变化。孢子是许多真菌产生的繁殖细胞,具有高度的环境耐受性,能够在不利条件下存活。当环境条件不适宜时,孢子停止活动和生长,以保存能量和资源。当条件变得适宜,或出现新的食物来源(如腐烂的植物或动物)时,孢子会从休眠中激活,开始生长和繁殖。

12. Viruses

Text

Viruses, remarkably, span the non-living and the living. Unlike living organisms, a virus is a particle not made up of a cell. Viruses consist of only a short chain of genetic material (nucleic acid) containing relatively few genes and protected by an outer protein coat, or capsid. They are parasites that reproduce and develop only when infecting a living cell. Nearly all viruses are minute, roughly between 20 and 200 nm, 1 nm being equal to one-billionth of a meter, or 1,000 μm. They can be seen only with the most powerful microscopes. A giant marine virus (Figure Ⅱ-13) is an exception, being 440 nm in diameter and thus visible through an ordinary light microscope.

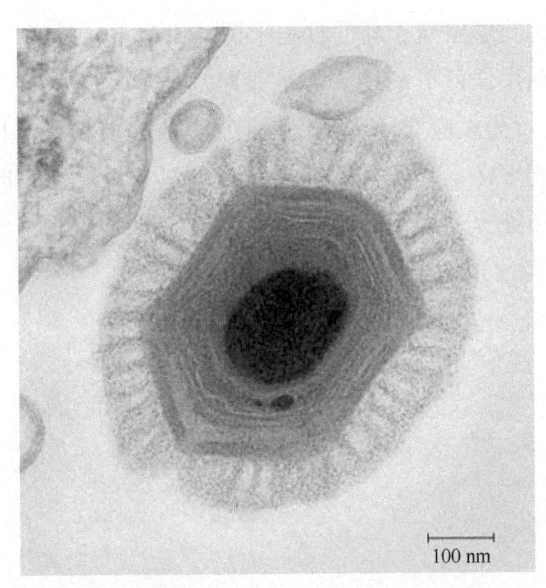

Figure Ⅱ-13: A giant marine virus. [Source: Chantal Abergel]

Several types of viruses have been described. Retroviruses store their genetic information in the form of single-stranded RNA, in contrast to the double-stranded DNA of most viruses and all forms of life[1]. Retroviruses are responsible for deadly human diseases, such as AIDS and some types of leukemia. Some viruses reproduce by attaching into a living cell and injecting their nucleic acid into the cell. The viral nucleic acid then commands the cell to produce, or replicate, copies of the virus, which are ultimately released as the infected cell bursts. The new generation of viruses then infects other cells or is released from the organism to infect new hosts. Another type, the lysogenic viruses, reproduce by inducing their nucleic acid to become part of the complete genetic information contained in the DNA, or genome, of the host cell. The infecting virus will then, sometimes long afterwards, direct the production not of host proteins but of new viruses. Many more viruses are then released as the infected cell is destroyed. Of particular importance in the ocean are the highly diverse bacteriophages, viruses that specifically target—and destroy—bacteria[2].

Until recently, marine viruses were practically ignored by marine scientists. They are nearly invisible, so it was not until the development of new techniques that viruses were shown to be extremely common in marine environments and to play an important role in the complex cycles of life in the ocean[3]. Though viruses are not referred to as living "organisms" in the strict sense of the word, viruses constitute the most abundant and diverse life-like particles in the ocean. Virus numbers in water samples can be determined by staining them with chemicals and observing them under a microscope, something that can be done on a research vessel. Their numbers in seawater samples can be staggering. Viruses are extremely abundant in deep-sea sediments.

The abundance of viruses in the water is directly related to the abundance of microbial life, particularly bacteria, which they infect and destroy in order to reproduce. Viruses also infect other marine microbes that form part of the phytoplankton, the very important primary producers that are part of the drifting plankton. The bursting of the cells of bacteria and phytoplankton as the result of viral infections also releases into the water large amounts of organic molecules and cell debris, which make up dissolved organic matter (DOM). DOM cannot be utilized as such by most organisms. It is, however, readily taken in by bacteria and other microbes. These microbes are eaten by small zooplankton, the non-photosynthetic components of the plankton, which are then eaten by larger zooplankton, so that the energy contained in DOM, a significant part of primary production by the phytoplankton, is not lost but ultimately is available to all kinds of animals, from small fishes to whales[4]. The bursting of cells as a result of viral infections also releases some essential nutrients, which can then be made available to primary producers. Some viruses infecting phytoplankton "hijack" the process of photosynthesis and shift some of the captured solar energy to their own reproduction.

Viruses are also responsible for diseases affecting marine life other than microbes. Some viruses affect molluscs and crustaceans of commercial importance, while others cause serious diseases in other invertebrates, fishes, sea turtles, and marine mammals. Humans are also vulnerable. Oysters and mussels that filter hepatitis viruses from sewage-contaminated water commonly infect humans eating them raw or improperly cooked.

Vocabulary

diameter：(n) 直径

retrovirus：(n) 逆转入酶病毒

double-stranded：(adj) 双链的

leukemia：(n) 白血病

lysogenic：(adj) 溶原性

bacteriophage：(n) 噬菌体

stain：(vt/vi) 留下污渍，给……染色

sediment：(n) 沉积物，沉积

debris：(n) 碎片，残骸

utilized：(adj) 被利用的

vt/vi：utilize

zooplankton：(n) 浮游生物

invertebrate：(adj) 无脊椎的；(n) 无脊椎动物

hepatitis：(n) 肝炎

sewage-contaminated：(adj) 污水污染的

Notes

[1] 逆转录病毒以单链 RNA 的形式储存其遗传信息，与大多数病毒和所有生命形式使用双链 DNA 形成对比。逆转录病毒（Retrovirus）是一类 RNA 病毒，其基因组由单链正义 RNA 构成。它们具有一种特殊的酶，称为逆转录酶，能够将病毒 RNA 逆转录为 DNA。这个过程与一般的 DNA 转录为 RNA 的过程相反，因此称为逆转录。

[2] 在海洋中，特别重要的是高度多样化的噬菌体，即专门感染并摧毁细菌的病毒。海洋噬菌体在海洋生态系统中起着关键作用，包括控制细菌和古菌的数量，影响海洋微生物群落的结构和功能，以及参与碳循环和能量流动。它们是海洋微生物生态系统中不可或缺的组成部分。

[3] 由于海洋病毒几乎不可见，直到新技术的发展，才发现它们在海洋环境中极为普遍，并在海洋中复杂的生命循环中发挥着重要作用。

[4] 这些微生物被小型浮游动物——浮游动物中非光合组分的微生物——所摄食，随后被更大的浮游动物捕食。因此，浮游植物初级生产的重要组分 DOM 所含的能量并未损失，而是通过微生物环，最终为各种动物所用，从小型鱼类到鲸类。

13. Bacteria

Text

Bacteria (domain Bacteria) appear to have branched out very early on the tree of life and are genetically distant from both archaea and eukaryotes. Because they are structurally simple, bacteria (singular, bacterium) are classified as prokaryotes. Bacteria are microbes with a cell structure simpler than that of many other organisms. Nevertheless, they have evolved a great range of metabolic abilities. They are abundant in all parts of the ocean.

Bacteria are classified into five groups according to their basic shapes: spherical (cocci), rod (bacilli), spiral (spirilla), comma (vibrios) or corkscrew (spirochaetes)[1]. They can exist as single cells, in pairs, chains or clusters. The unique chemistry of bacterial cell walls makes them rigid and strong. A stiff or slimy covering often surrounds the cell wall as additional protection or a means of attaching to surfaces[2]. The cells are very small, much smaller than those of single-celled eukaryotes. About 250,000 average bacterial cells would fit on the period at the end of this sentence. There are exceptions, however, such as a sediment bacterium discovered off southwest Africa, with cells as wide as 0.75 mm (0.03 in), large enough to be seen with the naked eye (Figure Ⅱ-14). Another giant bacterium (0.57 mm, or 0.02 in, long) is found inside the intestines of coral reef fishes. In large numbers, marine bacteria are sometimes visible as whitish hairs on rotting seaweed or iridescent or pink patches on the surface of stagnant pools in mudflats and salt marshes.

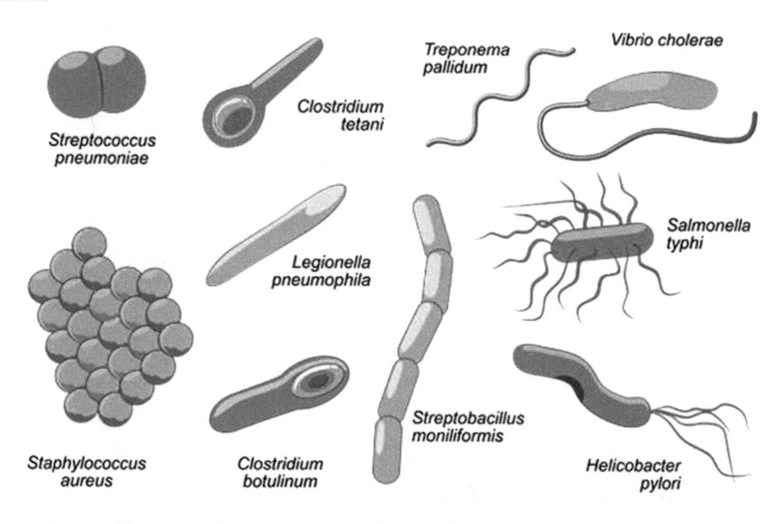

Figure Ⅱ-14: Different bacterial shapes. [Source: Microbiology Society]

Bacteria can grow to extremely high numbers in favorable environments such as detritus, particles of dead organic matter[3]. Decay bacteria break down waste products and dead organic matter and release nutrients into the environment. They are vital to life on Earth because they ensure the recycling of essential nutrients, as in the recycling of dissolved organic matter in oceanic food webs. Most organic matter is sooner or later broken down by decomposition, though in very deep, cold water the process is slower than elsewhere. Decay bacteria (or decomposers, those that carry out decomposition) play another crucial role because they constitute a major part of the organic matter that feeds countless bottom-dwelling animals. Even many organic particles in the water column are composed mostly of bacteria. Some marine bacteria are also involved in degrading oil and other toxic pollutants that find their way into the environment. The same process of decomposition is unfortunately involved in the spoilage of valuable fish and shellfish (molluscs and crustaceans) catches[4]. Other types of marine bacteria cause diseases in marine animals and humans.

Bacteria are found everywhere in the marine environment, on almost all surfaces and in the water column. In fact, the most abundant forms of life on the planet seem to be bacteria, found in surprisingly high numbers in open water. Their role, though probably highly significant due to their vast numbers, remains unknown. Bacteria have also been found in seemingly harsh environments, such as in sediment more than 3 km below the sea floor.

Vocabulary

archaea: (n pl) 古生菌
  sing: archaeum
metabolic: (n) 新陈代谢
cocci: (n) 球菌
rod: (n) 竿,杆子,棒
bacilli: (n) 杆菌
spirilla: (n) 螺旋状菌
vibrio: (n) 弧菌
corkscrew: (n) 螺旋形
spirochaete: (n)(微生物)螺旋体
rot: (n) 腐烂,腐朽;(vt/vi)(使)腐烂/腐朽
iridescent: (adj) 彩虹色的,闪光的
mudflats: (n) 泥滩

Notes

[1] 细菌根据其基本形状可分为五类:球形(球菌)、杆状(杆菌)、刚性螺旋状(螺旋菌)、逗号状(弧菌)和柔性螺旋状(螺旋体)。球菌可以单独存在,也可以形成链状或簇状聚集体。杆菌的形态有时呈直杆状或弯曲杆状。螺旋菌的螺旋形结构使其具备较高的运动能力,有利于在不同环境中的生存。弧菌呈逗号状,有些种类对人类和

其他生物具有致病性。螺旋体是一类细胞呈柔性螺旋形的细菌,包括一些与人类疾病相关的致病菌,如梅毒螺旋体和莱姆病螺旋体。

[2] 一层坚硬或黏稠的覆盖物通常包裹在细胞壁周围,作为额外的保护层或用于附着在表面上。

[3] 细菌可以在有利的环境中大量繁殖,例如,在碎屑和死亡有机物的颗粒上。这一过程在微生物环(the microbial loop)中起着重要作用(微生物环将在后续内容中详细介绍)。碎屑(detritus)是指枯枝落叶等死有机物的碎片,是分解者的重要营养来源。大量细菌生长在枯叶、腐木等环境中,将有机物质分解为更简单的化合物。

[4] 细菌的分解过程也涉及宝贵的鱼类和贝类(包括软体动物和甲壳类)渔获物的腐败。变质"spoilage"指食品因细菌或其他微生物作用而腐败的过程。在这种情况下,细菌导致有价值的鱼类和贝类渔获物腐败变质。

14. Archaea

Text

The Archaea (domain Archaea) are among the simplest, most primitive forms of life. Some look very similar to the oldest fossils ever found, cells estimated to be at least 3.8 billion years old[1]. Archaea are thought to have had an important role in the early evolution of life. Like bacteria, their cells are small and can be spherical, spiral, or rod-shaped. In fact, archaea were thought to be bacteria. The origins of Archaea remain unclear. Despite being prokaryotic, however, there is evidence that archaea are more closely related to eukaryotes than to bacteria.

Some groups of archaea were discovered only recently, first in extreme environments on land, such as hot sulfur springs, saline lakes, and highly acidic or basic environments[2]. Archaea were thus named "extremophiles," meaning "lovers of extremes." Archaea were subsequently found in extreme marine environments, such as in very deep water, where they survive at pressures of 300 to 800 atmospheres. Some archaea live at the high temperatures of hydrothermalvents[3]. Some of these cannot grow in temperatures less than 70–80 ℃, and one hydrothermal vent archaeum can live at 121 ℃, the highest of any known organism. Other archaea depend on extremely salty environments, such as coastal salt pans and deep ocean basins[4].

Because the first groups of microorganisms to be classified as archaea were extremophiles, it was thought that all archaea were extremophiles; indeed, the two terms became almost synonymous. New techniques based on detecting nucleic acid sequences, however, have shown that archaea are common in many marine environments, not only in the water column but also in sediments. Some archaea are also known to live symbiotically in sponges, sea cucumbers, and fishes. Thus, the hypothesis that archaea are restricted to extreme environments was proven false.

Archaea are prokaryotic microorganisms once thought to be bacteria but more closely related to eukaryotes. They were first known from extreme environments, but they are now known to be common in the marine environment.

Vocabulary

extremophile: (n) 极端微生物
hydrothermal vent: (n) 热泉
basin: (n) 盆地,盆,水池,流域
synonymous: (adj) 同义的
sponge: (n) 海绵
sea cucumber: (n) 海参

Notes

[1] 一些古菌与目前已知最古老的化石非常相似,这些细胞估计至少有 38 亿年的历史。

[2] 某些古菌群是最近才被发现的,最初是在陆地上的极端环境中发现,如热硫泉、盐湖以及高酸性或碱性的环境。

[3] 热液喷口 (hydrothermal vents) 是海洋底部的一种地质结构,常出现在洋中脊和地壳板块交界处。热液喷口周围存在高温、高压和丰富的化学物质,是一些特殊微生物生存的重要场所。

[4] 其他古菌依赖于极咸的环境,如盐沼和深海盆地。滨海盐湖 (coastal salt pans) 是位于海岸线附近的高盐度湖泊或盐田,通常由海水在陆地上蒸发形成。滨海盐湖是一些极端嗜盐微生物的理想栖息地,它们能够在高盐度环境中生存和繁殖。

15. The Microbial Loop

Text

The dissolved organic matter (DOM) in the ocean isn't just sitting there, but is the basis of a major energy pathway called the microbial loop (Figure Ⅱ-15). The DOM is not lost to the food web, but is used by heterotrophic bacteria and archaea. The most abundant plankton in the epipelagic are a group of heterotrophic bacteria collectively known as SAR11. SAR11 bacteria were first discovered by genomics: Marine biologists found their DNA without ever observing the cells. Eventually some cells were collected and biologists learned how to culture them—these have been named *Pelagibacter*, but there are many other strains of SAR11, sometimes referred to as *Pelagibacterales*, that haven't yet been collected or cultured[1]. Many other microbes in addition to SAR11 process DOM through the microbial loop.

The microbial loop heterotrophs are so small that most animals can't eat them, but tiny protozoans in the nanoplankton can. The energy from DOM is finally passed up the rest of the food chain when net zooplankton eat the protozoans. The microbial loop channels not only DOM but also the energy in living pico- and nanoplankton up the food chain. Photosynthetic cyanobacteria like *Prochlorococcus* and *Synechococcus* are consumed directly by protozoan grazers that are able to eat such small cells. Biologists have discovered that the cyanobacteria live in close association, perhaps even symbiosis, with heterotrophic bacteria, so that the photosynthetic production is transferred to the heterotrophic bacteria before moving up the food chain through the microbial loop[2].

Detritus—particulate rather than dissolved organic matter—is also important in the epipelagic, as in other marine food webs. Two important sources of detritus have already been mentioned: fecal pellets and abandoned larvacean houses. Larvacean houses and other accumulations of mucus can be very abundant and support rich populations of bacteria. They are often called marine snow because underwater they look like snowflakes.

Many zooplankton and fishes eat marine snow. Much of the detritus, however, sinks out of the epipelagic zone to deeper water before epipelagic organisms can get to it[3]. As previously noted, the epipelagic does not get much detritus from other systems.

The "microbial loop" in simplified form, refers to the flow of energy through the series phytoplankton → DOM → bacteria → protozoans → zooplankton[4]. Without the microbial loop, the energy in DOM would largely go unused. As much as half the primary production in the epipelagic is channeled through the microbial loop.

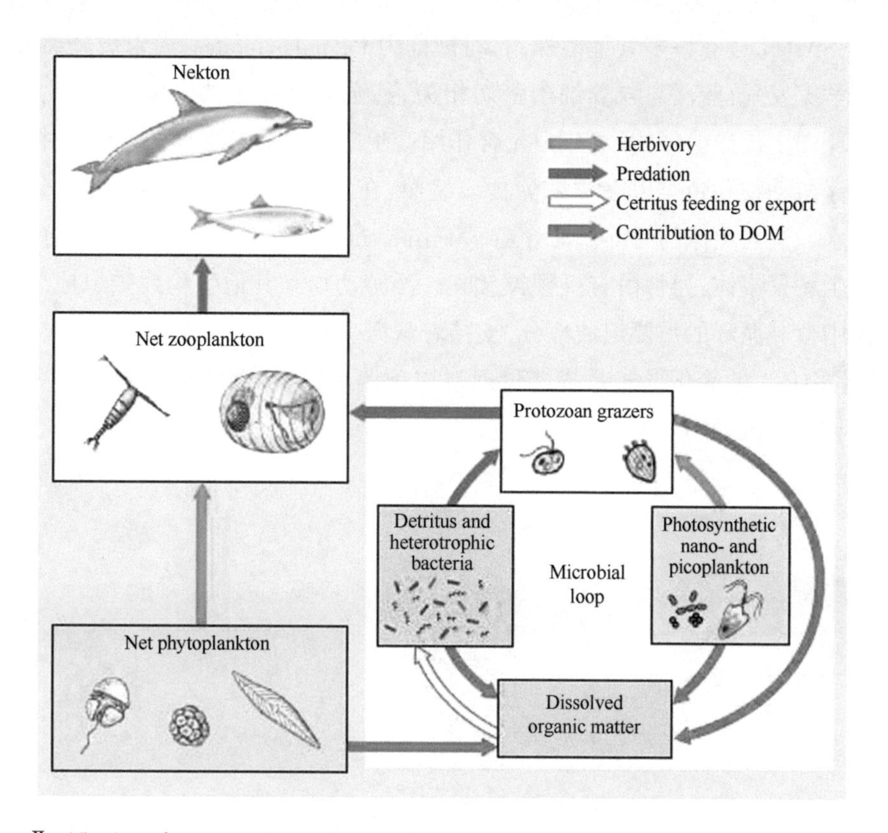

Figure Ⅱ-15: A modern view of epipelagic food webs. The left side shows the flow of energy from the net phytoplankton to the nekton. This part of epipelagic food webs has been known for decades. The right side of the diagram shows the more recently discovered microbial loop. Much of the flow from organisms into dissolved organic matter (blue arrows) is caused by viruses. [Source: Bill Ober]

Vocabulary

dissolved organic matter: 溶解性有机质
epipelagic: (adj) 上层的,光合作用带的
genomics: (n) 基因组学,基因体学
Pelagibacter: (n) 远洋杆菌属
loop: (n) 圈,环,环状物,回路,循环
Prochlorococcus: (n) 原绿球藻
Synechococcus: (n) 聚球藻
protozoan grazers: (n) 食草性原生动物
symbiosis: (n) 共生(关系)
heterotrophic bacteria: 异养细菌
larvacean: (n) 幼形纲
snowflake: (n) 雪花

Notes

[1] SAR11 细菌最初是通过基因组学发现的,即海洋生物学家在未观察到细胞的情况下检测到了其 DNA。最终,一些细胞被收集并培养,这些细胞被命名为 Pelagibacter。

然而,SAR11 还有许多其他菌株,有时被称为 Pelagibacterales,尚未被分离或培养。

[2] 生物学家发现,蓝藻与异养细菌密切相关,甚至可能存在共生关系,因此,在通过微生物环向上传递能量的过程中,光合作用产生的能量被转移到异养细菌中。

[3] 许多浮游动物和鱼类以海洋雪为食。然而,在上层生物到达之前,大部分碎屑会从上层水域沉入更深的海洋。海洋雪(marine snow)是指海洋中大量悬浮的有机碎片和微生物聚集体,通常由有机颗粒、细菌、浮游动物和其他生物残骸组成。海洋雪是海洋中营养循环的重要组成部分,支持着深海生物的生存和繁衍。

[4] "微生物环"的简化形式是指能量通过以下路径流动:浮游植物→DOM→细菌→原生动物→浮游动物。

16. Seaweeds

Text

The most familiar types of marine algae are those popularly known as seaweeds. "Algae" (singular "alga") and "seaweeds" are interchangeably used, except that, to complicate our lives, unicellular "algae" are not seaweeds but all seaweeds are algae[1]. Just remember that "red algae" and "red seaweeds" mean the same thing. "Seaweed" is a rather unfortunate name. First of all, the word "weeds" does not do justice to these conspicuous and often elegant inhabitants of rocky shores and other marine environments. Some biologists opt for the more formal name of macroalgae. On the other hand, the term "seaweeds" is useful in distinguishing them from the unicellular algae. By definition, seaweeds are all multicellular; unicellular green and red algae are therefore not considered seaweeds. All seaweeds and unicellular green and red algae are nevertheless considered protists (formerly kingdom Protista) whereas seagrasses, saltmarsh grasses, and mangroves are true plants (kingdom Plantae)[2].

Like unicellular algae, seaweeds are eukaryotic[3]. The structures of seaweeds, however, are far more complex than those of unicellular algae. Reproduction is also more elaborate. Though more complex than unicellular algae, seaweeds lack the highly specialized structures and reproductive mechanisms of most terrestrial true plants, especially the flowering plants. Some seaweeds are microscopic, such as the phytoplankton that live suspended in the watercolumn and provide the base for most marine food chains. Some are enormous, like the giant kelp that grow in abundant "forests" and tower like underwater redwoods from their roots at the bottom of the sea. Most are medium-sized, come in colors of red, green, brown, and black, and randomly wash up on beaches and shorelines just about everywhere.

Seaweeds any of the red, green, or brown marine algae that grow along seashores. Seaweeds are generally anchored to the sea bottom or other solid structures by rootlike "holdfasts," which perform the sole function of attachment and do not extract nutrients as do the roots of higher plants[4]. A number of seaweed species are edible, and many are also of commercial importance to humans. Some are used as fertilizers or as sources of polysaccharides.

The range of variation observed among seaweeds is spectacular. Those we see on rocky shores at low tide are usually small and sturdy as an adaptation to withstand waves. Some small, delicate ones live on other seaweeds. Kelps found offshore in cold waters are true giants that form dense underwater forests (Figure Ⅱ-16). The multicellular condition of seaweeds allows many adaptations not available to unicellular forms. The ability of seaweeds to grow tall and rise off the bottom provides new opportunities as well as challenges, particularly that of wave action.

Figure Ⅱ-16: Giant kelp (Macrocystis pyrifera) near Catalina Island, California. Giant kelp is a brown alga (Phaeophyceae) that can form extensive "kelp forests," which are an important marine habitat. [Source: hotshotsworldwide/Fotolia]

Seaweeds are autotrophs that play an important role in many coastal environments[5]. They transform solar energy into chemical energy in the form of organic matter and make it available to a long list of hungry creatures, which can include humans. There are always exceptions, and there are a few seaweeds that are not primary producers but parasites of other seaweeds. Other organisms live on or even within the tissues of seaweeds, and more importantly, seaweeds also take carbon dioxide from their environment and release oxygen to be used by organisms both in the ocean and on land. It has been estimated that seaweeds, together with the unicellular algae in the water, produce 90% of all the oxygen in the atmosphere.

Vocabulary

alga:(n) 水藻,藻类
　pl: algae
interchangeably:(adv) 互换地
opt:(vi) 选择,挑选
macroalgae:(n) 大型藻类
kingdom Protista:(n) 原生生物界
kingdom Plantae:(n) 植物界
reproduce:(vi) 繁殖,复制
　n: reproduction 繁殖,复制品
　adj: reproductive 繁殖的,复制的
terrestrial:(n) 陆地的,地球的
water column:(n) 水体
anchor:(vt) 使固定,扎根
extract:(vt) 提取,摄取
edible:(adj) 可食用的
polysaccharide:(n) 多糖
variation:(n) 变异
sturdy:(adj) 坚固的
kelp:(n) 海带,巨藻
autotroph:(n) 自养生物

Notes

[1] “藻类” 和 “海藻” 常被互换使用,但实际上,单细胞的 “藻类”并非海藻,而所有海藻都是藻类。海藻是生活在海洋中的大型藻类的总称。肉眼不可见的蓝藻等微型藻类不属于海藻;能构成海底“海藻森林”的红藻 (red algae)、褐藻 (brown algae) 和绿藻 (green algae) 等大型藻类,才是海藻家族的成员。

[2] 原生生物 (protist) 是最简单的真核生物,全部生活在水中,没有角质。可分为三大类:藻类、原生动物类和原生菌类。它们的细胞内具有细胞核和膜结合的细胞器。

[3] 真核生物 (eukaryote) 是由真核细胞构成的生物,包括原生生物界、真菌界、植物界和动物界,真核生物是所有单细胞或多细胞、细胞内具有细胞核的生物的总称。真核生物与原核生物 (prokaryote) 的根本区别在于前者的细胞内有以核膜为边界的细胞核,因此以 “真核” 命名。

[4] 海藻通常通过根状的 “固着器” (holdfast) 固定在海底或其他固体结构上,这些 “固着器” 仅执行附着功能,不像高等植物的根那样吸收养分。固着器的本义为钩子或锚,这里指海藻的根状结构像钉子一样固定在固体表面。

[5] 自养生物 (autotroph),也称作独立营养生物,指能够利用无机物质(如水、二氧化碳、无机盐等),通过光合作用 (photosynthesis) 或化学反应,合成有机物质以维持生命活动的生物。利用化学反应获得能量进行碳素同化的生物称为化能自养生物 (chemoautotroph);利用光能进行光合作用的生物称为光能自养生物 (photoautotroph)。与自养生物相对应的是异养生物 (heterotroph),它们无法通过光合作用或化学合成合成有机物,需通过摄取其他生物或有机物质来维持生命活动。异养生物包括动物、真菌和大部分细菌。

17. Seagrasses

Text

Seagrasses superficially resemble grass but actually are not grasses at all. The closest relatives of certain seagrasses seem to be members of the lily family, so we know that seagrasses evolved from land plants. Seagrasses have adapted to life in the marine environment. They have horizontal stems called rhizomes that commonly grow beneath the sediment (Figure Ⅱ - 17). Roots and erect leaves grow from the rhizomes.

The sexual reproduction of seagrasses involves tiny flowers [Figure Ⅱ -17(b)]. The Pollengrains, Which contain the sperm, are produced by male flowers that typically develop in plants that are separate from plants producing female flowers [Figure Ⅱ - 17(e)][1]. Some seagrasses produce long, thread-like pollen grains instead of the minute and round pollen of land flowering plants. Pollen, which is often produced in sticky strands, is carried to female flowers for fertilization, or pollination, of the egg. Water currents are involved in the dispersal of pollen, but marine animals, such as small crustaceans, could also be involved[2]. It has been suggested that seagrass flowers are inconspicuous because insects are not needed for pollination, as in most land plants.

Tiny seeds, which in some species develop inside small fruit within the flower or in floating fruit, result from successful fertilization. Seeds and fruit are dispersed by water currents and perhaps in the feces of the sea turtles, fishes, ducks, and other animals that graze on the plants. Seagrasses readily reproduce asexually by extending rhizomes under the substrate (Figure Ⅱ - 17). Some seagrass meadows actually consist of a single gigantic, long-living, genetically identical clone.

About 70 species of seagrasses are known. Most are tropical and subtropical, but several species are common in colder waters. Most seagrasses are restricted to muddy and sandy areas in shallow water, but some, like surf grass [Figure Ⅱ -17(e)], live on rocky shores. Different species of seagrasses vary in their maximum depth, but all are limited by the penetration of light through the water.

Seagrasses form extensive meadows that have a most important ecological role in shallow-water environments. They provide shelter to a variety of organisms including some of considerable economic importance, their roots and leaves decrease water turbulence that helps stabilize soft bottoms, and are highly productive, providing a rich source of detritus that directly or indirectly feeds many animals[3].

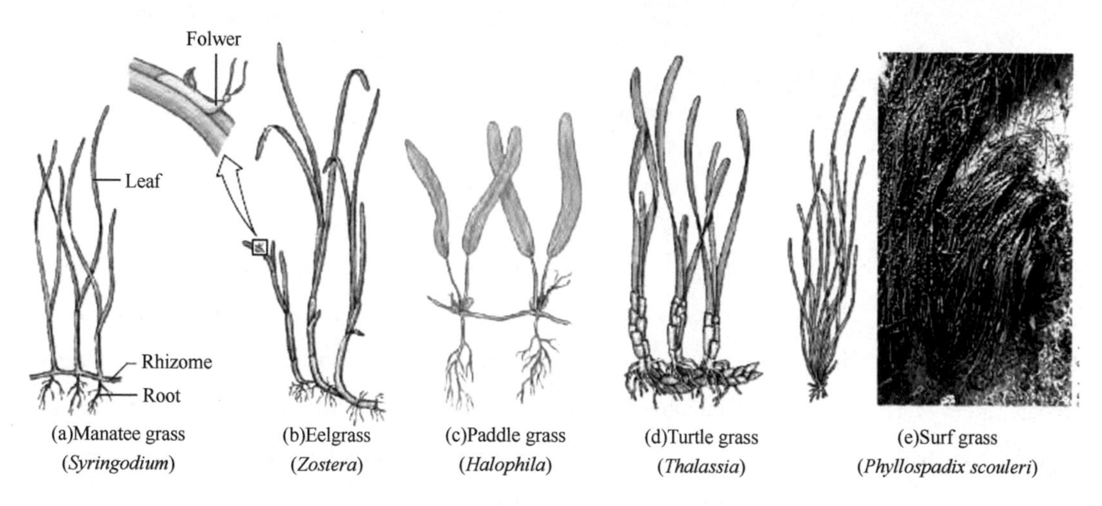

(a)Manatee grass (*Syringodium*)　(b)Eelgrass (*Zostera*)　(c)Paddle grass (*Halophila*)　(d)Turtle grass (*Thalassia*)　(e)Surf grass (*Phyllospadix scouleri*)

Figure Ⅱ-17: Some common seagrasses. [Source: Bill Ober]

Vocabulary

stem: (n) 茎

rhizome: (n) 根茎

pollen grain: (n) 花粉

thread: (n) 线,线状物

sticky strand: (n) 黏性线

pollination: (n) 授粉

feces: (n) 粪便

meadow: (n) 草地

n: meadows(河边的)低洼地

tropical: (adj) 热带的

subtropical: (adj) 亚热带的

penetration: (n) 穿透

turbulence: (n) 湍流

Notes

[1] 花粉粒(含有精子的雄性配子)由雄花产生。雄花通常生长在与产生雌花的植株分开的植物上。

[2] 甲壳动物(crustaceans)是节肢动物门中的一个大型类群,通常被视为一个亚门。它们包括常见的物种,如螃蟹(crab)、虾(shrimp)、龙虾(lobster)、淡水龙虾(freshwater lobster)、磷虾(krill)和藤壶(barnacle)等。

[3] 海草生态系统为许多经济物种提供庇护所、育幼场和食物来源。通过碎屑循环和直接摄食,海草床构成了食物网的基础。此外,海草床还能抵御海浪侵蚀,稳定沉积物。海草的叶片具有过滤作用,能够捕获悬浮颗粒(suspended particles)和细小沉积物(fine sediments),从而净化水体。

18. Salt-Marsh Plants

Text

Cordgrasses (Spartina; Figure Ⅱ-18) are true members of the grass family. At least 14 species are known. They are not really marine species but, rather, land plants tolerant of salt. Unlike seagrasses, which are true marine species, cordgrasses do not tolerate total submergence by seawater[1]. They live in salt marshes and other soft-bottom coastal areas throughout temperate regions worldwide (Figure Ⅱ-19). Cordgrass salt marshes have high primary production and provide habitat and breeding grounds for many species important to fisheries. They also offer protection against erosion, the result of the network of horizontal stems that extends under the sediment, and provide natural water purification systems (Figure Ⅱ-20)[2]. In detail, sewage is given preliminary treatment, and chlorine is added to kill harmful bacteria. It is then pumped into the marshes, where mud bacteria break down the organic matter. The released nutrients fertilize the marsh plants. The marshes prevent the pollution of the ocean by sewage and attract many species of birds and other wildlife. Flowers, and to a much lesser extent leaves, provide food to insects and other grazers[3]. Most organic matter is made available to other organisms in the form of detritus.

Figure Ⅱ-18: Cordgrass (*Spartina*) is a flowering plant that dominates the seaward margin of salt marshes in temperate and subarctic regions around the world. Salt marshes develop in soft-bottoms that are protected from wave action and are partially flooded at high tide. [Source: USGS]

Cordgrasses inhabit the zone above mudflats that becomes submerged by seawater only at high tide, so their leaves are always partly exposed to air. Salt glands in the leaves excrete excess salt. Other salt-tolerant plants, or halophytes, such as pickleweed, salt grasses (*Distichlis*), and rushes (*Juncus*) can be found at higher levels on the marsh.

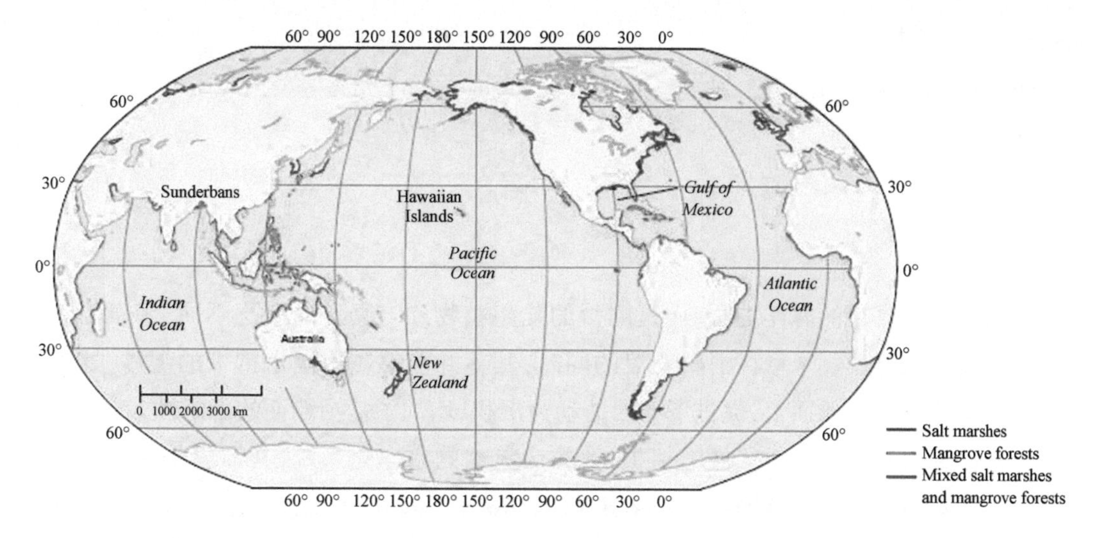

Figure Ⅱ-19: The world distribution of salt marshes and mangrove forests. [Source: Bill Ober]

Figure Ⅱ-20: This lush artificial marsh is part of a network of marshes where domestic sewage from the town of Arcata in Northern California is naturally purified before it is piped into the ocean. [Source: Gary Crabbe/age fotostock/Superstock]

The ecological significance of salt-marsh plants as a result of their high productivity, a source of detritus, shelter for many marine and terrestrial species, and protection against erosion is highly significant. Salt marshes are nevertheless rapidly disappearing as victims of coastal development[4].

Vocabulary

cordgrass: (n) 米草
submergence: (n) 浸没,淹没
marsh: (n) 沼泽,湿地
temperate region: 温带
erosion: (n) 侵蚀,腐蚀
horizontal stems: 水平茎
purification: (n) 净化
vt: purify 净化

gland:(n) 腺体
halophytes:(n) 盐生植物
pickleweed:(n) 海滨草
Distichlis:(n) 盐草属
rushes:(n) 灯芯草

Notes

[1] 米草并非真正的海洋物种,而是耐盐的陆生植物。与真正的海草不同,米草不能完全被海水淹没。米草在消减波浪的同时,其茎叶能够黏附潮水携带的泥沙,这些泥沙最终沉积在滩面上,促进了滩面的动力沉积,增加了盐沼滩面的稳定性。

[2] 盐沼湿地还提供防止侵蚀的保护,并作为天然的水净化系统,其形成是由于延伸到沉积物下方的水平茎网络(network of horizontal stems)。生活污水经过人工沼泽进一步自然净化后排入海洋。这里 sewage 指生活污水,经过初步处理,并添加氯消毒后,可被泵入人工沼泽。沼泽中的泥浆细菌分解有机物,释放的养分促进沼泽植物生长。沼泽不仅防止污水直接污染海洋,并吸引多种鸟类和其他野生动物。

[3] graze 意为提供所需食物、喂养。与 nourish 和 feed 相比,graze 更侧重于用正在生长的青草喂养牲畜。在这里,grazer 指的是食草动物,近义词有 herbivore 和 vegetarian。

[4] 盐沼湿地作为沿海开发的主要受害者,其面积正在迅速消失。这里的 as 作为介词,修饰盐沼湿地。此外,全球气候异常、城市围垦(urban reclamation)加剧和海平面上升等多种因素也导致了盐沼湿地的退化。

19. Mangroves

Text

A mangrove is a shrub or tree that grows mainly in coastal saline or brackish water. Mangroves grow in an equatorial climate, typically along coastlines and tidal rivers. They have special adaptations to take in extra oxygen and to remove salt, which allow them to tolerate conditions that would kill most plants. The term is also used for tropical coastal vegetation consisting of such species. Mangroves are taxonomically diverse, as a result of convergent evolution in several plant families[1]. They occur worldwide in the tropics and subtropics and even some temperate coastal areas, mainly between latitudes 30° N and 30° S, with the greatest mangrove area within 5° of the equator. Mangrove plant families first appeared during the Late Cretaceous to Paleocene epochs, and became widely distributed in part due to the movement of tectonic plates.

Mangroves are trees and shrubs adapted to live along tropical and subtropical shores around the world (Figure Ⅱ-19). They are essentially land plants that can tolerate salt to varying degrees. They contain a complex salt filtration system and a complex root system to cope with saltwater immersion and wave action (Figure Ⅱ-21). They are adapted to the low-oxygen conditions of waterlogged mud, but are most likely to thrive in the upper half of the intertidal zone[2].

Figure Ⅱ-21: The red mangrove (Rhizophora mangle) forms lush forests along shores in Florida, the Caribbean, the Gulf of California, and other tropical regions of the Western Hemisphere and West Africa. Notice the long roots extending into the mud, exposed here at low tide. [Source: Castro and Huber, 2018]

Mangroves include around 70 mostly unrelated species of flowering plants. They are adapted in various ways to survive in a salty environment where water loss from leaves is high and sediments are soft and poor in oxygen. Adaptations become more crucial in mangroves living right on the shore, such as species of the red mangrove (Rhizophora; Figure Ⅱ-21 and Figure Ⅱ-22), which are found throughout the tropics and subtropics. The extreme northern and southern limits of the red mangrove are those areas in which killing frosts begin (Figure Ⅱ-19). Salt marshes replace red mangrove forests in areas exposed to frosts. Other important species of mangroves include the black (*Avicennia*) and white (*Laguncularia*) mangroves. The largest number of species of these and other mangroves are found along the shores of the tropical Indian and Pacific oceans. The world's largest mangrove forest is the Sunderbans, some 10,000 km^2 along the coast of Bangladesh and western India. Mangroves are nevertheless abundant in the Atlantic Ocean and Gulf of Mexico.

(a) (b)

Figure Ⅱ-22: A seedling of the red mangrove (Rhizophora mangle) (a) as it appears in the tree and (b) one that has taken root in the soft sediment. [Source: Charlie Arneson]

The leaves of the red mangrove are thick, an adaptation to reduce water loss. As in several other mangroves, seeds germinate while still attached to the parent tree [Figure Ⅱ-22(a)]. They develop into elongated, pencil-shaped seedlings as long as 30 cm before falling from the parent. Successful seedlings stick in the soft, muddy sediment like a knife thrown into a lawn [Figure Ⅱ-22(b)], or float in the water to be carried by currents to new locations[3]. Many species of marine and terrestrial organisms live among the roots or branches of mangroves. Like seagrasses and salt-marsh plants, mangroves have high primary production, but relatively few animals graze on the hard-to-digest leaves, with most organic matter being consumed as detritus.

Vocabulary

shrub：(n) 灌木
saline：(adj) 咸的
brackish：(adj) 有盐味的
equatorial：(adj) 赤道的
tidal：(n) 潮汐的,潮水的
convergent：(adj) 趋同的
equator：(n) 赤道
Cretaceous：(n) 白垩纪
Paleocene：(n) 古新世
tectonic：(adj) 地壳(质)构造的
filtration：(n) 过滤
　vt：filter 过滤,渗透
immersion：(n) 浸没,浸泡
waterlog：(n) 涝灾
　adj：waterlogged 水涝的
germinate：(vi) 发芽
elongate：(vt) 拉长

Notes

[1] 红树林在分类学上多种多样,是多个植物科趋同(convergent)进化的结果。由于要适应恶劣的赤道气候,这些植物进化出相似的适应能力,统称为红树林(mangrove)。

[2] waterlog 意为涝灾,指长期阴雨或暴雨后,地势低洼、排水不畅的地区,由于地表积水无法及时排出,农田积水超过作物的耐淹能力,造成农业减产的灾害。这里的 waterlogged mud 指的是泥涝。

[3] 红树林的叶子很厚,以减少水分流失。与其他一些红树林物种一样,种子在仍附着于母树上时就开始发芽。在从母树脱落前,它们会发育成细长的、铅笔状的幼苗,最长可达 30 cm。成功的幼苗会像刀子插入草坪一样扎入松软泥泞的沉积物,或者漂浮在水面上,被水流带到新的地方。这说明了红树林由于其强大的耐受性和特殊的繁殖特性,得以广泛分布。

20. Primary Producers and Estuaries

Text

Seaweeds seagrasses, salt-marsh grasses, and mangroves are both primary producers capable of using light energy to perform photosynthesis[1]. Estuaries include plant-dominated communities with very high primary production. Much of the food manufactured by these plants is made available to consumers by way of detritus. Though estuaries are low in biodiversity when compared to rocky shores, they reap the benefits of living in a very productive ecosystem. The generalized food webs shown in Figure Ⅱ - 23 summarize the feeding relationships among different organisms in estuarine ecosystems.

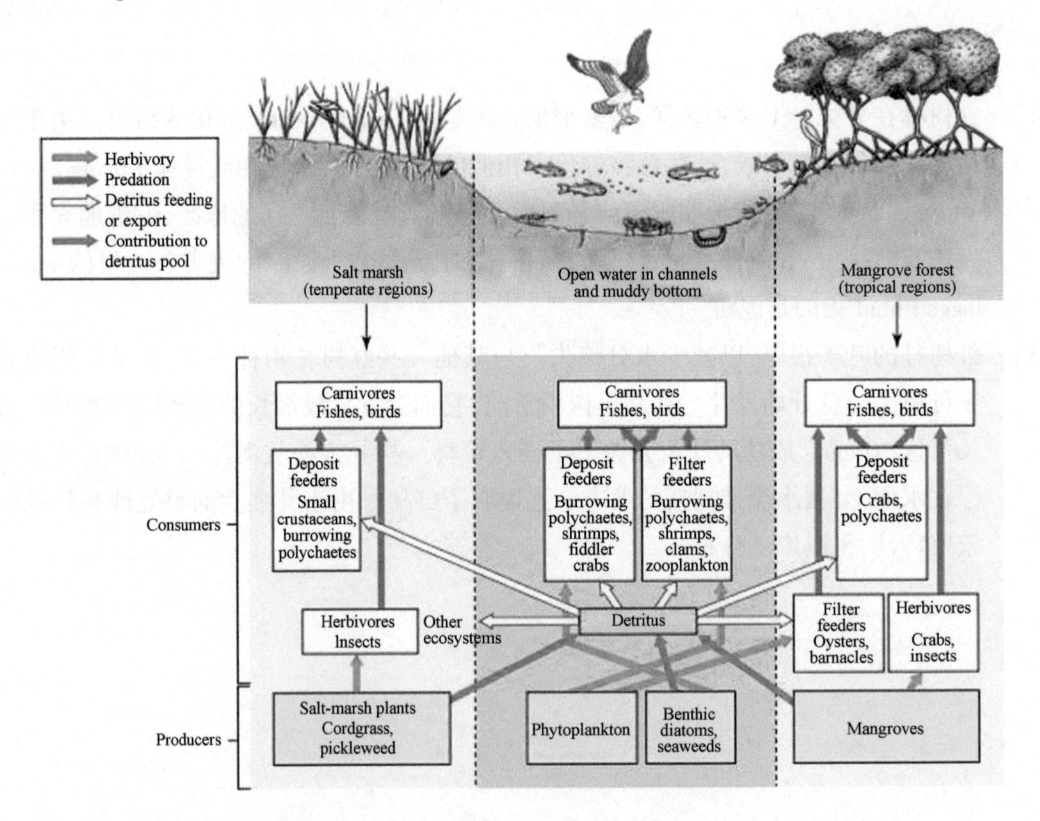

Figure Ⅱ - 23: Generalized food webs in estuarine ecosystems. Salt marshes (left) occur in temperate regions, mangroves (right) in the tropics. [Source: Bill Ober]

Why do estuaries have a high primary production? There are several reasons. Nutrients brought in by the tide and rivers, together with those generated by nitrogen-fixing organisms and the decomposition of detritus, are used by plants, algae, and bacteria, the primary producers[2].

Primary production is especially high in the communities that surround estuaries. The high biomass of cordgrass and other salt-marsh plants (or mangroves in the tropics) are particularly adapted to live on the mud and thus take advantage of the high concentration of nutrients in the sediments[3]. The diatoms and bacteria in the mud and the phytoplankton in the water also contribute significantly to primary production.

Detritus also tends to sink to the bottom. Bottom water, which has a higher salinity and density than shallower water, thus acts as a nutrient trap in deep estuaries. Some phytoplankton are known to migrate to deep water at night to take in nutrients and move up to shallow, sunlit water the next day to carry out photosynthesis[4].

Primary production by estuary plants and other organisms varies geographically and seasonally, as does their relative contribution to the ecosystem as a whole. Estimates of primary production range from 130 to nearly 6,000 g dry weight/m^2/year for cordgrasses in salt marshes on the Atlantic coast of the United States. For a summary of the typical rates of primary production for salt marshes, mangroves, and seagrass meadows.

The organic material manufactured by primary producers is made available to consumers mainly in the form of detritus. A distinctive feature of estuarine ecosystems is that most of the animals feed on dead organic matter. Except for insects, geese, and some land animals on the fringes, relatively few herbivores actually graze on salt-marsh plants. Many detritus feeders obtain more energy from the bacteria and other decomposers in the detritus than from the dead organic matter itself. They excrete any detritus that remains undigested, however, returning it to the detritus pool. The surplus detritus is exported to the open ocean and neighboring ecosystems in a process known as outwelling[5]. The exported detritus serves as a valuable source of food and nutrients to other ecosystems. The amount of exported detritus varies among estuaries, and some are actually net importers. Nonetheless, outwelling is an important role of estuaries, an additional reason that they should be preserved and protected.

Vocabulary

estuary: (n) 河口
  pl: estuaries
  adj: estuarine
decomposition: (n) 分解
  n: decomposer 分解者
biomass: (n) 生物量
migrate: (vt) 迁移,洄游
fringe: (n) 边缘,外围
herbivore: (n) 食草动物
surplus: (adj) 多余的,剩下的
outwelling: (n) 外涌,流出

Notes

[1] 在海洋生态系统中,海藻、盐沼草和红树林都是能够利用光能进行光合作用的海洋生物,统称为初级生产者,即 Text 16~Text 19 中所讲述的内容。

[2] 潮汐和河流带来的营养物质,以及固氮生物(nitrogen-fixing organisms)和碎屑分解产生的营养物质,被植物、藻类和细菌等初级生产者所利用,这导致了河口地区的初级生产力特别高。

[3] 具有高生物量的米草和其他盐沼植物(或热带地区的红树林)特别适合生活在淤泥上,从而利用淤泥中高浓度的营养物质。这里的 sediments 本义为沉积物,此处指淤泥。河口是大江大河与海洋系统对接的生态过渡带,具有丰富的生物多样性,因此,河口环境非常适宜米草的生长。

[4] migrate 意为迁徙,在这里表示垂直洄游,一些浮游植物在夜间洄游到深水中吸收养分,白天再洄游到阳光照射的浅水中进行光合作用。

[5] outwelling 意为流出,是一种假设的过程。沿海的盐沼和红树林作为生产“热点”,每年产生过量的碳,并将这些有机营养物和碎屑“流出”到周围的沿海海湾或海洋,提高了海洋的生产力,从而对当地渔业或其他沿海生物产生一定影响。流出还会滋养浮游生物群落,导致其活动激增。

Text Sources

Baron, 1996. Medical Microbiology. Galveston: University of Texas Medical Branch at Galveston.

Biology Dictionary. https://biologydictionary.net.

Castro, Huber, 2018. Marine Biology. New York: McGraw-Hill Education.

Dos Santos Severiano, et al., 2018. Effects of Increased Zooplankton Biomass on Phytoplankton and Cyanotoxins: A Tropical Mesocosm Study. Harmful Algae, (71): 10–18.

Encyclopedia Britannica. https://www.britannica.com.

Lalli, Parsons, 1997. Biological Oceanography: An Introduction. Oxford: Butterworth-Heinemann.

Lexico UK English Dictionary. https://www.dictionary.com.

Lumen Learning. https://courses.lumenlearning.com.

Microbiology Society. https://www.microbiologysociety.org.

Miller, Wheeler, 2012. Biological Oceanography. New Jersey: Wiley-Blackwell.

National Geographic. https://education.nationalgeographic.org.

Weisse, 2017. Functional Diversity of Aquatic Ciliates. European Journal of Protistology, 61: 331–358.

PART III

MARINE GEOLOGY AND GEOPHYSICS

1.Dating Rocks and Fossils Using Geologic Methods

Text

Accurate dating is of fundamental importance to geoscience. There are three general approaches that allow scientists to date geological materials and answer the question: "How old is this fossil/rock?" First, the relative age of a fossil can be determined. Relative dating puts geologic events in chronological order without requiring that a specific numerical age be assigned to each event[1]. Second, it is possible to determine the numerical age for fossils or earth materials. Numerical ages estimate the date of a geological event and can sometimes reveal quite precisely when a fossil species existed in time. Third, magnetism in rocks can be used to estimate the age of a fossil site. This method uses the orientation of the Earth's magnetic field, which has changed through time, to determine ages for fossils and rocks.

Relative Dating to Determine the Age of Rocks and Fossils

Geologists have established a set of principles that can be applied to sedimentary and volcanic rocks that are exposed at the Earth's surface to determine the relative ages of geological events preserved in the rock record. For example, in the rocks exposed in the walls of the Grand Canyon there are many horizontal layers, which are called strata. The study of strata is called stratigraphy, and using a few basic principles, it is possible to work out the relative ages of rocks.

In the Grand Canyon, the layers of strata are nearly horizontal. Most sediment is either laid down horizontally in bodies of water like the oceans, or on land on the margins of streams and rivers. Each time a new layer of sediment is deposited it is laid down horizontally on top of an older layer. This is the principle of original horizontality: layers of strata are deposited horizontally or nearly horizontally[2]. Thus, any deformations of strata must have occurred after the rock was deposited.

The principle of superposition builds on the principle of original horizontality. The principle of superposition states that in an undeformed sequence of sedimentary rocks, each layer of rock is older than the one above it and younger than the one below it. Accordingly, the oldest rocks in a sequence are at the bottom and the youngest rocks are at the top.

Sometimes sedimentary rocks are disturbed by events, such as fault movements, that cut across layers after the rocks were deposited. This is the principle of cross-cutting relationships. The principle states that any geologic features that cut across strata must have formed after the rocks they cut through.

The principles of original horizontality, superposition, and cross-cutting relationships allow events to be ordered at a single location. However, they do not reveal the relative ages of rocks preserved in two different areas. In this case, fossils can be useful tools for understanding the relative ages of rocks. Each fossil species reflects a unique period of time in Earth's history. The principle of faunal succession states that different fossil species always appear and disappear in the same order, and that once a fossil species goes extinct, it disappears and cannot reappear in younger rocks[3].

Fossil species that are used to distinguish one layer from another are called index fossils. Index fossils occur for a limited interval of time. Usually index fossils are fossil organisms that are common, easily identified, and found across a large area. Because they are often rare, primate fossils are not usually good index fossils. Organisms like pigs and rodents are more typically used because they are more common, widely distributed, and evolve relatively rapidly.

Using the principle of faunal succession, if an unidentified fossil is found in the same rock layer as an index fossil, the two species must have existed during the same period of time. If the same index fossil is found in different areas, the strata in each area were likely deposited at the same time[4]. Thus, the principle of faunal succession makes it possible to determine the relative age of unknown fossils and correlate fossil sites across large discontinuous areas.

Determining the Numerical Age of Rocks and Fossils

Unlike relative dating methods, absolute dating methods provide chronological estimates of the age of certain geological materials associated with fossils, and even direct age measurements of the fossil material itself. To establish the age of a rock or a fossil, researchers use some type of clock to determine the date it was formed. Geologists commonly use radiometric dating methods, based on the natural radioactive decay of certain elements such as potassium and carbon, as reliable clocks to date ancient events. Geologists also use other methods—such as electron spin resonance and thermoluminescence, which assess the effects of radioactivity on the accumulation of electrons in imperfections, or "traps", in the crystal structure of a mineral—to determine the age of the rocks or fossils.

All elements contain protons and neutrons, located in the atomic nucleus, and electrons that orbit around the nucleus. In each element, the number of protons is constant while the number of neutrons and electrons can vary. Atoms of the same element but with different number of neutrons are called isotopes of that element. Each isotope is identified by its atomic mass, which is the number of protons plus neutrons. For example, the element carbon has six protons, but can have six, seven, or eight neutrons. Thus, carbon has three isotopes: carbon 12 (^{12}C), carbon 13 (^{13}C), and carbon 14 (^{14}C).

Most isotopes found on Earth are generally stable and do not change. However, some isotopes, like ^{14}C, have an unstable nucleus and are radioactive. This means that occasionally the unstable isotope will change its number of protons, neutrons, or both. This change is called radioactive decay. For example, unstable ^{14}C transforms to stable nitrogen (^{14}N). The atomic nucleus that decays is called the parent isotope. The product of the decay is called the daughter isotope. In the example, ^{14}C is the parent and ^{14}N is the daughter.

Some minerals in rocks and organic matter (e.g., wood, bones, and shells) can contain radioactive isotopes. The abundances of parent and daughter isotopes in a sample can be measured and used to determine their age[5]. This method is known as radiometric dating.

The rate of decay for many radioactive isotopes has been measured and does not change over time. Thus, each radioactive isotope has been decaying at the same rate since it was formed, ticking along regularly like a clock. For example, when potassium is incorporated into a mineral that forms when lava cools, there is no argon from previous decay (argon, a gas, escapes into the atmosphere while the lava is still molten). When that mineral forms and the rock cools enough that argon can no longer escape, the "radiometric clock" starts. Over time, the radioactive isotope of potassium decays slowly into stable argon, which accumulates in the mineral.

The amount of time that it takes for half of the parent isotope to decay into daughter isotopes is called the half-life of an isotope[6]. When the quantities of the parent and daughter isotopes are equal, one half-life has occurred. If the half-life of an isotope is known, the abundance of the parent and daughter isotopes can be measured and the amount of time that has elapsed since the "radiometric clock" started can be calculated.

For example, if the measured abundance of ^{14}C and ^{14}N in a bone are equal, one half-life has passed and the bone is 5,730 years old (an amount equal to the half-life of ^{14}C). If there is three times less ^{14}C than ^{14}N in the bone, two half-lives have passed and the sample is 11,460 years old. However, if the bone is 70,000 years or older the amount of ^{14}C left in the bone will be too small to measure accurately. Thus, radiocarbon dating is only useful for measuring things that were formed in the relatively recent geologic past[7]. Luckily, there are methods, such as the commonly used potassium-argon (K-Ar) method, that allows dating of materials that are beyond the limit of radiocarbon dating.

Radiation, which is a byproduct of radioactive decay, causes electrons to dislodge from their normal position in atoms and become trapped in imperfections in the crystal structure of the material. Dating methods like thermoluminescence, optical stimulating luminescence and electron spin resonance, measure the accumulation of electrons in these imperfections, or "traps", in the crystal structure of the material. If the amount of radiation to which an object is exposed remains constant, the amount of electrons trapped in the imperfections in the crystal structure of the material will be proportional to the age of the material[8]. These methods are applicable to materials that are up to about 100,000 years old. However, once rocks or fossils become much older than that, all of the "traps" in the crystal structures become full and no more electrons can accumulate, even if they are dislodged.

Using Paleomagnetism to Date Rocks and Fossils

The Earth is like a gigantic magnet. It has a magnetic north and south pole and its magnetic field is everywhere. Just as the magnetic needle in a compass will point toward magnetic north, small magnetic minerals that occur naturally in rocks point toward magnetic north, approximately parallel to the Earth's magnetic field. Because of this, magnetic minerals in rocks are excellent recorders of the orientation, or polarity, of the Earth's magnetic field.

Through geologic time, the polarity of the Earth's magnetic field has switched, causing reversals in polarity. The Earth's magnetic field is generated by electrical currents that are produced by convection in the Earth's core. During magnetic reversals, there are probably changes in convection in the Earth's core leading to changes in the magnetic field. The Earth's magnetic field has reversed many times during its history. When the magnetic north pole is close to the geographic north pole (as it is today), it is called normal polarity. Reversed polarity is when the magnetic "north" is near the geographic south pole. Using radiometric dates and measurements of the ancient magnetic polarity in volcanic and sedimentary rocks (termed paleomagnetism), geologists have been able to determine precisely when magnetic reversals occurred in the past. Combined observations of this type have led to the development of the geomagnetic polarity time scale (GPTS). The GPTS is divided into periods of normal polarity and reversed polarity.

Geologists can measure the paleomagnetism of rocks at a site to reveal its record of ancient magnetic reversals. Every reversal looks the same in the rock record, so other lines of evidence are needed to correlate the site to the GPTS. Information such as index fossils or radiometric dates can be used to correlate a particular paleomagnetic reversal to a known reversal in the GPTS[9]. Once one reversal has been related to the GPTS, the numerical age of the entire sequence can be determined.

Summary

Using a variety of methods, geologists are able to determine the age of geological materials to answer the question: "how old is this fossil?" Relative dating methods are used to describe a sequence of events. These methods use the principles of stratigraphy to place events recorded in rocks from oldest to youngest. Absolute dating methods determine how much time has passed since rocks formed by measuring the radioactive decay of isotopes or the effects of radiation on the crystal structure of minerals. Paleomagnetism measures the ancient orientation of the Earth's magnetic field to help determine the age of rocks.

Vocabulary

dating:(n) 定年,(岩石或地层)年龄的确定

relative dating:相对定年

岩石或构造在地质年代框架中的位置可以确定其相对的年龄大小

magnetism:(n) 磁性,磁力

magnetic field:(n) 磁场

地磁场是地球内部存在天然磁性现象

stratum:(n) 层,地层

pl:strata 地层

stratigraphy:(n) 地层学

principle of original horizontality:原始水平定律,即地层沉积时是近于水平的

principle of superposition:地层叠覆律,原始地层具有下老上新的规律

superposition:(n) 地层叠加

fault:(n) 断层

principle of cross-cutting relationships:切割穿插定律,可以确定侵入体和围岩之间、断层之间的先后关系

faunal:(adj) 动物群的,动物区系的

index fossil:标志化石

principle of faunal succession:化石层序律,化石的出现具有先后的规律,古老的化石消失后不再出现

absolute dating:绝对定年,确定岩石或事件

发生的年代
chronological:(adj)按年代顺序的
radiometric dating:放射性测量定年,指用放射性同位素的方法确定岩石矿物的年龄
radioactive decay:放射性衰变,指不稳定同位素的质子数和中子数发生衰变而生成另一种同位素的过程
element:(n)元素
electron spin resonance:电子自旋共振
thermoluminescence:(n)热释光测年法
radioactivity:(n)放射性,放射能力
proton:(n)质子
neutron:(n)中子
atomic nucleus:原子核
nucleus:(n)核心
pl:nuclei
electron:(n)电子
isotope:(n)同位素
atomic mass:原子量
parent isotope:母体同位素
daughter isotope:子体同位素
half-life:(n)半衰期
radiocarbon dating:碳 14 测年法
potassium-argon (K-Ar) method:钾(^{39}K)-氩(^{40}Ar)定年法
optical stimulating luminescence:光释光测年法
paleomagnetism:(n)古地磁
polarity:极性
reversal:反转
normal polarity:(地磁)正极性
reversed polarity:(地磁)反向极性
geomagnetic polarity time scale:地磁极性年代表

Notes

[1] 确定地质年代方法主要有两大类,这句话主要是阐述了相对定年方法。相对定年方法主要是无须通过精确测量岩石或地质事件的发生年代,而只需把地质事件放在地质年代框架中确定其先后顺序。后文中提到的绝对定年方法则是指需要通过同位素等手段测量岩石或地质事件发生的精确年代。

[2] 此句主要说明沉积学的原始水平沉积定律,即地层沉积时是近于水平的,所有的地层也近似平行于这个水平面。

[3] 在两个不同区域,地层沉积定律不适用于判断地层年代的先后顺序,化石层序律是相对定年的主要方法。依据生物演化规律,生物灭绝后不会再次出现,地层中的生物化石也就具有新旧演替序列,据此可以建立化石层序律。

[4] 这句话主要阐述即使在不同区域,当有相同的标志化石出现时,也可以根据该现象判断两个区域的地层处于同一年代。

[5] 放射性同位素定年方法主要依据岩石、贝壳等物体中含有的放射性同位素经衰变后其母体同位素和子体同位素含量的比值来确定该物体的精确年龄。该方法的前提是样品形成以来对母、子体同位素保持体系封闭,只有在地质历史时期体系一直保

持封闭,测得的年龄才是样品的形成年龄。若体系封闭状态受构造事件、热事件等影响而发生改变,那么所获得的年龄并非样品的真实年龄。

[6] 此句主要解释了放射性元素半衰期的概念。放射性元素的半衰期是母体数量减少到原来的一半,即母体和子体同位素个数相等的时候所需要的时间。

[7] 这部分主要用放射性^{14}C举例说明放射性同位素测年的年代限制,当放射性同位素^{14}C经过长时间衰变后,其母体同位素^{14}C的剩余数量极低时,低于现代技术的检测标准后,测试出来的年龄结果较不准确,因此一般放射性^{14}C定年局限于近7万年。

[8] 光释光、热释光测年主要是利用光和加热的方式来将物体晶格缺陷中储存的自由电子以光子能量形式释放出来。这句话说明的是,当物体在被加热或者埋藏后,其周围环境中的辐射稳定时,晶格缺陷中储存的自由电子数量与年龄成正比。

[9] 古地磁在地质历史上经历过多次极性反转变化,每次的反转变化相似。一般结合化石年代或放射性同位素测年方法建立古地磁极性年代表,之后各个区域的古地磁极性信息可以与其对比而确定地层年龄。

2. Deep Carbon Cycle

Text

As one of the basic elements of life and the main component of fossil fuels, carbon has played a key role in the birth of human civilization. Furthermore, greenhouse gases such as CO_2 and CH_4 in the Earth's atmosphere affect the global climate. Given the current significant problems regarding global warming and carbon emission, the cycle of carbon among the Earth's spheres has become one of the most important topics in scientific research on the Earth's system. The carbon cycle of Earth has been divided into the shallow carbon cycle, which affects the atmosphere, hydrosphere, biosphere, and pedosphere, and the deep carbon cycle, affecting Earth's surface and deep systems[1]. Although more than 90% of carbon may be stored in the deep Earth, the form and behavior of carbon in this region are poorly understood. The petrological, geochemical, and geodynamical factors influencing the deep carbon cycle have become topics of widespread concern in the past decade. The research program *Deep Carbon Observatory* (*DCO*) was launched in 2009, which marked the beginning of systematic research on the origin, composition, distribution, and recycling of deep carbon in the Earth by multiple institutions across multiple disciplines. DCO is structurally divided into four research communities (Extreme Physics and Chemistry, Reservoirs and Fluxes, Deep Energy, and Deep Life), each focused on a different area of deep carbon research. The Extreme Physics and Chemistry Community aims to improve our understanding of the physical and chemical behavior of carbon at extreme conditions, as found in the deep interiors of Earth and other planets. The Reservoirs and Fluxes Community is dedicated to identifying the principal deep carbon reservoirs, to determining the mechanisms and rates by which carbon moves among those reservoirs, and to assessing the total carbon budget of Earth[2]. The Deep Energy Community is focused on quantifying the environmental conditions and processes from the molecular to the global scale that control the origins, forms, quantities and movements of reduced carbon compounds derived from deep carbon through deep geologic time. The Deep Life Community explores the evolutionary and functional diversity of Earth's deep biosphere and its interaction with the carbon cycle.

In the long-term geological history, the shallow and deep carbon cycle are intimately connected to each other through plate subduction. Carbon that enters subduction zones includes carbonate minerals and reduced organic carbon that exist within the serpentinized lithospheric mantle, altered oceanic crust (AOC), and seafloor sediments of the incoming oceanic plate[3]. During subduction, carbon may be liberated from the slab through metamorphic decarbonation, carbonate dissolution into COH fluids, or formation of carbonate melts at sub-arc depths[4]. Upward migration of the slab-derived carbon-bearing fluids/melts can induce melting of the overlying mantle wedge and thus provide a mechanism for carbon release through arc volcanism. Residual carbonate phases could have been carried into the deep mantle, possibly to the convecting upper mantle and transition zone or even to the lower mantle, reduced into diamond that may reach the surface via kimberlite eruptions or trigger redox melting of mantle and subsequently liberate carbon through volcanic degassing in intraplate settings[5] (Figure Ⅲ-1).

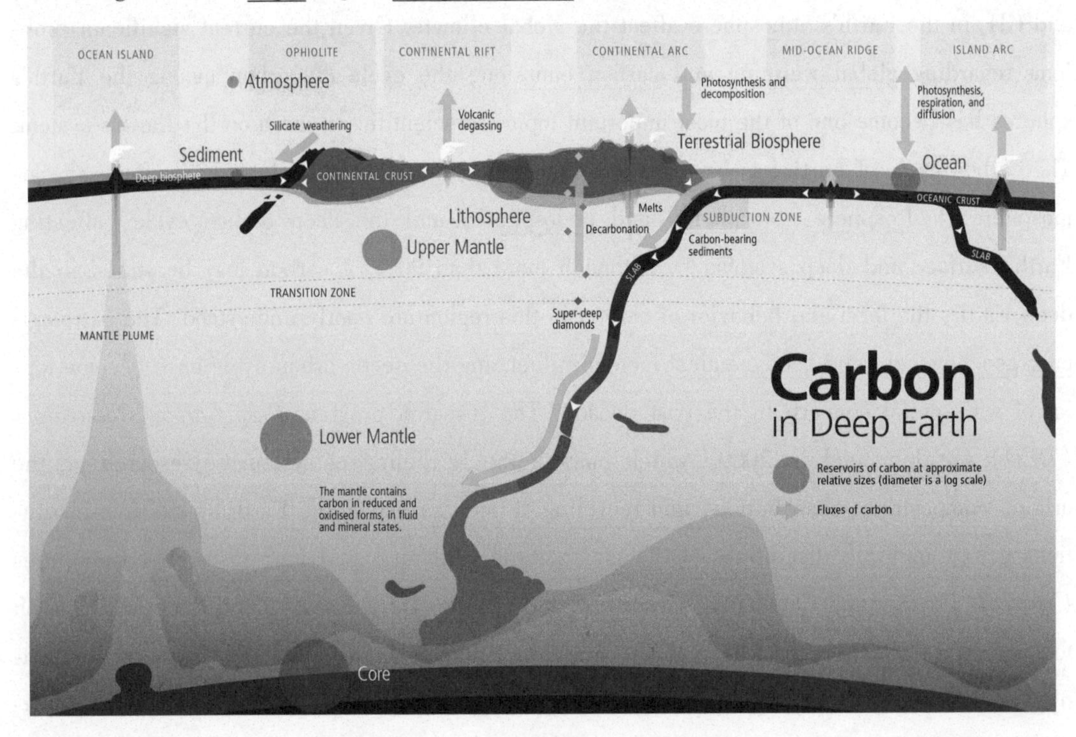

Figure Ⅲ-1: Schematic for deep carbon cycle.

Subducting Carbon

Mantle peridotites that form the bulk of the subducting lithosphere readily hydrate and carbonate if exposed to seawater, forming carbonated serpentinites. However, most peridotites usually reside at least 6 km beneath the seafloor and are not in direct contact with seawater. Faulting and fracturing are necessary to bring mantle rocks to the seafloor or seawater to the mantle. Carbonated serpentinites may form near spreading centers or near trenches. Near spreading centers, extensional faulting is linked to hydration and carbonation reactions, as well as to the precipitation of carbonate veins in mantle peridotites. Serpentinization is most pronounced at slow-spreading mid-ocean ridges, but most subducting lithosphere is not formed there, because of the w plate production rate[6]. Oceanic carbonated serpentinites therefore make a minor contribution to the global carbon input flux to subduction zones, although they may be locally important. Near trenches, the subducting plate deforms and fractures because of bending. However, the extent of hydration and carbonation in outer-rise serpentinites is unknown, and fluid pathways longer than 5 km may lead to low fluid-to-rock ratios and low carbon transport into the mantle section of the downgoing plate.

In contrast to peridotite carbonation, there is abundant evidence for pervasive carbonation of the near-ridge oceanic crust. Seawater-derived fluids circulate predominantly in the higher-permeability upper-crustal volcanic section, leading to precipitation of carbonate minerals through water-rock interactions. The dominant form of carbon in altered oceanic crust is calcium carbonate (calcite and aragonite) that precipitates in veins and vugs, as has been found in samples from the *International Ocean Drilling Program* (*IODP*)[7]. Carbon uptake occurs near the ridge axis in crust that is 20 Myr old, but surprisingly, AOC older than 80 Myr has higher carbonate content. This may be due to higher bottom-water temperatures in the Cretaceous period promoting greater abiogenic carbonate precipitation. Although low in carbonates, young AOC (less than 10 Myr old) is isotopically light owing to the intense bio-alteration of young crust. Thus, an important prediction for carbon inputs is that old plates (for example, Marianas and Tonga) will have greater AOC carbonate concentrations and higher average $\delta^{13}C$, whereas young plates (for example, Cascadia and Central America) will have fewer AOC carbonates with lower $\delta^{13}C$.

The sedimentary layer that is deposited on top of the oceanic crust contains dramatically different forms of carbon than the largely inorganic precipitates of the oceanic crust and peridotite. A significant proportion of seafloor sediments are carbon-bearing and are derived from the weathering of continental crust and accumulation of the skeletal components of marine organisms. Organisms that grow a carbonate shell, such as nannoplankton coccoliths and bottom-dwelling foraminifera, are the richest source of carbon deposits on the seafloor. Although sedimentary carbon has the potential to dominate global input fluxes, it may be entirely absent from some subducting sections. The ocean's cold and corrosive bottom water is particularly challenging to carbonate survival. The calcite compensation depth (CCD), which marks the transition between carbonate-bearing and carbonate-absent sediments, is about 4,500 m deep in today's oceans[8]. The depth of the CCD is dependent on several factors, including the rate and composition of terrestrial sediment flux, biological productivity, and sea-surface temperature and salinity. Organic carbon is also consumed in the oxic ocean and in the sediments themselves by microbially mediated reactions, so its preservation in sediments requires rapid supply and burial. These conditions are met beneath regions of high biological productivity and in deep-sea fans, where rivers deliver high fluxes of carbon-bearing sediments to the ocean from regions of active uplift and erosion.

Carbon Returned at the Arc

Explosive volcanic eruptions—common within the volcanic arcs that form above subduction zones—are driven by the dramatic exsolution of volatiles, including CO_2, as magmas rise to the surface. This is due to a strong drop in solubility with decreasing pressures. The standard method used to measure the concentration of magmatic volatiles is to analyze melt inclusions trapped in early-formed minerals[9]. This approach has been successful in estimating the concentrations of H_2O, S, Cl and F, but not CO_2, owing to its much lower solubility at crustal pressures. Thus, a major outstanding challenge is to determine the CO_2 concentration of primary arc magmas. A recent promising approach for the estimation of volcanic CO_2 fluxes has been to make use of the ratio of CO_2 to S. Integrated CO_2/S correlates in some volcanic arcs with ratios of non-volatile trace elements (for example, Sr/Nd or Ba/La) linked to subduction recycling of carbonates and slab fluids. Arcs with low CO_2/S ratio (for example, the Western Pacific) are associated with a lack of subducting sedimentary carbon, whereas those with higher CO_2/S ratios are associated with subducting sedimentary carbonates. Another useful tool is C isotopic ratios in volcanic gases, which may be lower or higher than mantle values for individual volcanoes. Low-$\delta^{13}C$ volcanic carbon occurs at margins where low-$\delta^{13}C$ organic carbon is subducted (for example, the Aleutians), whereas high $\delta^{13}C$ values occur where high-$\delta^{13}C$ carbonates are subducted (for example, Central America). Globally, volcanic arcs with the highest C fluxes have the highest $\delta^{13}C$, suggesting a dominance of carbonate sources. This could reflect preferential recycling of carbonates in subduction zones. Volcanic-arc data are still sparse, however, and portable mass spectrometers mounted on helicopters and field vehicles are providing new methods for carbon isotope analysis in remote regions.

Carbon beyond the Arc

Carbon that survives beyond sub-arc depths is exceptionally challenging to investigate. In the absence of a fluid phase, even trace carbonates remaining in slab lithologies can trigger melting that may be small in volume but is geochemically important. Most slab geotherms will eventually intersect conditions at which carbonated crust melts, producing a magma highly enriched in carbonate known as carbonatite, which has extremely high mobility[10]. Such melts will quickly migrate into the surrounding mantle, where vastly different chemical conditions lead to reactions that transform carbon into new and more resistant forms such as diamond. There are two geodynamically distinct reservoirs of subducted carbon in the mantle: the deep lithosphere beneath continents and the more voluminous, convecting upper mantle, transition zone and upper part of the lower mantle. Although most diamonds are derived from the volumetrically minor lithosphere, rare sub-lithospheric diamonds (SLD) appear to be more directly linked to deeper subduction processes extending into the transition zone and beyond. Key to the origin of SLD is the strong chemical contrast between the subducting crust and the ambient convecting mantle. This difference is not just in bulk composition but also, more importantly, in oxidation state. Whereas most crustal rocks are derived from Earth's oxidized surface reservoirs, the mantle is relatively reducing, and becomes more so with depth. Driven chiefly by changes in the stability of iron-rich minerals with depth, this leads to conditions that stabilize an iron-rich metallic phase and, at depths greater than about 140 km, produce diamond as the stable form of carbon. SLD are identified by virtue of the mineral inclusions that they hold, which signal a depth greater than that of the deepest lithosphere (>200 km). SLD are younger than lithospheric diamonds, show highly complex growth histories and display strong chemical links to subducted crust. Carbon isotope data for SLD are quite variable, extending to very low $\delta^{13}C$ values consistent with carbon derivation from subducted sediments or altered basalts. As noted earlier, the vast majority of diamonds come from the subcontinental mantle lithosphere. Like SLD, they are best interpreted as metasomatic in origin, and some show evidence for the involvement of subducted carbon, but they differ from SLD in their great age (extending to more than 2.5 Gyr). Overall, the total amount of carbon that exists in the form of diamond is unknown. It seems likely that the more common return path for carbon stored in diamond is as a component of magma derived from the mantle.

Vocabulary

fossil fuel:(n) 化石燃料
greenhouse gas:(n) 温室气体
atmosphere:(n) 大气,大气圈
  [-sphere]:圈层
global warming:(n) 全球变暖
carbon emission:(n) 碳排放
carbon cycle:(n) 碳循环
hydrosphere:(n) 水圈
biosphere:(n) 生物圈
pedosphere:(n) 土壤圈
petrological:(adj) 岩石学的
  n:petrology 岩石学
  [petro-]:石,岩石
geochemical:(adj) 地球化学的
  n:geochemistry 地球化学
  [geo-]:地球
geodynamical:(adj) 地球动力学的
  n:geodynamics 地球动力学
flux:(n) 通量,流,涌流
  pl:fluxes
deep carbon reservoir:(n) 深部碳储库
carbon budget:(n) 碳收支
molecular:(adj) 分子的
  n:molecule 分子
geologic:(adj) 地质(学)的
  syn:geological
plate subduction:(n) 板块俯冲
carbonate mineral:(n) 碳酸盐矿物

organic：（adj）有机的
  ant：inorganic 无机的
serpentinized：（adj）蛇纹石化的
  vt/vi：serpentinize 蛇纹石化
  n：serpentinization 蛇纹石化作用
  n：serpentinite 蛇纹岩
lithospheric：（adj）岩石圈的
  n：lithosphere 岩石圈
  ~asthenosphere：（n）软流圈
mantle：（n）地幔
  ~upper mantle：（n）上地幔
  ~lower mantle：（n）下地幔
  ~mantle wedge：（n）地幔楔
oceanic crust：（n）洋壳
  ~continental crust：（n）陆壳
seafloor sediment：（n）海底沉积物
  ~sedimentary：（adj）沉积的
metamorphic：（adj）变质的
  n：metamorphism 变质作用
  [meta-]：变化
decarbonation：（n）脱二氧化碳
  ant：carbonation 碳酸盐化
dissolution：（n）溶解
  vt/vi：dissolve 溶解
carbonate melt：（n）碳酸盐熔体
sub-arc depth：（n）弧下深度
  ~fore-arc depth：（n）弧前深度
arc volcanism：（n）弧火山作用
  ~arc volcano：（n）弧火山
  ~volcanic arc：（n）火山弧
transition zone：（n）转换带，过渡带
diamond：（n）金刚石，钻石
degas：（vi）脱气，去气
  syn：outgas
intraplate setting：（n）板内（构造）环境

peridotite：（n）橄榄岩
hydrate：（vt/vi）水化，水合；（n）水合物
  ant：dehydrate：脱水
faulting：（n）断层作用
spreading center：（n）扩张中心
trench：（n）海沟
precipitation：（n）沉淀
  precipitate：（vt/vi）沉淀；（n）沉淀物
slow-spreading：（adj）慢速扩张的
  ~fast-spreading：（adj）快速扩张的
mid-ocean ridge：（n）洋中脊
fluid-to-rock ratio：（n）流体/岩石比率
permeability：（n）渗透性
  vt/vi：permeate 渗透
  adj：permeable 可渗透的
water-rock interaction：（n）水岩相互作用
calcite：（n）方解石
aragonite：（n）文石，霰石
vug：（n）晶簇，晶洞
International Ocean Drilling Program：国际大洋钻探计划
Cretaceous：（n/adj）白垩纪（的）
abiogenic：（adj）非生物成因的
  syn：abiotic
isotopically：（adv）同位素地
  n：isotope 同位素
  adj：isotopic 同位素的
  ~isotopic analysis：（n）同位素分析
weathering：（n）（地质）风化作用，（水利）泄水斜度，挡风雨材料；（vt/vi）风干，使褪色，经受住（weather 的-ing 形式）
skeletal：（adj）骨骼的
  n：skeleton 骨骼
marine organism：（n）海洋生物
nanoplankton：（n）微型浮游生物

=nannoplankton

coccolith：(n) 颗石藻

=coccolithophore

bottom-dwelling：(n) 底栖生物

foraminifer：(n) 有孔虫

pl：foraminifera 或 foraminifers

calcite compensation depth：(n) 方解石补偿深度

terrestrial：(adj) 地球的，陆生的

~terrestrial planet：(n) 类地行星

syn：terrigenous：(adj) 陆源的

oxic：(adj) 好氧的，含氧的

ant：anoxic 厌氧的，缺氧的

deep-sea fan：(n) 深海扇

uplift：(n/vt)(地壳)隆起，抬升

erosion：(n) 侵蚀，腐蚀

vt/vi：erode 侵蚀，腐蚀

syn：corrosion

volcanic eruption：(n) 火山喷发

exsolution：(n) 出溶作用

volatile：(n) 挥发分

magma：(n) 岩浆

adj：magmatic 岩浆的

~magmatic rock 岩浆岩

solubility：(n) 溶解度，可解决性

adj：soluble 可溶解的，可解决的

n：solution 溶液，解决方法

melt inclusion：(n) 熔体包裹体

~fluid inclusion：(n) 流体包裹体

trace element：(n) 微量元素

~major element：(n) 主量元素

mass spectrometer：(n) 质谱仪

lithology：(n) 岩性

pl：lithologies

geotherm：(n) 地热等温线

carbonatite：(n) 碳酸岩

oxidation state：(n) 氧化状态

metallic：(adj) 含金属的

n：metal 金属

basalt：(n) 玄武岩

adj：basaltic 玄武岩/质的

metasomatic：(adj) 交代的

n：metasomatism 交代作用

Notes

[1] 此句反映了全球碳循环的多尺度特征。地球系统中的碳循环可分为“地表碳循环”和“深部碳循环”：前者指碳在固体地球外部大气圈、水圈、生物圈和土壤圈之间周期相对较短（万年尺度）的循环过程；后者指碳在地球表层系统与深部系统（如地幔、地核）之间周期漫长（百万年尺度）的循环过程。

[2] 此句介绍了“深部碳观测（DCO）”计划中的“储库与通量”子课题的研究目标。be dedicated to 后面通常加名词或动名词，意为“致力于、献身于”。by which 作定语从句引导词来修饰先行词 mechanism 和 rate，这里表示“碳在不同储库中运移的机制和速率”。

[3] 典型的俯冲大洋岩石圈板块自下而上分别包含蛇纹石化地幔橄榄岩、蚀变洋壳（辉长岩和玄武岩）和海底沉积物。这些物质在俯冲之前与海水经历了不同程度的水岩

相互作用,因而或多或少都含有一些碳酸盐矿物和/或有机碳。“蛇纹石化”是指基性、超基性岩中的橄榄石和斜方辉石等富镁矿物发生水化形成蛇纹石的过程,多发生于洋中脊、海沟和地幔楔等构造环境中。

[4] 此句介绍了俯冲板片中的碳在弧下深度常见的三种释放机制。板片沿不同地温梯度俯冲时所对应的弧下深度也略有差异,一般地,弧下深度在 80~150 km 范围内,而 80 km 以浅部位则称为弧前深度。变质脱碳是指含碳相通过变质反应释放 CO_2 的过程;溶解脱碳是指流体诱发含碳相溶解为 CO_3^{2-}、HCO_3^- 和/或 CO_2, aq 的过程;熔融脱碳是指含碳相发生高温熔融形成易迁移的含碳酸盐成分熔体的过程。

[5] 此句反映了俯冲板块中的碳除了在弧下深度发生部分释放,残留的含碳相可能通过俯冲作用进入地球更深部,如上地幔、地幔转换带/过渡带(深度约 410~660 km)、甚至是下地幔。由于地球内部环境相对还原,这些含碳相可转变为金刚石这一稳定的含碳相。这里,“that” 引导的定语从句进一步解释了金刚石可被金伯利岩浆作用携带至地表或诱发地幔熔融并通过板内岩浆作用(如大陆裂谷、洋岛等)发生碳的释放。

[6] 全球洋中脊根据其扩张速率可分为快速扩张洋中脊、中速扩张洋中脊、慢速扩张洋中脊和超慢速扩张洋中脊,反映了洋中脊下方的岩浆供给程度。其中,快速扩张洋中脊下方的岩浆供给充足,其半扩张速率一般大于 8~10 cm/a,如东太平洋洋隆(East Pacific Rise);而慢速-超慢速扩张的洋中脊由于岩浆供给不足(对应半扩张速率一般小于 3~4 cm/a),通常发育大型低角度正断层(即拆离断层)将地幔橄榄岩拆离至海底并发生强烈的蛇纹石化作用,如大西洋洋中脊 (Mid-Atlantic Ridge)。

[7] 国际大洋钻探计划是一项致力于推进对地球的科学了解,进行海底监测和采样的国际地球科学合作计划。该计划起始于 1968 年,不同阶段具有不同的名称:1968—1983 年,深海钻探计划 (Deep Sea Drilling Project, DSDP);1985—2003 年,大洋钻探计划 (Ocean Drilling Program, ODP);2003—2013 年,综合大洋钻探计划 (Integrated Ocean Drilling Program, IODP);2013—2023 年,国际大洋发现计划 (International O-cean Discovery Program, IODP)。现在 IODP 主要是指国际大洋发现计划 (Interna-tional Ocean Discovery Program)。

[8] 方解石补偿深度又称碳酸钙补偿深度、碳酸盐补偿深度 (CCD),是指海洋中碳酸钙输入海底的补给速率与溶解速率相等的深度面。CCD 在海底沉积物的分布特征上具有明显反映,浅于这一临界深度的海底,广布白色碳酸钙沉积;而在这一深度之下,则缺失钙质沉积(多为硅质沉积)。因此,CCD 犹如海底雪线,是海底沉积物最重要的相界面。由于碳酸钙溶解度随温度升高而降低,故 CCD 自赤道向两极升高。现代海洋中 CCD 平均约 4500 m,其中大西洋最深,平均约 5300 m;太平洋最浅,平均只有 4400 m。

[9] 熔体包裹体也称岩浆包裹体、熔融包裹体或玻璃包裹体,是指存在于岩浆矿物中、通

常粒径大小为 1~300 μm 的微小硅酸盐熔体,是矿物在岩浆结晶生长过程中捕获的周围熔体。熔体包裹体在常温下形态多样,有的保存为成分均匀的玻璃质;有的则含结晶相,表明其被捕获后冷却速率较慢,有足够时间形成子晶;还有一些因冷凝收缩而出现气泡。熔体包裹体多出现于火山岩矿物中,也有少量出现在侵入岩或脉体矿物中。

[10] 此句介绍了碳酸岩的成因及性质。at which 作定语从句引导词来修饰先行词 condition,这里表示“在俯冲板片地温梯度线与碳酸盐化洋壳固相线相交的条件下,后者便发生熔融”。注意区分碳酸岩（carbonatite）与碳酸盐岩（carbonate rock）,前者指主要由碳酸盐矿物（>50%）组成的火成岩,也含有一定量的原生岩浆矿物及稀土元素矿物,通常也称之为“火成碳酸岩”;后者指由碳酸盐矿物组成的沉积岩,因而也称之为“沉积碳酸盐岩”。

3. Estuarine Delta

Text

A deltaic system can typically be defined as a sedimentary body that progrades basinward into a standing body of water, built by fluvial processes in combination with a more or less pronounced reworking through waves and tides[1]. Although it is common to classify deltas as river-dominated, wave-dominated and tide-dominated, the analysis of many examples shows that the basic types can be considered merely as relatively unusual end-members of a continuum[2]. Mixed-energy deltas (Figure Ⅲ-2) should therefore be much more common in the stratigraphic record than currently recognized. The study of mixed-energy deltas usually involves very complex systems, where all three processes are active at the same time, but the strength and preservation potential of each individual process changes in space, even over short distances (few hundreds of meters), as well as in time[3]. Therefore, it becomes important to identify and separate the signals of the three main processes (rivers, waves and tides) through the stratigraphy, and to understand what controls their latero-vertical partitioning[4]. The high degree of variability can occur in modern deltas and other coastal systems but research has yet to form a clear methodology tested on ancient cases. An important point to make is that the classical ternary diagram for delta classification does not fully reflect or explain the great complexity and variability in mixed-energy deltaic systems.

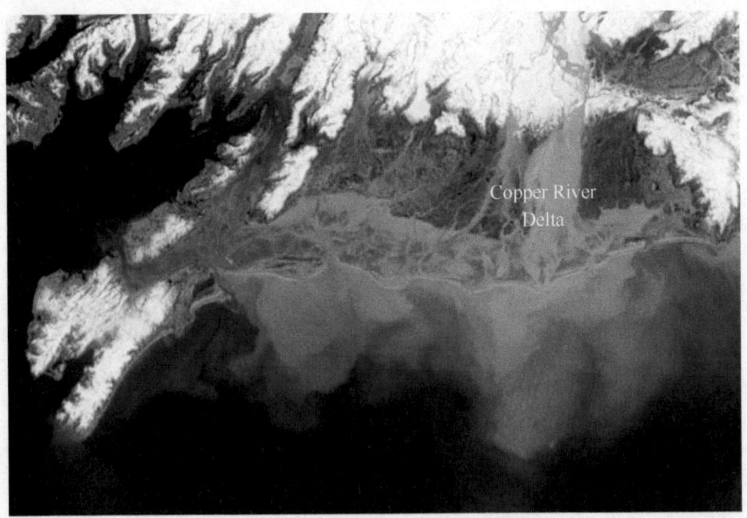

Figure Ⅲ-2: Satellite image of the Copper River delta (a mixed-energy delta) in Alaska. [Source: NASA]

In the Lajas system (Middle Jurassic, Argentina) there was a broad spatial and temporal partitioning of the three processes. Fluvial signals in the stratigraphy are the strongest landward, whereas wave signals are the strongest in the most distal strata or in lateral areas with respect to the main river input. Tidal signals are the least predictable, but they show the strongest influence in the deltafront and subaqueous platform deposits. An additional complexity in the partitioning of wave, river and tidal signals is that they not only change spatially, but also temporally. Through time, landward and seaward movements of the shoreline, accompanied by changes in coastal morphology, enhanced or reduced the influence of any of the three processes, thus further complicating process partitioning in the stratigraphy[5]. Lateral facies transitions related to process changes can be extremely abrupt in mixed energy systems. Vertical process (and facies) variability is related to time and spatial changes in process dominance in different sub-environments, and can be caused by allogenic or autogenic processes. Instead, lateral process variability is more likely to be related to autogenic causes[6]. For example, the change between tidal channels and wave-influenced lateral shorefaces is probably related to autogenic processes, i.e. it occurs without any obvious change in external forcing[7]. On the contrary, the change from tide-reworked delta-front deposits to amalgamated, tidally modulated braided distributary channels is more likely to be related to an allogenic change in relative sea level[8]. This process partitioning is important because it controls the distribution of sand, which in the Lajas system is concentrated in areas of strong tides and river currents (delta front and tide-influenced subaqueous channels). As a general rule, sediment sorting is indicative of the degree of marine reworking: better sorted sediments are more likely to be the product of wave and/or tidal reworking, whereas more poorly sorted sediments are likely to be river-borne[9]. Similar process distributions (river influence stronger in proximal and distal distributary channels, tidal influence stronger in the subaqueous platform and delta-front areas, wave influence stronger in more distal settings and areas lateral to the main river mouth) have also been recognized in the Tilje Formation (Early Jurassic, Norway).

In the modern, there are few well-described examples of mixed-energy deltas. The Mahakam delta shows strong river and tidal influence, with minor wave reworking. The Mekong delta shows an evolution from strongly tide-influenced to wave-influenced and tide-influenced through time, as the delta prograded and moved out of a protected embayment[10]. In the Holocene Mekong River delta deposits, subaqueous platform facies show tidal influence accompanied in places by wave influence. Another possible good analogue for the Lajas system is the north-west Borneo coast. Here, transtensional and transpressional tectonics affected the shelf and slope areas and they influenced deltaic deposition as well as sedimentary processes. As pointed out before, many of these modern systems occur in the Indo-Pacific zone and tend to be mud-rich, whereas the ancient example of the Lajas deltas is extremely sand-rich. Possible additional modern analogues for such a system can be found in Canada and Alaska, in the Fraser and Copper river deltas, respectively. Both of these systems are developed close to the sediment source (coastal range) and are quite sandy. The Fraser River delta shows a very sandy delta front and development of many subtidal channels. The Copper River delta shows clear influence of waves (barrier islands and sand spits), tides (tidal channels) and rivers (braided system) (Figure Ⅲ-2)[11].

The ternary diagram based on Galloway (1975) does not encompass the great complexity and variability in mixed-energy deltaic systems. Mixed-energy deltas show a high degree of process variability both in space and time. When thinking about changes in process regime or depositional systems through time, there is a tendency to associate these events with base-level changes associated with the formation of incised valleys, estuaries and deltas. However, when dealing with mixed-energy deltas, changes in dominant processes can easily occur at shorter time scales due to autogenic processes (for example, the Mekong delta). These changes are very important, because they control the distribution of architectural elements and their heterogeneities, sediment partitioning and also delta morphology.

Overall, wave influence decreased and river influence increased in the succession upward through time, due to the overall progradational trend of the succession. In other words, wave energy was stronger in the distal and lateral reaches of the delta system, whereas river currents were stronger in more proximal areas[12]. Tidal energy was stronger in the delta front and subaqueous platform areas. Strong tidal currents played a key role in the reworking of sediments in the delta front and subaqueous platform, producing tidal bars and compound dunes. The final stratigraphic record is strongly influenced by this reworking, so that these deltaic deposits are very different from the simple coarsening-upward trend typical of river-dominated and wave-dominated deltas. The Lajas system at Lohan Mahuida was probably influenced by syndepositional tectonic activity, that controlled the locus of strong tidal reworking and the development of sequence boundaries associated with abrupt increase in grain sizes.

Vocabulary

deltaic system:(n) 三角洲系统
sedimentary body:(n) 沉积体
prograde:(vt) 进积
  ant: backstep (vt) 退积
basinward:(adv) 向盆地方向
fluvial process:(n) 河流过程
reworking:(n) 再作用过程
wave:(n) 波浪
tide:(n) 潮汐
end-member:(n) 端元
mixed-energy delta:(n) 混合(河流、潮汐、波浪)能量控制的三角洲

stratigraphic record：（n）地层记录
latero-vertical partitioning：（n）侧向-垂向空间分配
spatial and temporal partitioning：（n）时空分配
landward：（adv）向陆地方向
delta front：（n）三角洲前缘
coastal morphology：（n）海岸形貌
allogenic：（adj）外生的
autogenic：（adj）自生的
tidal channel：（n）潮汐水道
shoreface：（n）临滨
braided：（adj）辫状的
distributary channel：（n）分流河道
relative sea level：（n）相对海平面
sediment sorting：（n）沉积物分选
river mouth：（n）河口
embayment：（n）海湾
transtensional：（adj）张扭的
transpressional：（adj）压扭的
shelf：（n）陆架
slope：（n）陆坡
deltaic deposition：（n）三角洲沉积
sedimentary process：（n）沉积过程
mud-rich：（adj）富泥的
sand-rich：（adj）富砂的
sediment source：（n）沉积物物源
barrier island：（n）障壁岛
sand spit：（n）沙嘴
ternary diagram：（n）三元图解
at shorter time scale：（adv）在较短时间尺度上
ant：at longer time scale：（adv）在较长时间尺度上
sediment partitioning：（n）沉积物分配
delta morphology：（n）三角洲形貌
succession：（n）（地层）层序
progradational trend：（n）进积趋势
ant：backstepping trend（n）退积趋势
proximal area：（n）近端区域
ant：distal area（n）远端区域
coarsening-upward：（adj）向上变粗
ant：fining-upward（adj）向上变细
syndepositional tectonic activity：（n）同沉积构造活动
sequence boundary：（n）层序边界
grain size：（n）颗粒粒度

Notes

[1] 三角洲体系是由河流输送沙体向盆地内部进积形成的沉积体。在进入盆地之后，三角洲可以受河流作用、波浪作用、潮汐作用所控制。其中，主要受河流作用控制的三角洲称为河控三角洲，受波浪作用控制的三角洲称为浪控三角洲，受潮汐作用控制的三角洲称为潮控三角洲。

[2] 尽管通常根据能量控制作用类型，将三角洲划分为河控三角洲、浪控三角洲、潮控三角洲等，然而诸多实际案例分析表明，这些基础分类往往是在特殊条件下才能发育的端元产物。然而，现实中的三角洲往往受到河流、波浪、潮汐的联合作用，只不过在部分环境条件下，某种作用占主导地位。

[3] 河流、波浪、潮汐联合控制的三角洲包含了非常复杂的内容,这三个能量单元可以同时发生作用,但是它们之间的相对作用大小可以在很短的空间范围(几百米)内发生变化。

[4] 识别河流、波浪、潮汐对于三角洲的作用,并理解三种作用在空间上的分配,即明确河流、波浪、潮汐在何处以何种方式为主。

[5] 随着时间的推移,滨线会发生迁移、海岸线会有所变化、河流与波浪以及潮汐的强度会发生变化,这些因素的变化进一步使得河流与波浪以及潮汐对三角洲的控制作用复杂化。

[6] 在河流、波浪、潮汐联合作用的沉积体系之中,沉积相的侧向变化会很剧烈。垂向上沉积相的时空变化往往与外生因素或者自生因素有关,而侧向上沉积相的变化更多是与自生因素有关。

[7] 潮汐水道和波浪改造的临滨沉积之间的转换(水深相似)可能与内生作用有关,也就是说这两种沉积物之间的转变没有受到外部因素的控制。

[8] 由潮汐改造的三角洲前缘沉积向潮汐改造的辫状河分流河道沉积之间的改变(水深发生了变化)更有可能与外生作用(如相对海平面的变化)有关。

[9] 沉积物分选指示了海洋沉积物改造的程度,分选更好的沉积物可能与波浪、潮汐的改造作用有关,而分选较差的沉积物可能与河流的改造作用有关(尚未受到明显的波浪、潮汐作用改造)。

[10] 湄公河三角洲显示了由强烈潮控向浪控-潮控转变的过程,主要由于三角洲由海湾向外海逐渐进积,受到了更多的波浪控制作用。

[11] 铜河三角洲显示出典型的波浪控制作用(表现为障壁岛、沙嘴)、潮汐控制作用(表现为潮汐水道)、河流控制作用(表现为辫状河)。

[12] 波浪的能量在远离岸线的区域更强,而河流的能量在靠近岸线的区域更强。

4. Igneous Rocks

Text

What are Igneous Rocks?

Igneous rocks are those that form from molten products of the Earth's interior[1]. Petrologists use two words for molten rock. Magma is the more general term that embraces mixtures of melt and any crystals that may be suspended in it[2]. A good example would be flowing lava which contains crystals suspended in the melt: the term magma refers to the entire assemblage, embracing both solid and liquid states of matter present in the lava[3]. Melt, on the other hand, refers to the molten state on its own, excluding any solid material which might be suspended in or associated with it. The difference becomes clearer if one considers how one would chemically analysis the distinct chemical compositions of the magma and melt, once the lava flow had solidified. The magma composition could be estimated by crushing up a sample of the solidified lava, including both phenocrysts and groundmass (ensuring they are present in representative proportions). Analysing the melt composition, however, would require the groundmass or glassy matrix—the solidified equivalent of the melt between the phenocrysts—to be physically separated out and analysed on its own.

The word magma, from the Greek μαγμα ("paste"), was actually introduced in a Neptunist context, not igneous, by Dolomieu in 1794, in the belief that the rocks originating from it were reduced to paste by evaporation. In fact, "magma" may be used in a still broader sense. An ascending magma body, as it approaches the surface, commonly contains gas bubbles as well as phenocrysts, bubbles formed by gas that has escaped from the melt due to the fall in pressure that accompanies ascent. The term "magma" is generally understood to embrace melt, crystals and any gas bubbles present. Once erupted on the surface, on the other hand, and having lost some of its gas content to the atmosphere, the molten material is more appropriately called "lava"[4]. Determining a representative chemical analysis of the original magma composition, including the gaseous component, would however be difficult: as the melt solidified and contracted on cooling, the gaseous contents of the vesicles would escape to the atmosphere (and they would in any case be lost during crushing of the rock prior to analysis). Determining the concentrations of these volatile magma constituents—from the solid rock that the magma eventually becomes—therefore requires a different analytical approach that will be discussed later.

Magmas are originally formed by melting deep within the Earth. The initial melting event most commonly takes place in the mantle, though passage of hot magma into or through the continental crust may cause additional melting to occur there as well, adding to the chemical and petrological complexity of continental magmatic rocks[5]. In oceanic and continental areas, mantle-derived magmas are liable to undergo cooling and partial crystallization in storage reservoirs (magma chambers) within the crust, and such processes widen considerably the diversity of magma compositions that eventually erupt at the surface[6] (Figure Ⅲ-3).

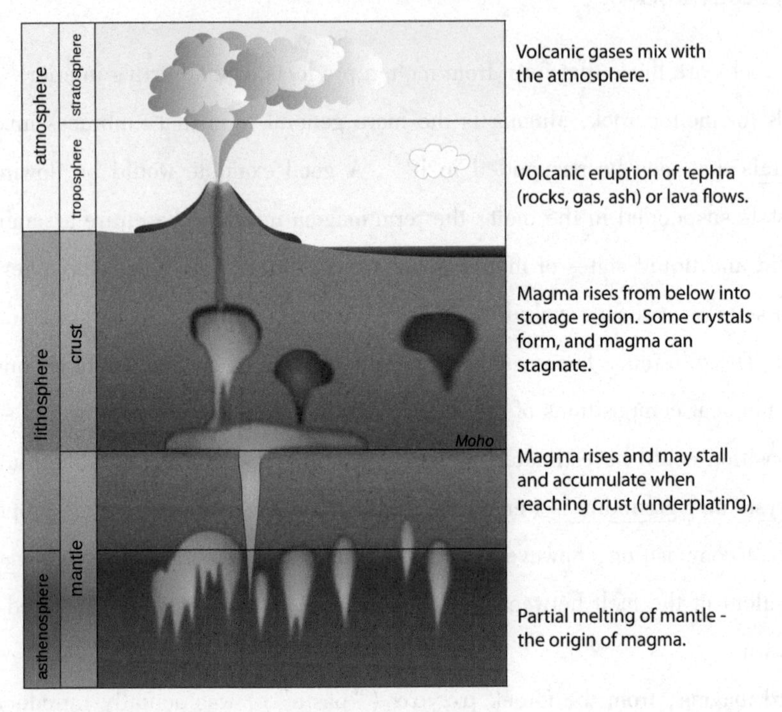

Figure Ⅲ-3: Illustration of the basic process of magma formation, movement to the surface, and eruption through a volcanic vent. [Source: USGS]

Igneous rocks can tell us not only about processes taking place on the Earth's surface at the present time, but also about processes that have taken place earlier in Earth history, and about processes that operate in parts of the Earth that are not directly accessible to us, for example in a magma chamber that originally lay 5 km below an active volcano (but whose contents—or erupted products—are now exposed at the surface). Today, anyone working with igneous rocks must have a range of skills, including the analysis of field relationships, hand-specimen identification in the field, the description and interpretation of thin sections, the allocation of informative rock names, the quantitative interpretation of rock and mineral analyses (often including trace elements and isotope ratios), and the interpretation of experimental equilibria and phase diagrams.

How to Classify Igneous Rocks?

Igneous petrologists need a consistent nomenclature for academic communication. Modern igneous nomenclature rests on three types of observation, each of which may influence the name given to a rock:

(1) Classification by qualitative criteria—grain size

Igneous rocks are divided into coarse (>3 mm), medium (1~3 mm) and fine (<1 mm) grained categories, based on a qualitative (or semiquantitative) estimate of the average grainsize of the groundmass of the rock[7]. This estimate may be based on hand-specimen observation or, more reliably, on thin section examination. According to the grain size category in which it falls (fine, medium or coarse).

(2) Classification by mineral proportions—color index

Familiar adjectives like "ultramafic" and "leucocratic" refer to the relative proportions of dark and light minerals in an igneous rock, where "dark" and "light" relate to the appearance of the minerals in hand-specimen[8]. Dark minerals are known alternatively as mafic or ferromagnesian minerals (e.g., olivine, pyroxene, amphibole, biotite, opaques); light minerals are also known as felsic minerals (e.g., quartz, feldspars, feldspathoids, muscovite). The percentage of dark minerals is known as the color index of the rock.

(3) Classification by chemical composition—acidic versus basic

The first classification of igneous rocks that most students encounter is the one that divides rocks into ultrabasic (<45 wt.% SiO_2), basic (45~52 wt.% SiO_2), intermediate (52~63 wt.% SiO_2) and acid (>63 wt.% SiO_2) categories[9]. This classification is based on the SiO_2 content of the rock. The SiO_2 content, cannot be known until the sample has undergone laboratory analysis, and herein lies the main disadvantage of this classification: it cannot be used to describe rocks as they are being collected in the field or examined under the microscope.

Vocabulary

petrologist: (n) 岩石学家

melt: (n) 熔体

crystal: (n) 晶体

phenocryst: (n) 斑晶

groundmass: (n) 基质

glassy matrix: (n) 玻璃基质

gas bubble: (n) 气泡

lava: (n) 熔岩

cooling: (n) 冷却

content: (n) 含量

escape: (vt/vi) 逃离,逃逸

analysis: (vt) 分析,测试

constituent: (n) 成分

approach: (n) 方法

mantle-derived：地幔来源的
magma chamber：岩浆房
crust：(n) 地壳
active volcano：(n) 活火山
field relationship：野外关系
hand-specimen：(n) 手标本
thin section：(n) 薄片
phase diagram：(n) 相图
nomenclature：(n) 命名法
observation：(n) 观察
classification：(n) 分类
coarse：(adj) 粗粒的
medium：(adj) 中粒的
fine：(adj) 细粒的
qualitative：(adj) 定性的
semiquantitative：(adj) 半定量的
color index：(n) 色率
ultramafic：(adj) 超镁铁质
mineral：(n) 矿,矿物;(adj) 矿物的
olivine：(n) 橄榄石
pyroxene：(n) 辉石
amphibole：(n) 角闪石
biotite：(n) 黑云母
opaque：(adj) 不透明的
quartz：(n) 石英
feldspar：(n) 长石
feldspathoid：(n) 似长石
muscovite：(n) 白云母
chemical composition：化学组成
ultrabasic：(adj) 超基性的
basic：(adj) 基性的
intermediate：(adj) 中性的
acid：(adj) 酸性的
laboratory：(n) 实验室
disadvantage：(n) 不利条件
microscope：(n) 显微镜

Notes

[1] 此句反映了火成岩的形成背景。火成岩是地球内部物质熔融形成的产物。

[2] 此句阐述了岩浆的内涵。岩浆是一个更笼统的术语,它包含了熔体和任何可能悬浮在其中的晶体混合物。

[3] 此句阐述了熔岩的特点及定义。一个很好的例子就是流动的熔岩,它含有悬浮在熔体中的晶体;岩浆一词指的是整个组合,包括熔岩中存在的固态和液态物质。

[4] 此句阐述了熔岩的形成过程。另一方面,一旦在地表喷发,并失去了一些气体到大气中,熔融物质更合适被称为“熔岩”。

[5] 此句阐述了地球内部熔体产生及上升的过程。最初的熔融事件通常发生在地幔中,尽管炙热岩浆进入或穿过大陆地壳也可能导致那里发生额外的熔融,增加了大陆岩浆岩的化学和岩石学复杂性。

[6] 此句阐述了地表出露的岩浆岩成分多样性的原因。在海洋和大陆地区,地幔来源岩浆容易在地壳内的储层(岩浆房)中经历冷却和部分结晶,这一过程扩大了最终在地表喷发的岩浆成分的多样性。

[7] 此句阐述了火成岩分类的一种方案,该方案一般适用于具有等粒结构的火成岩。基

于对岩石中矿物平均粒度的定性(或半定量)估计,分为粗粒(>3 mm),中粒(1~3 mm)和细粒(<1 mm)火成岩。有些火成岩中矿物粒度具有显著差异,呈斑状结构,则该分类方案不适用。

[8] 此句介绍了火成岩石中超镁铁质和淡色体的概念。我们熟悉的形容词,如“超镁铁”和“淡色”,指的是火成岩中深色和浅色矿物的相对比例,其中“深色”和“浅色”与手工标本中矿物的外观有关。

[9] 此句介绍了火成岩的另一种分类方案。大多数学生遇到的火成岩的第一种分类是将岩石分为超基性(< 45 wt.% SiO_2),基性(45~52 wt.% SiO_2),中性(52~63 wt.% SiO_2)和酸性(> 63 wt.% SiO_2)类别。

5. Intraplate Magmatism

Text

Intraplate magmatism constitutes igneous activity distal from the boundaries of the tectonic plates and is thus considered to be unrelated to the processes of seafloor spreading, subduction, and transform faulting[1].

Intraplate magmatism is pervasive within both oceanic and continental crust. Expansive spatially and across geologic time, it is complex because of both its causative magmatic processes as well as its interplay with Earth's mantle geodynamics and plate tectonic processes. The scale of intraplate magmatism ranges from large igneous provinces (LIPs) encompassing millions of cubic kilometers of igneous rock to small individual volcanoes. Similarly, compositions of intraplate igneous rock span the spectrum of extrusive compositions between highly mafic and highly silicic[2]. However, intraplate volcanic rocks are dominantly basaltic and have compositions distinctly different from those found at "normal" mid-ocean ridges and in arc-trench systems.

Intraplate magmatism both present and past in the oceans has occurred predominantly within oceanic crust. The geological record of oceanic crust flooring much of the world's oceans extends back to ~200 million years ago, or slightly less than 5%, of Earth history.

Currently, 25 tectonic plates collectively occupy 97% of Earth's surface. Boundaries of the majority of these are relatively distinct; however, diffuse boundaries characterize portions of the Capricorn, Caribbean, Lwandle, Macquarie, and Sur plates (Figure Ⅲ-4). Forebulges, in this case flexural lithospheric bulges seaward of subduction zone trenches, are common features throughout the global ocean and are considered to be intraplate rather than part of a plate boundary. Intraplate stress can result in zones of extension and convergence, examples of which are found in the global ocean; these may or may not develop into plate boundaries.

The scale of intraplate volcanic features in the oceans ranges over many orders of magnitude, from small volcanoes on the seafloor that rise <100 m above surrounding seafloor to massive oceanic plateaus with areas exceeding 10^6 km^2 and volumes of igneous rock greater than 10^7 km^3. Within this continuum are volcanoes rising hundreds to thousands of meters above the ambient seafloor, linked chains of volcanoes, and submarine ridges, all of which can have subaerial expressions.

Individual undersea volcanoes, known as seamounts[3], sea knolls, and/or guyots, are among the most ubiquitous landforms on Earth and are found across the entire range of ages of oceanic crust, i.e., from zero age to approaching 200 million years. Excluding subduction-related

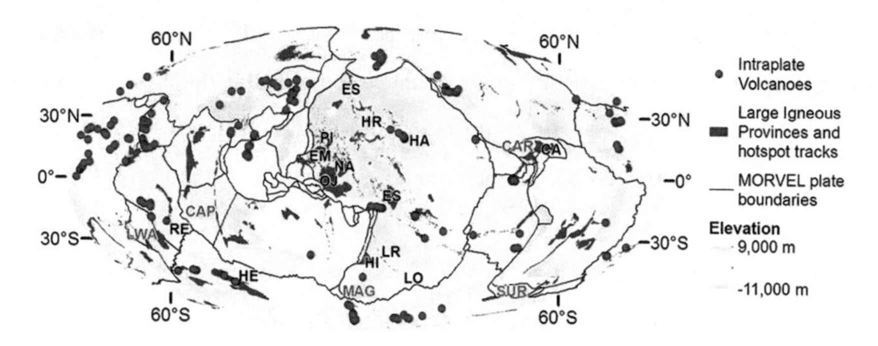

Figure Ⅲ-4: Global distribution of intraplate magmatism. Abbreviations: CA: Caribbean, EM: East Mariana, ES: Emperor Seamount Chain, HA: Hawaii (island), HR: Hawaiian Ridge, HE: Heard Island, HI: Hikurangi, LO: Louisville, LR: Louisville Ridge, MA: Manihiki, NA: Nauru, OJ: Ontong Java, PI: Pigafetta, RE: Réunion Island. Plate abbreviations: CAP: Capricorn, CAR: Caribbean, LWA: Lwandle, MAC: Macquarie, SUR: Sur. [Source: Siebert et al., 2010]

island arc volcanoes but including volcanoes formed near seafloor spreading centers.

As with mid-ocean ridge basalt (MORB), the dominant magma type for oceanic intraplate volcanism is basalt (commonly called ocean island basalt (OIB), and suggesting a mantle origin. As we might expect, intraplate volcanism is collectively much less voluminous than volcanism at plate margins. It is estimated the volume ratio of MORB to OIB at 9:1. If we consider convergent margin igneous activity as well, the proportion of OIB falls to a few percent of all oceanic-related volcanism. Nonetheless, OIB volcanism is estimated at about 1.5 km^3/a, which is far from trivial[4].

The chemical character of oceanic intraplate volcanism is distinct from that of constructive or consumptive plate margins, permitting us to treat the products as a single petrogenetic province. The chemistry and petrography are highly variable, however, more so than for MORB. It is beyond the scope of this book to attempt a comprehensive survey of the varied products of oceanic intraplate volcanism. Space limitations dictate that we address the major themes by way of some well-studied and constrained examples, encompassing a reasonable compositional diversity of OIBs and their differentiation products. These examples serve as a basis for later speculation on petrogenetic processes and source regions of oceanic intraplate magmas.

The most familiar products of oceanic intraplate volcanism are the numerous islands that dot the world oceans. Seamounts (eroded or sunken islands or accumulations that never rose above sea level) also compose a significant (although poorly studied) proportion of the total intraplate igneous activity. It is estimated that there are between 22,000 and 55,000 seamounts dotting the ocean floor, of which only about 2000 are presently active or dormant.

Seamounts appear to be concentrated along fracture zones, which supply convenient shallow conduits for magma rising toward the surface. There are also about 15 oceanic plateaus: massive outpourings of basalt on the ocean floor, most similar to continental flood basalts on land. All of the basalts from oceanic intraplate settings are referred as OIBs, regardless of whether the accumulations rise above sea-level, risking the occasional misnomer ("island") for the sake of simplicity[5].

The concept of intraplate magmatism developed at the same time as the plate tectonic paradigm. Such magmatism requires an explanation other than the plate tectonic processes of seafloor spreading and subduction. Age-progressive volcanism along the Hawaiian island chain provided the first insights into potential causes of intraplate magmatism. The age of volcanism systematically increases to the northwest from the island of Hawaii, a "hot spot"[6] generating significant magmatism that is distal from any current plate boundary. Another explanation for intraplate magmatism is mantle ascending and partially melting via fissures and faults where the lithosphere is extending. The trend of the Hawaiian-Emperor seamount chain, however, is oblique to preexisting fracture zones, which show no evidence for extension. Observations of multiple age-progressive volcanic island chains led to the interpretation that hot spots are surface expressions of deep mantle plumes[7], to which large igneous provinces have also been attributed, both on the continents and in the oceans.

Vocabulary

intraplate: (adj) 板块内的
magmatism: (n) 岩浆作用
tectonic: (adj) 区域构造的
plate: (n) 板块
  ~plate tectonics (n) 板块构造(学)
seafloor spreading: (n) 海底扩张
subduction: (n) 俯冲
transform faulting: (n) 转换断层
oceanic: (adj) 大洋的
continental: (adj) 大陆的
geologic time: (n) 地质时代
magmatic: (adj) 岩浆的
  adv: magmatically
  syn: igneous
geodynamics: (n) 地球动力学
large igneous province: (n) 大火成岩省
volcano: (n) 火山
  adj: volcanic 火山的
mafic: (adj) 镁铁质的
silicic: (adj) 酸性的
basaltic: (adj) 玄武质的
arc-trench system: (n) (岛) 弧-(海) 沟体系
surface: (n) 表面
subduction zone: (n) 俯冲带
plate boundary: (n) 板块边界
extension: (n) 拉张
convergence: (n) 汇聚
seafloor: (n) 海底
oceanic plateau: (n) 洋底高原

seamount:(n) 海山
sea knoll:(n) 海丘
guyot:(n) 平顶海山
landform:(n) 地形,地貌
island arc:(n) 岛弧
mid-ocean ridge basalt:(n) 大洋中脊玄武岩
ocean island basalt:(n) 洋岛玄武岩
plate margin:(n) 板块边缘
convergent margin:(n) 汇聚边缘
petrogenetic:(adj) 岩石成因的
petrography:(n) 岩相学
eroded:(adj) 被侵蚀的(erode 的过去分词)
sunken:(adj) 沉没的(sink 的过去分词)
fracture zone:(n) 断裂带
hot spot:(n) 热点
partially melting:(n) 部分熔融
lithosphere:(n) 岩石圈
seamount chain:(n) 海山链
mantle plume:(n) 地幔柱
plume:(n) 羽流,羽状柱

Notes

[1] intraplate magmatism 意为板内岩浆作用。它主要包括远离板块边界的岩浆活动,与大洋扩张、俯冲作用和转换断层无关。大洋内部广泛发育该类型的岩浆活动。

[2] 此句主要强调板内岩浆岩的成分变化较大,存在基性-酸性的喷出岩。除了 mafic 和 silicic 外,还可用 ultramafic(超基性的)和 intermediate(中性的)来描述岩石的成分。

[3] seamount 意为海山,是大洋板内岩浆活动的重要产物之一。海底火山有多种形态,包括海山以及下文提到的 sea knoll(海丘)和 guyot(平顶海山)。海山呈线状或者簇状分布分别形成海山链(seamount chain)和海山群(seamount group)。

[4] OIB 为洋岛玄武岩,尽管与 MORB 相比,其在大洋中所占的比例非常小,但由于其独特的成因,一直是地幔地球化学研究的热点,是了解地幔组成最重要的探针和窗口之一。

[5] 为了使用方便,所有的大洋板内玄武岩都被称为 OIB(不管其是否出露于海平面),但可能会导致“岛”这个词的误用。

[6] hot spot 意为热点,是地幔柱在地表的表现形式。

[7] mantle plume 即地幔柱,一般认为它起源于核幔边界,由蘑菇状的头和细长的尾组成。大陆和大洋中的大火成岩省一般也被认为与地幔柱活动有关。

6. Marine Gas Hydrate Resources

Text

Gas hydrates represent crystalline solid formations comprising water and gas[1]. While visually and functionally reminiscent of ice, they possess vast quantities of methane, occurring ubiquitously across every continent. These remarkable compounds are abundant within marine sediments, forming a substantial stratum spanning several hundred meters directly beneath the sea floor, while also exhibiting a close association with permafrost in Arctic regions (Figure Ⅲ-5). However, due to their inherent instability under customary sea-level pressures and temperatures, comprehending their intricate nature presents a formidable challenge to scientific inquiry.

Gas hydrates are important for three primary reasons:

- They possess the potential to harbor a substantial reservoir of energy.
- Their presence poses a noteworthy hazard, as it engenders alterations in the stability of seafloor sediment, thereby exerting influence over the occurrence of collapse and landslides.
- The reservoir of hydrates wields considerable sway over the environment and climate, given that methane, a potent greenhouse gas, assumes a consequential role.

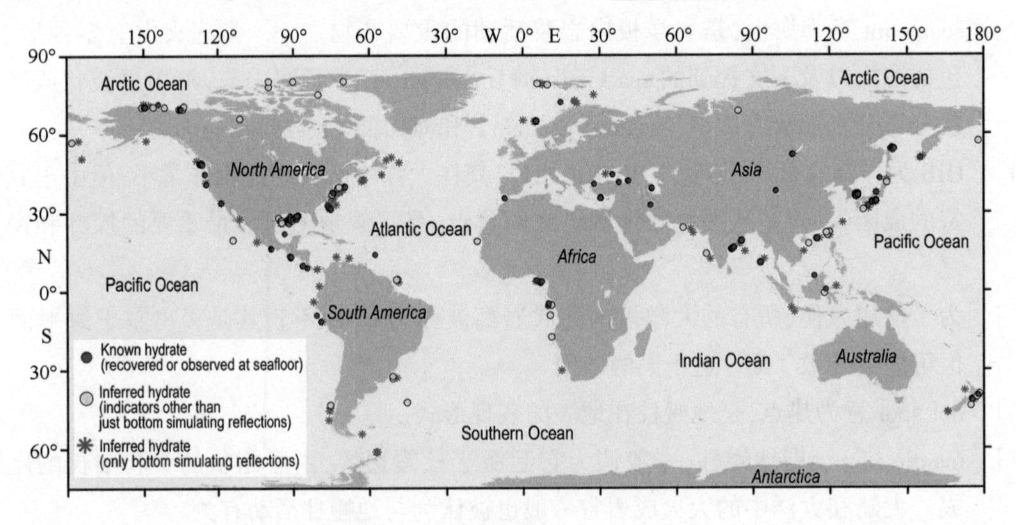

Figure Ⅲ-5: The global distribution of observed and inferred gas hydrates in marine and permafrost-associated settings, which have been the focal point of drilling initiatives, is depicted. The color coding refers to the primary sediment type in each location, thereby delineating the likely type of gas hydrate reservoir at each site. [Source: USGS]

Gas hydrate materializes through the amalgamation of water and natural gas, catalyzed by precise pressure and temperature thresholds, giving rise to a solid akin to crystalline ice. Within this intricate framework, the water molecules craft a lattice-like confinement, ensnaring the gas in a manner that permits substantial gas retention. Gas hydrates are found in sub-oceanic sediments in the polar regions (shallow water) and in continental slope sediments (deep water)[2], where pressure and temperature conditions combine to make them stable.

On a planetary magnitude, gas hydrate reservoirs harbor vast quantities of methane, residing within depths that are relatively close to the Earth's surface, rendering them susceptible to the temperature fluctuations coinciding with the phenomenon of climate change. Notably, methane itself assumes the role of a formidable greenhouse gas, and discerning scholars posit the possibility that the liberation of methane consequent to the disintegration of gas hydrates during prior climatic occurrences may have indeed accentuated the intensification of global warming[3].

Methane, the primary component of natural gas, stands as the preeminent among the gases that congregate to shape the enigmatic entity known as gas hydrate. In truth, the vastness of natural gas contained within the global reservoirs of gas hydrate is presumed to transcend the combined volume of all conventionally acknowledged gas reserves. It is this extraordinary potential that has spurred a concerted endeavor spanning several decades, wherein a collective assemblage of erudite scholars, governmental entities, and private industry pioneers have diligently delved into the intricate pursuit of unraveling the methodologies by which natural gas extraction from hydrate formations may be realized.

The latent propensity of water to engender ethereal bonds with gaseous companions was initially unveiled in the early 1800s, yet the fortuitous revelation of naturally occurring gas hydrates remained shrouded until the 1960s when Siberia unveiled its enigmatic treasures. As the 1980s dawned, the United States Geological Survey (USGS) and the United States Department of Energy (DOE) collectively embarking upon an odyssey to explore the prospect of procuring copious quantities of natural gas from the hydrate deposits.

Subsequent to those discoveries, domestic and international entities orchestrated an expanding array of field trials, meticulous laboratory investigations, and comprehensive geophysical analyses across diverse corners of the globe. Their collective pursuit, resolute in purpose, sought to unravel the intricate mechanism of gas hydrate distribution while deciphering the precise conditions that would bestow upon humanity the possibility of unlocking the coveted elixir of natural gas from these ethereal formations.

Gas hydrates, those ethereal entities of scientific inquiry, beckon to be examined, both within the hallowed confines of the laboratory, where a contrivance of intricate machinery orchestrates the precise interplay of pressure and temperature conditions for hydrate formation. Alternatively, one may venture into the realms of in situ exploration, where seismic data, meticulously collected aboard ships, intertwines with geophysical models, unraveling the enigmatic fabric of gas hydrates in their natural abode.

Presently, groups of distinguished scientists from the United States, China, Canada, Norway, Great Britain, Japan, and India are diligently engaged in unraveling the profound mysteries enshrouding gas hydrates[4]. With unwavering resolve, these intellectual trailblazers strive to decipher the intricate interplay of these enigmatic formations, meticulously exploring their pivotal role within the intricate tapestry of our planet's climate. Moreover, they embark upon a visionary voyage, contemplating the profound implications that gas hydrates hold for the future of fuels, as humanity grapples with the relentless pursuit of sustainable energy sources.

Vocabulary

crystalline:(adj) 结晶的,晶状的
reminiscent:(adj) 使人联想的,回忆的
permafrost:(n) 永久冻土,永冻层
intricate:(adj) 错综复杂的
formidable:(n) 巨大的,艰难的,可怕的,令人敬畏的
harbor:(vt) 怀有
hazard:(n) 危害,危险
engender:(vt) 产生,引起
consequential:(adj) 重要的
materialize:(vi) 发生,实现
amalgamation:(n) 合并,混合
akin:(adj) 类似的,相似的
craft:(vt) 制造,精心制作
confinement:(n) 限制,约束
retention:(n) 保留,保持
fluctuation:(n) 波动,起伏,涨落
posit:(vt) 假设,认定,安置,放置
liberation:(n) 释放,摆脱束缚
disintegration:(n) 分解
preeminent:(adj) 卓越的,优秀的
congregate:(vi) 聚集,集合
enigmatic:(adj) 神秘的
transcend:(vt) 超越,超出
spur:(vt) 刺激,促进
endeavor:(n/vi) 努力,尽力
erudite:(adj) 博学,有学问的
diligent:(adj) 刻苦的,孜孜不倦的
pursuit:(n) 追求,事业
propensity:(n) 倾向,倾向性
unveil:(vt) 使公之于众,揭开
fortuitous:(adj) 偶然的,意外的
revelation:(n) 启示,新发现
embark:(vi) 着手,从事
odyssey:(n) 艰苦的旅程
procure:(vt/vi) 获得,取得
copious:(adj) 大量的,充裕的
orchestrate:(vt) 策划,精心安排
meticulous:(adj) 细心的,小心翼翼的
bestow:(vt) 赠予,授予
coveted:(adj) 垂涎的,梦寐以求的
elixir:(n) 灵丹妙药,万能药
ethereal:(adj) 优雅的,缥缈的
beckon:(vi/vt) 召唤,吸引
hallowed:(adj) 神圣的
contrivance:(n) 发明,发明物
machinery:(n) 机器,机械
venture:(vi) 冒险,敢于去做

intertwine:(vi) 紧密相连,交织
abode:(n) 居所
enshroud:(vt) 笼罩,遮蔽,掩盖
unwavering:(adj) 未察觉到的,不注意的
intellectual:(adj) 智力的,聪明的
trailblazer:(n) 先驱,开拓者
strive:(vi) 努力,奋斗
voyage:(n) 航行,远行
contemplate:(vi/vt) 考虑,沉思,思忖
grapple:(vi) 努力设法解决,抓住
relentless:(adj) 不停的,持续强烈的

Notes

[1] 天然气水合物是天然气与水在高压低温条件下形成的类冰状结晶物质,其气体分子被束缚在由水分子通过氢键连接而构成的多面体笼子里,在低温(<10 ℃)高压(>100 bar)条件下稳定存在,化学式为 $8CH_4 \cdot 46H_2O$,因其外观像冰,且遇火即可燃烧,所以也被称为"可燃冰"。

[2] 已发现的天然气水合物主要存在于北极地区的永久冻土区和世界范围内的海底、陆坡、陆基及海沟中。

[3] 甲烷是一种很强的温室气体,对大气的暖化威力比二氧化碳强 23 倍。因此,天然气水合物开采过程中如果不能很好地对甲烷气体进行控制,就必然会加剧全球温室效应。

[4] 截至目前,我国海域天然气水合物已完成二次试采工作,均取得了巨大的成功。2017 年,我国海域天然气水合物第一轮试采成功,实现连续稳定产气 60 d,累计产气 30.9×10^4 m^3。2020 年,我国海域天然气水合物第二轮试采成功,实现连续稳定产气 30 d,累计产气 86.14×10^4 m^3,日均产气 $2.87\times10^4 m^3$,创造了产气总量、日均产气量两项新的世界纪录。通过两次试采,我国实现了从探索性试采向试验性试采的重大跨越,使天然气水合物开发利用成为现实可能,向天然气水合物勘查开采产业化迈出了极为关键的一步。

7.Marine Geohazards

Text

Marine geohazards, or "dangers in the deep" include earthquakes, volcanic eruptions, submarine landslides, and tsunamis, as well as dissociation of gas hydrates—which can cause seafloor collapse—and oil spills or toxic seeps that affect deep sea life or change the physical characteristics of ocean environments.

Earthquakes, Landslides, and Tsunamis

Underwater earthquakes and landslides can generate tsunamis that cause hazards for coastal communities[1]. USGS scientists study the subduction zone and the recent history of marine hazards and evaluate the future potential and probable impacts of such events on a regional basis. Quantifying these various hazards (e.g., earthquakes, landslides, tsunamis, and volcanoes) using geological and geophysical data, interpretations, and models improves understanding of the underlying processes of marine geohazards to assess the threats they pose. The USGS develops reliable deterministic and probabilistic estimates of the hazards that are used by engineers and policymakers to help reduce risk.

One barrier to measuring seismic risk has been the scarcity of high-resolution maps of the ocean floor. To fill these gaps, USGS scientists conduct high-resolution mapping offshore, especially near urban regions such as Southern California and the Pacific Northwest that are particularly at risk from seismic hazards. Creating three-dimensional views of the seafloor has given scientists remarkable ways to examine how a fault works, or how fluids may follow underground paths and potentially trigger submarine landslides. These landslides threaten offshore structures such as seafloor pipelines, cables, and equipment for oil and gas exploration. They can also trigger tsunamis that endanger coastal communities worldwide.

To discover a fault's structure, scientists go to sea to collect streams of data that they turn into comprehensive underwater maps. This type of imaging, along with knowing the age of the sediment along faults and measuring other factors such as magnetics and density, can help tell the story of when the fault last ruptured or how fast it is moving. USGS incorporates these data, which have historically been challenging to collect, into earthquake models to estimate their actual hazard risk. Using high-resolution mapping and seismic technology to gather detailed seafloor data can directly impact human life and cities by improving earthquake and tsunami forecasts.

Research results are used in evaluations of earthquake risk zoning, public disaster education and preparedness, and engineering and building codes. Additionally, reassessing the threat of earthquake, tsunami, and landslide hazards to ports and nuclear power plants can directly impact facility management, emergency-management planning, and plant re-licensing.

The Atlantic and Gulf of Mexico margins are heavily urbanized, support extensive port and industrial/resource facilities, and host 10 nuclear power plants. The USGS completed quantitative assessments of submarine landslides for the U.S. Atlantic coast from Maine to Florida and throughout the Gulf of Mexico to better comprehend the risk of potential submarine landslides and tsunamis to these areas and associated infrastructure.

Gas Hydrates and Seafloor Collapse

Naturally occurring gas hydrates are ice-like combinations of water and (usually methane) gas that form in sediment below the sea floor and in areas of continuous permafrost when pressure and temperature conditions are appropriate (Figure Ⅲ-6).

Figure Ⅲ-6: Ice-like gas hydrates under capping rock encrusted with mussels on the seafloor of the northern Gulf of Mexico. [Source: USGS]

In deep water marine settings where warm fluids are pumped from great depths below the seafloor for extraction of conventional[2] oil and gas. Heating of sediments near a well could lead to breakdown of gas hydrates and release gas and water. Intact gas hydrates generally strengthen marine sediments, and dissociation of gas hydrates could lead to subsidence or collapse of the seafloor near the well[3]. Features associated with natural failure of the seafloor (landslides) have also been linked to gas hydrates in some cases. USGS scientists support submarine geohazards research through field-based surveys that refine understanding of the hydrates-slope failure association and through geotechnical studies that evaluate the response of sediments to dissociation or dissolution of gas hydrates.

Vocabulary

geohazard：（n）地质灾害
earthquake：（n）地震
submarine landslide：（n）海底滑坡
tsunami：（n）海啸
dissociation：（n）分解
gas hydrate：（n）天然气水合物
seafloor collapse：（n）海底崩塌
oil spill：（n）石油泄漏
seep：（n）渗出
characteristic：（n）特征，特性
coastal：（adj）近岸的，沿海的
on a regional basis：（adj）基于区域上的
risk：（n）风险
seismic：（adj）地震的
high-resolution：（adj）高分辨率的
offshore：（adj）海上的，近海的，离岸的，向海的
urban region：（n）城市区域
three-dimensional：（adj）三维的
fluid：（n）流体；（adj）流体的
underground path：（n）地下通道
offshore structure：（n）近海设施
pipeline：（n）管道
cable：（n）光缆
oil and gas exploration：（n）油气勘探
underwater：（adj）水下的
sediment：（n）沉积物
magnetics：（n）电磁
density：（n）密度
rupture：（vi）断裂
historically：（adv）历史上地
earthquake model：（n）地震模型
forecast：（n）预报
zoning：（n）区划
disaster：（n）灾难
preparedness：（n）整备
code：（n）准则
port：（n）港口
nuclear power plant：（n）核电站
planning：（n）规划
licensing：（n）许可
Atlantic：（adj）大西洋的
Gulf of Mexico：（n）墨西哥湾
margin：（n）沿岸
urbanize：（vi）城市化
resource：（n）资源
infrastructure：（n）基础设施
capping rock：（n）上覆岩石
encrust：（vi）结壳
mussel：（n）贻贝
naturally occurring：（adj）自然形成的
methane：（n）甲烷
extraction：（n）提取
conventional：（adj）常规的
well：（n）油气井
intact：（adj）结构完整的
subsidence：（n）下沉
natural failure：（n）自然崩溃
field-based survey：（n）野外调查
hydrates-slope failure association：（n）水合物-陆坡崩溃体系
geotechnical：（adj）地质技术的

Notes

[1] communities 意为群体、群落。这里泛指海岸附近的人类活动设施和社群。

[2] conventional 意为常规的,这里指的是传统油气资源,从技术和经济角度界定,常规油气是指应用现有技术方法能够经济有效开采的油气资源。非常规(unconventional)油气是指现今无法用常规技术方法进行经济性勘探开发的油气资源,资源规模大,储层物性差。

[3] 天然气水合物主要赋存于低温高压下的海底沉积物层中,当周围环境的温度或者压力发生变化时,其稳定性会受到破坏,诱发坍塌、海底滑坡等地质灾害,对钻井平台、海底电缆等造成巨大破坏。

8. Marine Geologic Survey

Text

The research direction and content of conventional marine geology include coastal and submarine landforms[1], geological structures[2], rocks[3], sediments[4], mineral resources and ocean geological history. In recent years, more and more attention has been paid to natural gas hydrates, hydrothermal deposits, marine climate, environment, ecosystems, hydrodynamics and other resources and environmental issues, and extended to the interaction between the lithosphere and hydrosphere, biosphere, and atmosphere. Some scholars have proposed that the study of geotectonic geology must go beyond plate tectonics, take the earth as a whole, and use global dynamics and even the dynamics of heaven and earth. In order to solve the problem of coastal zones, we cannot rely on the narrow coastal zone itself in isolation. We need to expand to the inland areas that affect the coastal zone and extend to the ocean to form an integrated science. China's coastal marine geological survey needs to accurately meet the needs of economic and social development and solve problems in the process of marine development from the perspective of earth system and land and sea overall planning.

The research direction and content of marine geologic investigation[5] are included in the overall framework of Earth system science. In the next 15–20 years, China will focus on offshore China, especially the South China Sea; The hot spots will be the western Pacific Ocean, the North and South poles; The frontier areas are natural gas hydrates in the sea, Mesozoic deep oil and gas in offshore basins, the mechanism and evolution of plate tectonics in the Western Pacific, and the comprehensive geological survey of sea areas along the Maritime Silk Road; Interdisciplinary research covers mid-ocean ridges[6], seafloor sediments, mineral-geochemical characteristics, deep-sea mining, plate subduction processes, seismic tsunamis, seafloor hydrothermal[7] and mining, and seafloor environment. Comprehensive geo-geophysical exploration methods can be used to explore deep structures, such as the Yap subduction zone in the western Pacific Ocean. High-resolution seismic stations have been used in China to carry out wide-angle reflection-refraction sea-land correlation surveys to study the impact of plate subduction in the Western Pacific Ocean, revealing the mechanism and process of lithosphere thinning in the east of the North China Craton and the Yangtze Craton, reflecting the Earth system science concept of integrating land and sea.

From shallow water to deep water, the submarine geomorphic features are widely distributed and of many kinds. For example, there are many kinds of geomorphologic features such as submarine floor anomalies, sand waves, optical cable and submarine facilities, hard seabed, rugged terrain and landslide, fault, submarine hydrothermal solution, cold spring[8] and so on. At present, in the interpretation of the distribution characteristics, detailed changes and extraction of the relevant parameters of the submarine geomorphology and submarine structural features, attention should be paid to the investigation and comprehensive analysis of the marine environment and regional geological background data such as coastal zones[9], coral reefs[10], continental shelves[11], continental slopes[12], deep-sea basins[13] and seamounts[14] in combination with marine geological disasters[15], seabed routing geology[16] and marine engineering sites[17]. In order to improve the accuracy and reliability of identifying various meaningful geomorphic features. To find out all kinds of submarine geomorphic features of the target work area, evaluate their potential hazards to related submarine facilities and engineering construction, provide technical support for the safety and feasibility evaluation of various engineering construction and drilling operations, and help reduce engineering risks.

In the search for offshore oil, gas and mineral resources, we need to carry out ocean mineral survey[18], offshore oil and gas resources survey[19], natural gas hydrate survey and so on. The survey mainly uses marine geology, geophysics, geochemistry and other methods to conduct detailed investigation and research on the sediment, stratigraphic structure, geophysical field, etc. at the bottom of the ocean to determine the mineral, natural gas and oil and gas resource potential and reserves in the ocean area.

Vocabulary

natural gas hydrate：(n) 天然气水合物,可燃冰

hydrothermal deposit：(n) 热液矿床

ecosystem：(n) 生态系统

geotectonic：(adj) 大地构造的

deep-sea mining：(n) 深海采矿

seismic tsunami：(n) 地震海啸

anomaly：(n) 异常事物,反常现象

hard seabed：(n) 硬海底

rugged terrain：(n) 崎岖地形

landslide：(n) 滑坡

submarine hydrothermal solution：(n) 海底热液

Notes

[1] 海底地貌是指海洋底部的地形特征。海底地貌的形成受到多种因素的影响,包括地质构造、海洋动力作用、沉积作用等。常见的海底地貌包括海山、海沟、海岭等。

[2] 海底构造是指海洋底部的地质构造特征。海底构造的形成与地球内部的板块运动、

地壳变动、火山活动等密切相关。

[3] 岩石取样是指从地质体中采集岩石样品进行分析和研究的过程。岩石样品可以提供有关岩石的物理、化学和矿物学特性的信息,有助于了解岩石的成因、演化和地质历史等方面的知识。

[4] 沉积物取样是指从水体底部或沉积物层中采集样品进行分析和研究的过程。这些样品可以提供关于沉积物的物理、化学和生物特性的信息,有助于了解沉积物的成因、演化和环境变化等方面的知识。

[5] 海底地质调查是指对海洋底部地质构造、地貌、沉积物特征、地球物理性质等进行系统观测和研究的科学活动。

[6] 洋中脊调查是指对海洋中的洋中脊进行综合性的调查和评估,以了解洋中脊的地质构造、地貌特征、岩石成分、热液活动等情况。

[7] 海底热液是指在海底的特定地点,从海底底部喷发出的热水。与一般海水相比,海底热液的水温较高,通常为50~400℃。海水沿着洋壳裂隙下渗,在靠近岩浆房的过程中被不断加热,并与围岩发生反应,从中萃取成矿金属元素,最终演化为高温热液流体。热液活动主要发育于洋中脊、弧后扩张中心或岛弧火山附近。热液流体主要分为黑烟囱流体(富含金属元素)和白烟囱流体(富含硫酸盐),二者成因存在显著差异。

[8] 海底冷泉是海底之下天然气水合物分解后向上运移,喷出海底而形成的低温(约2~4℃)流体,流体组成主要为水和甲烷,喷出海底之后流体中甲烷被氧化常形成大量碳酸盐沉积。

[9] 海岸带调查是指对海岸带区域进行综合性的调查和评估,以了解海岸带的地貌特征、生态环境、资源利用等情况。

[10] 珊瑚礁调查是指对珊瑚礁生态系统进行综合性的调查和评估,以了解珊瑚礁的生物多样性、生态功能、健康状况等情况。

[11] 大陆架调查是指对海洋大陆架进行综合性的调查和评估,以了解大陆架的地质构造、地貌特征、地层结构、沉积物分布等情况。

[12] 大陆坡调查是指对海洋大陆坡进行综合性的调查和评估,以了解大陆坡的地质构造、地貌特征、沉积物分布、生物多样性等情况。

[13] 深海盆地调查是指对深海盆地进行综合性的调查和评估,以了解深海盆地的地质构造、地貌特征、沉积物分布、生物多样性等情况。

[14] 海山调查是指对海洋中的海山进行综合性的调查和评估,以了解海山的地质构造、地貌特征、岩石成分、生物多样性等情况。

[15] 海洋地质灾害调查是指对海洋地质灾害进行系统性的研究和调查,以了解其发生机制、影响范围和潜在风险。

[16] 海底路由地质勘查是指对海底地形和地质条件进行调查和评估,以确定适合敷设

海底光缆和管道的最佳路由。

[17] 海洋工程场址勘查是指对海洋区域进行调查和评估,以确定适合进行海洋工程建设的最佳场址。

[18] 大洋矿产调查是指对海洋底部的矿产资源进行勘探和评估的工作。

[19] 海洋油气资源调查是指对海洋地区进行的油气资源勘探和评估工作。这种调查主要通过海洋地质、地球物理、地球化学等方法,对海底沉积物、地层构造、地球物理场等进行详细的调查和研究,以确定海洋地区的油气资源潜力和储量。

9.Marine Geophysical Exploration

Text

Marine geophysical exploration is the process of using various geophysical techniques to study the geology and geophysical properties of the seafloor and subsurface. These techniques are crucial for understanding the oceanic crust and have diverse applications in scientific research and commercial ventures. Among these techniques are marine gravity surveys[1], which map seafloor topography by measuring variations in the Earth's gravity field, allowing identification of geological structures such as seamounts, subduction zones, and faults. Marine magnetic surveys[2] utilize measurements of the Earth's magnetic field variations to map magnetic anomalies, which can indicate subsurface rock formations and tectonic boundaries. Marine seismic exploration[3] involves controlled sound waves, generated by air guns or seismic sources, to image subsurface layers of the seafloor, enabling the identification of geological structures, oil and gas resources, and potential earthquake zones. Submarine heat flow surveys[4] measure the heat flowing out of the Earth's crust into the ocean, providing valuable insights into the Earth's thermal properties and tectonic processes.

One of the primary methods used in marine geophysical exploration is marine gravity surveying. This technique measures variations in the Earth's gravity field to map seafloor bathymetry and identify subsurface structures such as seamounts, subduction zones, and rifts. By understanding these features, scientists gain valuable insights into plate tectonics and the dynamic processes shaping the ocean floor. Another essential tool in marine geophysical exploration is marine magnetic surveying. This method detects variations in the Earth's magnetic field to map magnetic anomalies, which can indicate subsurface rock formations and tectonic boundaries. By analyzing these magnetic anomalies, researchers can reconstruct the Earth's magnetic history and better understand past geological events. Marine seismic exploration is another crucial aspect of marine geophysical exploration. It utilizes controlled sound waves, generated by air guns or seismic sources, to image the subsurface layers of the seafloor. This allows scientists to identify geological structures, potential hydrocarbon deposits, and even potential earthquake zones. Seismic data from marine exploration has been instrumental in improving our understanding of tectonic plate movements and seafloor spreading. A specialized technique used in marine geophysical exploration is the submarine heat flow survey. This method measures the amount of heat flowing out of the Earth's crust into the ocean, providing valuable information about the Earth's thermal properties and tectonic processes. It helps researchers understand the distribution of heat sources and the role of submarine volcanism in shaping the seafloor.

Marine electromagnetic exploration is another technique used in marine geophysical exploration. It measures electromagnetic fields to investigate subsurface electrical properties, identifying potential mineral deposits or hydrocarbon reservoirs. Seafloor in-situ exploration involves direct investigation and sampling of the seafloor to study geological, biological, and chemical processes. Researchers use specialized tools and equipment to collect samples and data from the seafloor, providing valuable information about marine ecosystems and geological formations. The study of seabed sediment acoustics focuses on the acoustic properties of seabed sediments. Understanding these properties is essential for characterizing sediment types, studying sediment transport processes, and assessing seafloor stability. Ocean acoustic detection utilizes acoustic signals to detect and locate underwater objects or marine life. This technique is widely used in marine biology and marine resource exploration, helping researchers locate underwater features and study marine biodiversity. Offshore drilling[5] is a common practice in marine geophysical exploration. It involves extracting geological samples or resources from beneath the seafloor for scientific research or commercial purposes. Offshore drilling provides direct access to subsurface materials, offering valuable information about geological processes and potential natural resources.

Additionally, to generate high-resolution bathymetric maps of the seafloor, marine geophysicists employ multibeam sounding[6]. This technique uses multiple sonar beams to collect data simultaneously, resulting in detailed images of the seafloor topography. These maps are crucial for studying underwater features, planning infrastructure development, and conducting marine ecological research. There are techniques like multibeam bathymetry, which uses multiple sonar beams to generate high-resolution seafloor maps; and side-scan sonar[7], using sonar technology to create detailed images of the seafloor, aiding in the identification of seafloor objects and geomorphic features; sub-bottom profiling[8] is employed to provide information about the layers and properties of sediment and rock beneath the seafloor. This technique helps in understanding sedimentary processes, identifying potential geohazards, and mapping subsurface geological structures. High-resolution multichannel seismic profiles[9], similar to marine seismic exploration but utilizing more recording channels to obtain more detailed subsurface images. Furthermore, there are in-situ seafloor heat flow[10] measurements, marine electronic exploration[11], marine electromagnetic exploration[12], seafloor in-situ exploration[13], seabed sediment acoustics[14], and ocean acoustic detection[15]. Side-scan sonar is another valuable tool used in marine geophysical exploration. By producing detailed images of the seafloor using sonar technology, it enables researchers to identify objects and features on the seafloor, such as shipwrecks or geological formations. Seismic tomography[16] is a sophisticated imaging technique that uses seismic waves to create 3D models of subsurface structures. By analyzing the travel times and amplitudes of seismic waves recorded at various locations, scientists can infer the properties and configurations of subsurface features. Resistivity tomography[17] is another imaging technique used in marine geophysical exploration. It measures variations in subsurface electrical resistivity to create detailed images of geological formations and identify potential resources.

To carry out these exploration tasks, various equipment and technologies are utilized. Marine detection equipment[18] encompasses various instruments and devices used in marine geophysical exploration. These tools include seafloor sensors, geophysical survey instruments, and data acquisition systems used to collect and process geophysical data, such as ocean bottom seismometers[19], underwater towed vehicles, fiber optic hydrophone arrays[20], and more. Additionally, unmanned ship vehicles (USVs), human-occupied vehicles (HOVs), remotely operated vehicles (ROVs), and autonomous underwater vehicles (AUVs) are employed, serving as unmanned or manned tools for deep-sea exploration and data collection.

Ocean bottom seismometers (OBS) are essential tools used in marine geophysical exploration. These instruments are placed on the seafloor to record seismic waves generated by earthquakes or controlled seismic sources. OBS data provide valuable information about seismic activity and the structure of the Earth's crust beneath the ocean. Underwater towed vehicles[21] are robotic vehicles towed behind ships to collect data or images from the seafloor or water column. These vehicles can carry various sensors and instruments to study marine environments and geological features. Fiber optic hydrophone arrays use fiber optic cables to detect underwater acoustic signals. These arrays are sensitive and can capture acoustic data over long distances, providing insights into underwater noise, marine life distribution, and geological processes. Unmanned Ship Vehicles[22] (USVs) are self-propelled vessels that operate without a crew on board. These autonomous vehicles can be equipped with various sensors and data collection systems, making them ideal for conducting long-duration marine geophysical surveys. Human Occupied Vehicles[23] (HOVs) are manned submersibles used for deep-sea exploration. These vehicles allow researchers to directly observe and interact with the underwater environment, making them invaluable tools for studying deep-sea ecosystems and conducting targeted research. Remote Operated Vehicles[24] (ROVs) are robotic vehicles controlled from the surface to perform tasks and collect data in the deep ocean. ROVs are widely used in marine geophysical exploration, underwater archaeology, and offshore industries, as they can access and work in extreme depths. Autonomous Underwater Vehicles[25] (AUVs) are self-contained, untethered robotic vehicles programmed to carry out specific tasks without direct human control. AUVs are widely used in marine geophysical exploration for mapping seafloor bathymetry, studying underwater environments, and collecting data in remote locations.

The International Ocean Drilling Program (IODP) is a collaborative research initiative that conducts scientific ocean drilling expeditions to study the Earth's history and structure. Marine geophysical exploration provides essential means and data sources for unveiling the mysteries of the seafloor and Earth's interior.

In conclusion, marine geophysical exploration is a multidisciplinary field that utilizes various advanced techniques and instruments to investigate the geological and geophysical properties of the seafloor and subsurface. These techniques play a crucial role in our understanding of Earth's geology, plate tectonics, marine ecosystems, and potential natural resources. The knowledge gained through marine geophysical exploration is not only essential for scientific research but also has significant implications for the sustainable management of marine environments and resources.

Vocabulary

marine geophysical exploration: (n) 海洋地球物理勘探
geophysical property: 地球物理性质
subsurface: (n) 地下
scientific research: (n) 科学研究
commercial venture: (n) 商业活动
seafloor topography: 海底地形
magnetic anomaly: 磁异常
subsurface rock formation: 地下岩石构造
tectonic boundary: 构造边界
controlled sound wave: 控制声波
air gun: (n) 气枪
seismic source: 地震源
oil and gas resource: (n) 油气资源
potential earthquake zone: (n) 潜在地震带
Earth's thermal property: (n) 地球的热性质
tectonic process: (n) 构造过程
Earth's gravity field: (n) 地球重力场
seafloor bathymetry: (n) 海底地形
subsurface structure: (n) 地下结构
rift: (n) 断裂,裂谷
dynamic process: (n) 动态过程
marine magnetic surveying: (n) 海洋磁测
Earth's magnetic history: (n) 地球的磁性历史
geological event: (n) 地质事件
marine seismic exploration: (n) 海洋地震勘探
geological structure: (n) 地质结构
hydrocarbon deposit: (n) 油气沉积物
seismic data: (n) 地震数据
tectonic plate movement: (n) 构造板块运动
submarine heat flow survey: (n) 海底热流测量
distribution of heat source: (n) 热源分布
submarine volcanism: (n) 海底火山活动
marine electromagnetic exploration: (n) 海洋电磁勘探
electromagnetic field: (n) 电磁场
subsurface electrical property: (n) 地下电

性性质
mineral deposit：(n) 矿物质沉积
hydrocarbon reservoir：(n) 油气储层
seafloor in-situ exploration：(n) 海底原位勘探
direct investigation：(n) 直接调查
acoustic property：(n) 声学性质
seafloor stability：(n) 海底稳定性
ocean acoustic detection：(n) 海洋声学探测
acoustic signal：(n) 声学信号
marine resource exploration：(n) 海洋资源勘探
underwater feature：(n) 水下特征
geological sample：(n) 地质样本
bathymetric map：(n) 海底地形图
sonar beam：(n) 声呐波束
sonar technology：(n) 声呐技术
seismic wave：(n) 地震波
seafloor sensor：(n) 海底传感器
geophysical survey instrument：(n) 地球物理勘测仪器
data acquisition system：(n) 数据采集系统
geophysical data：(n) 地球物理数据
deep-sea exploration：(n) 深海勘探
marine environments：(n) 海洋环境
underwater noise：(n) 水下噪声
collaborative：(adj) 合作的，协作的

Notes

[1] 海洋重力调查通过测量海洋上地球重力场的变化，以确定海底结构和地质特征。

[2] 海洋磁性调查通过测量地球磁场的变化，以确定与海底扩散、地壳磁性和潜在矿床有关的磁性异常。

[3] 海洋地震勘探是指利用地震调查来研究海底地质情况。通过将声波发送到海底并分析反射，地质学家可以创建地下图层的图像并识别地质结构。

[4] 海底热流调查通过测量来自海底的热流，以了解海底地壳的热特征。

[5] 海上钻探是指在海床进行钻探以开采石油、天然气或矿物的过程。

[6] 多波束探测是指由多波束声呐系统提供高分辨率的水深数据，允许研究人员创建详细的海底地图，其中包括关于深度变化和海底特征的信息。

[7] 侧扫声呐是指通过横向发射声学信号和捕获反射信号来创建海底的详细图像。

[8] 浅地层剖面是指利用低频声信号对沉积物层和海底地质进行成像的方法。

[9] 高分辨率多通道地震剖面是一种先进的地震测量技术，使用多个传感器提供更详细的地下图像。

[10] 海底原位热流是指使用专门的仪器直接测量来自海底的热流。

[11] 海洋电子勘探是指为海洋研究和勘探目的使用各种电子设备。

[12] 海洋电磁学勘探主要利用电磁学的方法来研究地下的电学特性，帮助识别潜在的资源和地质结构。

[13] 海底原位勘探是指使用各种专门的仪器对海底进行直接勘探和观测。

[14] 海底沉积物声学主要研究声波与海底沉积物的相互作用,以了解它们的特性。

[15] 海洋声学探测利用声学信号探测和研究海洋生物、水下结构或其他海洋现象。

[16] 地震层析成像是一种利用地震波创建地下结构图像并识别地壳和地幔异常的技术。

[17] 电阻率层析成像是指通过测量地下的电阻率来研究地质构造和识别潜在资源。

[18] 海洋探测设备是用于海洋研究和勘探的各种工具和仪器的统称。

[19] 海底地震仪是放置在海底的地震仪器,用来记录地震活动和研究地壳。

[20] 光纤水听器阵列是一种使用光纤电缆探测水下声音信号的水听器网络。

[21] 水下拖曳体是指拖在船只后面的车辆或传感器系统,以收集有关海底或水柱的数据。

[22] 无人船是指用于海洋研究和探索的自主或远程操作的船只。

[23] 载人潜水器是指用于深海勘探和研究的载人潜水器。

[24] 遥控潜水器是一种由水面控制的无人驾驶的水下运载器,用于各种海洋任务,包括探测和数据收集。

[25] 自治式潜水器是一种自行推进和无栓系的水下航行器,可以在无人干预的情况下执行任务。

10. Marine Sedimentation

Text

The sea floor, being the place of accumulation of solid detrital material of inorganic or organic origin, is virtually covered with unconsolidated sediments; therefore, the study of materials found on the sea bottom falls largely within the field of sedimentation, and the methods of investigation employed are those used in this branch of geology[1]. Sediment is produced by the weathering (the chemical and mechanical breaking down) of rock such as granite or basalt into particles that are then moved by air, water, or ice. Sediment can also form by the accumulation of shells of dead organisms. Therefore, sediment can consist of either mineral or fossil particles, and both types can be found in many places on the bottom of the sea.

Sediments can be subdivided on the basis of the size of the particles (grain size) or on the basis of their mode of formation. In the first case, the classification depends on a measurement of particle size; in the second case, the classification requires an interpretation of the origin of the deposit[2]. Both classifications are useful and are widely employed by geological oceanographers. The size of particles produced from the breakdown of rock ranges from enormous boulders to tiny grains of microscopic clay or even finer particles called colloids. From the largest to the smallest particles common in sediment, there are gravel, sand, silt, and clay (Table Ⅱ-1). Silt and clay particles are typically mixed together and form a deposit of mud. The most common sedimentary deposits in the ocean are mud and sand; gravel is very rare in the sea.

Table Ⅲ-1: Grain-size scale

Sediment	Type	Diameter (mm)
Gravel	Boulder	>256.0
	Cobble	65.0-256.0
	Pebble	4.0-64.0
	Granule	2.0-4.0
Sand	Very coarse	1.0-2.0
	Coarse	0.50-1.0
	Medium	0.25-0.50
	Fine	0.125-0.25
	Very fine	0.0625-0.125

Table Ⅲ-1

Sediment		Type	Diameter (mm)
Mud	Silt		0.0039-0.0625
	Clay		0.0002-0.0039
Colloid			<0.0002

Sediments by the way that they form can be subdivided into six categories: (1) detrital material, largely of immediate terrigenous origin, (2) products of subaerial and submarine volcanism, (3) skeletal remains of organisms and organic matter, (4) inorganic precipitates from seawater, (5) products of chemical transformation taking place in the sea, and (6) extraterrestrial materials[3].

Shelf Sedimentation

A continental shelf is a relatively broad, essentially flat platform 70 to 100 kilometers (~43 to 62 miles) wide that represents the submerged edge of a continent.

Water depths on the shelf are shallow, varying from zero at the shoreline to 120 to 150 meters (~396 to 495 feet) at the shelf break, where the gradient of the sea bottom steepens to about 4 degrees and marks the beginning of the continental slope. The sea floor of the shelf is nearly horizontal, with a regional slope that rarely exceeds 1 degree[4]. Energy for eroding and transporting sediment grains is provided by the tides and wind-generated waves and currents. Over most continental shelves, waves seem to be the dominant process affecting the sea bottom. Drawing from our swimming experience at the beach, we know that large waves contain more energy than do small waves. Diving beneath an unbroken wave, we will notice that the water becomes increasingly calmer with depth. In fact, if we go deep enough, the water will be relatively calm despite the surface agitation by waves. The bigger the waves, the deeper we must dive in order to escape the wave motion. We can infer that bottom energy induced by surface-water waves must diminish with distance offshore, because water depths increase seaward. This means that, if you began from the shore and walked seaward on the shelf bottom into deeper water, you first would be pushed around by the wave energy. As the bottom got deeper, you would feel less and less of an effect from the waves, until at some depth the water would be quite still.

The shoreline is affected by breaking waves, and these high-energy conditions suspend and remove all the fine sediment and allow mainly medium and coarse sand and gravel to be deposited on the beach and in the nearshore zone. Seaward from the nearshore zone, bottom energy induced by waves decreases because of increasing water depths. This decrease of bottom energy with water depth results in a systematic decrease in grain size with distance offshore. The beach, which is composed of coarse to medium sand and gravel, grades into fine sand farther offshore and, moving seaward, into muddy sand (sand with some mud), sandy mud (mud with some sand), and finally mud far offshore (Figure Ⅲ-7).

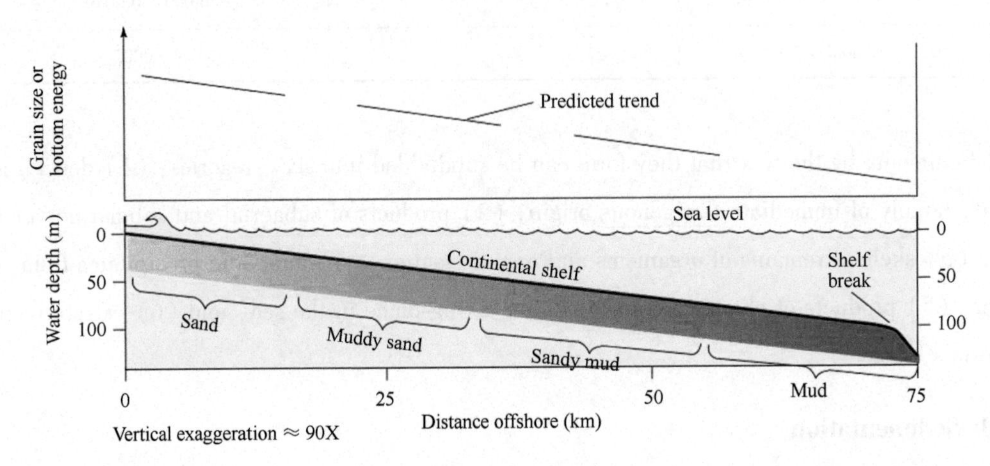

Figure Ⅲ-7: Shelf sedimentation. Assuming that bottom energy on the continental shelf is inversely proportional to water depth and, hence, distance offshore, the sedimentary cover should grade systematically from coarse sands and gravel onshore to mud at the shelf break. [Source: Pinet, 2019]

Deep-Sea Sedimentation

There are two main sources of sediment for the deep-ocean floor: (1) terrigenous mud and sand that bypass the shallow continental shelf and (2) the hard parts of surface-water microorganisms that settle to the deep-sea bottom[5].

Sediment that settles to the bottom of the deep sea is derived from either external or internal sources (Figure Ⅲ-8). External sources are the terrigenous rocks of the land. Weathering, the chemical and mechanical disintegration of rock at or near the Earth's surface, breaks down the bedrock of the land into small particles—mainly sand and mud—that are transported to the oceans by rivers and winds. The major sources of terrigenous sediment in the oceans are rivers that drain large mountain belts, such as the Himalayas of Asia. Internal sources of sediment furnish material that is produced largely by organisms and, to a lesser degree, by geochemical and biochemical precipitation of solids, such as ferromanganese nodules (hard pebbles enriched in metals). As a general rule, the proportion of deep-sea sediment derived from external sources (the terrigenous material) relative to that derived from internal sources (the biogenic material) decreases with distance offshore. In other words, the farther from the river supply, the greater tends to be the fraction of biogenic material in deep-sea deposits.

A simple classification of deep-sea deposits uses three broad categories based on the mode

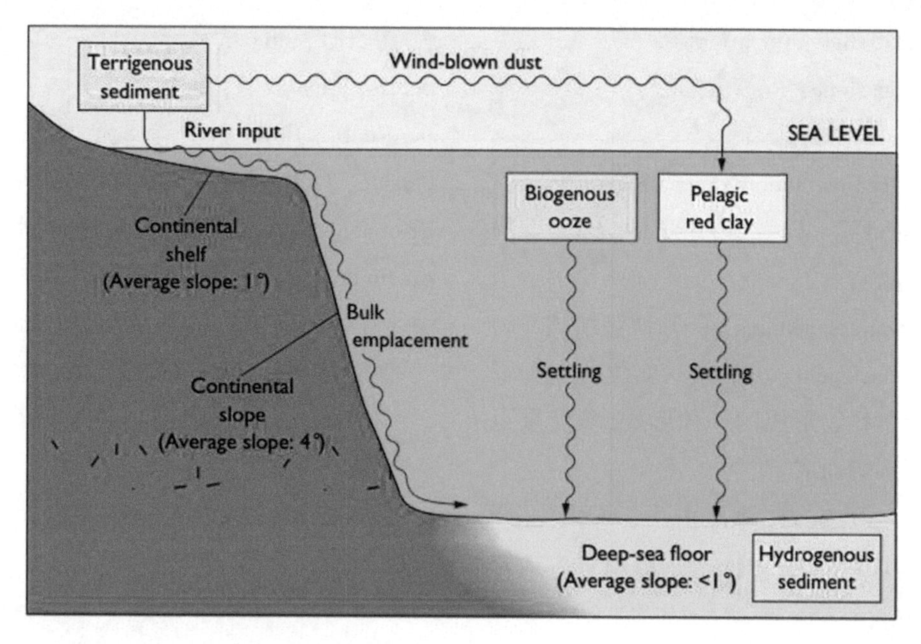

Figure Ⅲ-8: Sedimentation in the deep sea. [Source: Pinet, 2019]

of sedimentation. Bulk emplacement is the means by which large quantities of sediment are transported to the deep-sea floor as a mass rather than as individual grains. The processes of bulk emplacement are induced by gravity: material resting high up on a slope moves downward and comes to rest on the sea floor of the deep sea. All types of sedimentary debris—terrigenous and biogenic, both fine- and coarse-grained—can be swept seaward and dispersed across the deep-ocean floor by bulk emplacement. In contrast, pelagic sediment is the fine-grained fallout of terrigenous and biogenic material that settles through the water column, particle by particle, much as snowflakes fall out of the sky and accumulate as a snow cover on land. Hydrogenous sediment, as you may recall, consists largely of biochemical precipitates that form in situ (in place); that is, they originate at the site of deposition by geochemical and biochemical reactions.

Vocabulary

accumulation: (n) 积累,堆积,堆积物,堆积量

detrital: (adj) 碎屑的,由岩屑形成的

inorganic: (adj) 无机的,无生物的,不可分割的,(发生、发展过程)自然的,演进的,器官的,器质性的,仿自然形态的

sedimentation: (n) 沉积(作用)

particle size: (n) 颗粒尺寸
指物质中颗粒的大小,通常用直径或体积分布来描述

terrigenous: (adj)(尤其海中沉积)由泥土产生的,陆源的

subaerial: (adj) 接近地面的,地面上的,天空下的

continental shelf:(n) 大陆架
continental slope:(n) 大陆坡
current:(n) 海流
deep-sea sedimentation:(n) 深海沉积
biochemical:(adj) 生物化学的;(n) 生物化学物质
ferromanganese nodules:(n) 铁锰结核
bulk emplacement:(n) 块状就位
　海洋沉积物形成的一种形式,多受重力作用影响而成
gravity:(n) 重力,地心引力,严重性,严肃,庄严,重(量)
debris:(n) 残骸,碎片
　~debris flow:(n) 碎屑流
pelagic:(adj) 浮游的,远洋的,深海的
hydrogenous:(adj) 氢的,含氢的,水成的
precipitate:(vt) 使……突然降临,加速(坏事的发生),(使)淀析,使急落直下,使陡然下落,降水(如雨、雪、冰雹),使突然陷入(某种状态);(n) 沉淀物,析出物
　~precipitate sb./sth. into:(使)沉淀

Notes

[1] 此句讲述了海底作为沉积空间,通常接收了多种有机和无机物质,因此对于海洋沉积物的研究和方法涉及地质学中关于沉积物研究的方方面面。

[2] 此句对沉积物进行了分类,其中包括两种分类方案,一种是从粒径上来划分,另一种是从沉积物的来源上划分。

[3] 此句解释了海洋沉积物主要来源有六大类,分别包括陆源碎屑物质、陆地和海洋的火山物质、有机物的碎屑、海水中的化学沉积(如碳酸盐岩)、海洋中化学置换物质(如化学交代作用新生成的沉积物,如黏土矿物等),以及外来碎屑物质(如行星物质)。

[4] 此句解释了大陆架的基本特征:大陆架是一个相对宽阔、基本平坦的平台,是大陆向海的自然延伸,其基底由陆壳组成,宽 70~100 km。大陆架的水深较浅,从海岸线到转折端在 120~150 m 范围内,在转折端坡度陡峭,约为 4°,标志着大陆坡的开始。大陆架的海底几乎是水平的,区域坡度很少超过 1°。

[5] 此句解释了深海沉积物的来源主要包括两类:一种是陆源物质通过浅海大陆架后到达深海的泥沙;第二种是沉降到深海海底的表层微生物壳体,根据壳体组分可进一步分为硅质软泥和钙质软泥,钙质软泥在水深超过碳酸盐补偿深度(CCD)的区域几乎不会存在。

11. Offshore Oil and Gas Resources

Text

In the field of marine geology, studying the basin evolution of prototype basins is the key to understanding marine geological structures[1]. The application of seismic sedimentology provides us with a new insight to reveal the paleoenvironment, and by analyzing the reflection layers in seismic records, we can infer changes in the ancient marine depositional environment[2]. These seismic records reveal the sedimentary systems of different stages in the paleo-ocean, providing a strong basis for our understanding of the sedimentary facies of ancient basins. The presence of buried hills near the continental margin may affect the sedimentary system and basin evolution. Salt structures have a significant impact on marine geological structures, as they can form special structures and sedimentary facies within basins and may also play a crucial role in burial history[3]. By the analysis of sequence stratigraphy, we can infer the sea level changes and sedimentary system evolution during the marine Mesozoic-Paleozoic period.

When studying marine geological structures, the characteristics of sedimentary systems and variations in sedimentary facies are crucial for understanding basin evolution and geological processes. By the logging facies analysis[4], we can obtain information about lithology, grain size, and hydrocarbon properties within boreholes, thereby revealing the nature and distribution of stratigraphic layers. The buried history underground also serves as a critical clue in exploring marine geological structures. Analyzing burial history allows us to understand the compaction and deformation history of the strata, thereby inferring the process of geological structural evolution. Research in marine geological structures encompasses multiple aspects, ranging from seismic sedimentology that unveils paleoenvironments to geological structures' impacts on sedimentary systems and facies. By investigating elements like prototype basin evolution, continental margin characteristics, salt structures, and burial history, combined with seismic records and logging facies analysis, we can gain a deeper understanding of the formation and evolution of marine geological structures.

In recent years, significant progress has been made in the study of marine geological structures. Hydrocarbon-rich depressions have attracted widespread attention as potential oil and gas accumulation areas. By analyzing the kerogen content and organic facies characteristics within hydrocarbon-rich depressions[5], the maturity and type of organic matter can be understood, thereby assessing the potential oil and gas resources. However, biodegradation phenomena could potentially impact the oil and gas within reservoirs, which need to be taken into account in reservoir characterization and prediction.

The study of biomarkers holds significant importance in distinguishing different sources of oil and gas. Assessment of paleoproductivity provides clues for studying sedimentary environments and biodegradation. Reservoir characterization and prediction[6], achieved through petrophysical analysis and fluid inclusion research, reveal the properties and fluid characteristics of reservoirs. In the realm of deepwater petroleum, the mechanisms of hydrocarbon accumulation and reservoir evolution in deep-sea environments pose unique challenges. Exploration and development of deep petroleum and tight oil and gas have also become focal points, involving complex issues of hydrocarbon migration, subtle traps, and intricate transport systems.

Understanding the source-to-sink system is crucial in studying hydrocarbon accumulation[7]. The interactions between different source rocks and reservoirs influence the formation and migration of oil and gas. Oil and gas transfer stations play a significant role in the process of hydrocarbon migration, facilitating their movement between various reservoirs and creating subtle traps. Additionally, hydrocarbon detection techniques have played a vital role in identifying potential oil and gas resources. Analyzing subsurface hydrocarbon content through these techniques can indicate the presence of oil and gas.

From hydrocarbon-rich depressions to deepwater petroleum, from biodegradation to the source-sink system, a comprehensive overview of the marine geology and tectonics field has been depicted. Encompassing a wide range of topics, this field provides valuable scientific foundations for petroleum exploration and development. Furthermore, energy safety has always been of paramount importance. With the continuous growth in global energy demand, deepwater petroleum has become a critical sector for exploration and development. Modern geological exploration techniques such as marine earthquake exploration, marine gravity surveys, and marine magnetic surveys are widely employed to meet this demand. These technologies not only aid in identifying potential oil and gas reservoir locations but also assist in understanding hydrocarbon accumulation models.

However, once potential oil and gas reservoir sites are identified, drilling and completion operations are essential[8]. This constitutes a crucial phase in oil and gas exploration and production. Throughout this process, cementing technology plays a pivotal role in ensuring the stability and sealing of wellbores[9], preventing oil and gas leakage as well as groundwater contamination.

To maximize production, stimulation techniques like hydraulic fracturing are employed to enhance reservoir permeability. Simultaneously, reservoir reconstruction technology can be utilized to optimize reservoir production performance. In this entire process[10], a profound understanding of hydrocarbon migration is crucial, as it determines how oil and gas are transported from their accumulation sites to wellheads. Moreover, in-depth research into the hydrocarbon accumulation process can reveal oil and gas resources' origins, enrichment mechanisms, and reservoir characteristics.

Vocabulary

basin evolution:(n) 盆地演化
prototype:(n) 原型,雏形
seismic sedimentology:(n) 地震沉积学
paleoenvironment:(n) 古环境
  [paleo-]:古老的
buried hill:(n) 潜山
continental margin:(n) 大陆边缘
sedimentary system:(n) 沉积体系
burial history:(n) 埋藏史
sequence stratigraphy:(n) 层序地层学
Mesozoic-Paleozoic:(n/adj) 中生界-古生界(的)
hydrocarbon:(n) 碳氢化合物(烃类)
depression:(n) 坳陷
kerogen:(n) 干酪根
organic facies:(n) 有机相
biodegradation:(n) 生物降解
  [bio-]:生物的
biomarker:(n) 生物标志物
paleoproductivity:(n) 古生产力
petrophysical:(adj) 岩石物理的
  [-physical]:物理学的
fluid inclusion:(n) 流体包裹体
tight oil and gas:(n) 致密油气
subtle trap:(n) 隐蔽圈闭
deepwater petroleum:(n) 深水石油

Notes

[1] 在海洋地质领域,研究原型盆地的演化是理解海洋地质构造的关键。此句介绍了原型盆地的特征与性质。其中,原型盆地(prototype basin)是指在一定的地质历史时期内形成、后未经改造或改造甚微、基本保持了原来盆地的性质和分布范围的原始沉积盆地,与之相对应的是残留盆地(remnant basin)。原型盆地指示了相对单一的地球动力学系统或单旋回构造演化阶段所形成的具有特定沉积实体的盆地。

[2] reflection layers in seismic records 意指地震记录中的反射层,地震勘探使用声波或地震波来探测地下结构,当这些波在不同地层之间的界面上发生反射时,就形成了地震记录中的反射层。这些反射层的位置、强度和形态可以提供关于地下地层的信息,例如地层的厚度、分布和性质。paleo-意指一系列的代表"古"的单词前缀,例如古环境(paleoenvironment)、古生物学(paleontology)、古生代(paleozoic)、古气候学(paleoclimatology)。

[3] 盐构造(salt structures)是指由地下盐体的流动和塑性变形所形成的地质构造。这种类型的构造通常出现在盐岩等可塑性岩石中,其中盐岩的流动性质允许它在地壳的变形过程中形成各种复杂的形状。盐结构对于油气成藏和地质构造的研究非常

重要。在一些情况下,盐体的流动和变形可能会形成储藏油气的隐蔽圈闭,因此,在石油勘探中,盐构造被认为是有潜力的勘探目标。同时,盐构造的复杂性也可能影响地下地质构造的发展和演化,因此在地质研究中具有重要地位。沉积相(sedimentary facies)是指在地层中具有相似沉积特征和性质的一组岩石或地层单元。这些特征和性质可以包括岩石类型、颗粒大小、颜色、结构、化石含量等。

[4] 测井相(logging facies)是一种地球物理学技术,用于通过测量地下岩石的物理性质来识别和解释地层的性质和组成。测井相常常与地层的沉积相和岩石相相关联,但它是通过测量地下岩石的物理性质,如电阻率、自然伽马辐射、声波速度等,来进行分析。测井相可以提供有关地层组成、岩石类型、孔隙度、饱和度等信息。

[5] 干酪根(kerogen)是存在于沉积岩中的有机质的一种形式,通常是由植物和微生物的残骸经过生物降解和化学变化后形成的。它是一种含碳丰富的有机物质,是生烃(油和气)的潜在源头。干酪根存在于各种不同类型的沉积岩中,如页岩、泥岩和一些煤炭。它的形成过程通常发生在富含有机物的湿地、河流、湖泊等地方,当植物和微生物残骸逐渐埋藏在沉积物中,经过高压、高温和时间的作用,有机质会发生一系列复杂的化学反应,从而转化为干酪根。

[6] 储层表征(reservoir characterization)意指对岩石储层的性质、组成、孔隙结构、渗透性、饱和度等进行详细描述和分析的过程。在石油工业和地质学领域,储层表征旨在深入了解储层岩石的特性,以确定储层的石油和天然气储存能力、渗透性、可开发性以及生产潜力。储层表征通常涉及以下方面:岩石性质和组成、孔隙结构和孔隙度、渗透性、饱和度、岩石物理特性、岩石结构和构造。

[7] 源-汇系统(source-sink / source-to-sink system)是指剥蚀地貌形成的物源通过搬运路径到汇水盆地分散沉积下来的动力学系统,在地球科学领域具有重要研究意义。源-汇系统包含物源、搬运路径和沉积体系三个重要组成要素,必须把三个要素作为一个系统过程来研究,才能完整地认识地球表层的动力学过程及其演化。

[8] 钻完井(drilling and completion)是油气资源勘探与开采中的关键过程。它通常包括以下几个步骤:钻井、测井、套管、完井。"钻完井"包括创建功能性井筒的程序,该井筒可以从地下储层中提取油气。它涵盖了钻井井筒、评估地质层、套管、准备产出和最终提取用于商业用途的油气的过程。

[9] 固井技术(cementing technology)指的是一种用于油井和天然气井的技术,主要用于确保井筒的稳固性和封闭性。它的基本原理是在钻井井筒周围注入水泥浆,将井筒牢固地固定在地下岩层中,并密封井筒以防止油气泄漏到地表或污染地下水。固井是一项关键的技术,确保了油井和天然气井的安全性和有效性。它通过使用水泥浆在井筒周围创建坚固的屏障,以保护地下资源,防止泄漏和维护井筒的结构完整性。

[10] 储层改造技术(reservoir reconstruction technology)是一种用于油气勘探和生产的先进技术,旨在优化和改善油气储层的性能。这项技术的主要目标是提高油气的

采收率和产量,同时减少生产中的浪费,以更有效地利用地下储层中的油气资源。储层改造技术是为了更有效地开采地下油气资源而采取的一系列工程和技术措施。通过优化储层性能,可以提高采收率、产量和经济效益,从而更有效地满足能源需求。

12. Paleoceanography

Text

Introduction: The Relevance of Paleoceanography

Paleoceanography encompasses (as its name implies) the study of "old oceans", that is, the oceans as they were in the past. In this context, "the past" ranges from a few decades through centennia and millennia ago to the very deep past, millions to billions of years ago[1]. In reconstructing oceans of the past, paleoceanography needs to be highly interdisciplinary, encompassing aspects of all topics in this encyclopedia, from plate tectonics (positions of continents and oceanic gateways, determining surface and deep currents, thus influencing heat transport) through biology and ecology (knowledge of present-day organisms needed in order to understand ecosystems of the past), to geochemistry (using various properties of sediment, including fossil remains, in order to reconstruct properties of the ocean waters in which they were formed)[2].

Because paleoceanography uses properties of components of oceanic sediments (physical, chemical, and biological) in order to reconstruct various aspects of the environments in these "old oceans", it is limited in its scope and time resolution by the sedimentary record. Ephemeral ocean properties cannot be measured directly, but must be derived from proxies[3]. For instance, several types of proxy data make it possible to reconstruct such ephemeral properties as temperature (Sea Surface Temperatures, SST) and nutrient content of deep and surface waters at various locations in the world's oceans, and thus obtain insights into past thermohaline circulation patterns, as well as patterns of oceanic primary productivity. The information on ocean circulation can be combined with information on planktic and benthic microfossils, allowing a view of interactions between fluctuations in oceanic environments and oceanic biota, on short but also on evolutionary timescales.

Paleoceanography offers information that is available from no other field of study: a view of a world alternative to, and different from, our present world, including colder worlds (ice ages) as well as warmer worlds[4]. Paleoceanographic data thus serve climate modelers in providing data on boundary conditions of the ocean-atmosphere system very different from those in the present world. In addition, paleoceanography provides information not just on climate, but also on climate changes of the past, on their rates and directions, and possible linkages (or lack thereof) to such factors as atmospheric pCO_2 levels and the location of oceanic gateways and current patterns. Paleoceanography therefore enables us to gauge the limits of uniformitarianism[5]: in which aspects is the present world indeed a guide to the past, in which aspects is the ocean-atmosphere system of the present world with its present biota just a snapshot, proving information only on one possible, but certainly not the only, stable mode of the Earth system? How stable are such features as polar ice caps, on timescales varying from decades to millions of years? How different were oceanic biota in a world where deep-ocean temperatures were 10−12 C rather than the present (almost ubiquitous) temperatures close to freezing? Such information is relevant to understanding the climate variability of the Earth on different timescales, and modeling possible future climate change ("global warming"), as recognized by the incorporation of a paleoclimate chapter in the *Fourth Assessment Report of the Intergovernmental Panel on Climate Change* (2007).

Paleoceanography thus enables us to use the past in order to gain information on possible future climatic and biotic developments: the past is the key to the future, just as much and maybe more than the present is the key to the past[6].

Paleoceanographic Techniques

Over the last 30 years there has been an explosive development of techniques for obtaining information from oceanic sediments (Figure Ⅲ-9). One of the limitations of paleoceanographic research on samples from ocean cores is the limited size of each sample. However, a positive result of this limited availability is that researchers from different disciplines are forced to work closely together, leading to the generation of independent proxy records on the same sample set, thus integrating various aspects of chemical and biotic change over time. Proxies are commonly measured on carbonate shells of pelagic and benthic microorganisms, thus providing records from benthic and several planktonic environments (surface, deep thermocline)[7].

Methods used since the first paleoceanographic core studies include micropaleontology, with the most commonly studied fossil groups including pelagic calcareous (Planktonic Foraminifera) and siliceous-walled (Radiolarians) heterotroph protists, organic-walled cysts of heterotroph and autotroph dinoflagellates, siliceous-walled (Diatoms) and calcareous-walled autotroph protists (Coccolithophores), benthic protists (Benthic Foraminifera), and microscopic metazoa, Ostracods (Crustacea)[8]. Micropaleontology is useful in biostratigraphic correlation as well as in its own right, providing information on evolutionary processes and their linkage (or lack thereof) to climate change, as well as on changes in oceanic productivity[9].

Classic stable isotopes used widely and commonly in paleoceanographic studies include those of oxygen, e.g., in Cenozoic climate and carbon cycle models[10]. Carbon isotope records are of prime interest in investigations of deep oceanic circulation and of oceanic productivity. In addition to these classic paleoceanographic proxies, many new methods of investigation have been and are being developed using different geochemical (isotope, trace element, organic geochemical) proxies. Many more proxies are in development, including proxies on different organic compounds, to investigate aspects of global biogeochemical cycles, biotic evolution and productivity, and thermohaline circulation patterns (Figure Ⅲ-9).

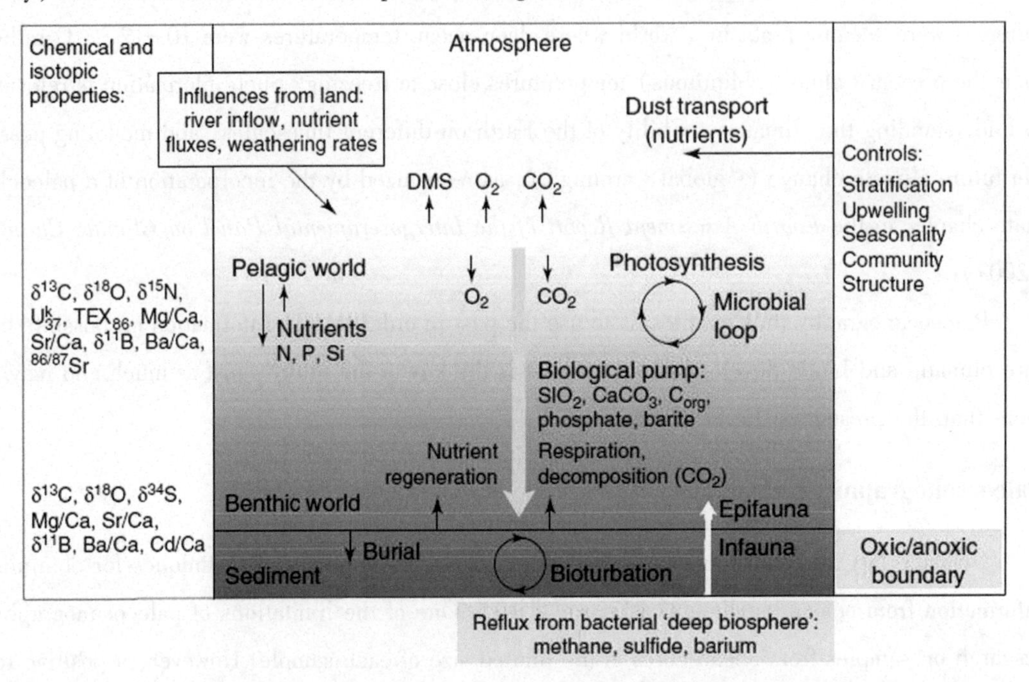

Figure Ⅲ-9: Linkages between the marine biosphere and global biogeochemical cycles, as well as to various proxies used in paleoceanographic studies. The proxies include isotope measurements (indicated by lower case deltas, followed by the elements and the heavier isotope), elemental ratios, and organic geochemical temperature proxies (e.g., U^K_{37}, an alkenone ratio in which the 37 refers to the carbon number of the alkenones; TEX_{86}, a proxy based on the number of cyclopentane rings in sedimentary membrane lipids derived from marine crenarchaeota[11]). DMS is dimethylsulfide, a sulfur compound produced by oceanic phytoplankton. The various proxies can be used to trace changes in ocean chemistry (including alkalinity), temperature, and productivity, as well as changes in the reservoirs of the carbon cycle. [Source: Steele, Thorpe, Turekian, 2011]

Techniques to correlate the age of features in sediment records recovered at different locations and to assign numerical ages to sediment samples are of the utmost importance to be able to estimate rates of deposition of sediments and their components. Correlation between sediment sections is commonly achieved by biostratigraphic techniques, which cannot directly provide numerical age estimates, and are commonly limited to a resolution of hundreds of thousands of years. Techniques used in numerical dating include the use of various radionuclides, the correlation of sediment records to the geomagnetic polarity timescale, and the more recently developed techniques of linking high-resolution records of variability in sediment character (e.g., color, magnetic susceptibility, density, and sediment composition) to variability in climate caused by changes in the Earth's orbit and thus energy supplied by the sun to the Earth's surface at specific latitudes[12]. Such an orbitally tuned timescale has been fully developed for the last 23 My of Earth history, with work in progress for the period of 65–23 Ma. Remote sensing techniques are being used increasingly in order to characterize sediment in situ in drill holes, even if these sediments have not been recovered, and to establish an orbital chronology even under conditions of poor sediment recovery.

Contributions of Paleoceanographic Studies

Paleoceanographic studies have contributed to a very large extent to our present understanding that the Earth's past environments were vastly different from today's, and that changes have occurred on many different timescales. Paleoceanographic studies have been instrumental in establishing the fact that climate change occurred rapidly and stepwise rather than gradually, whether in the establishment of the Antarctic ice cap on timescales of tens to hundreds of thousands of years rather than millions of years, or in the ending of a Pleistocene ice age on a timescale of decades rather than tens of thousands of years[13].

Unexpected paleoceanographic discoveries over the last few years include the presence of large amounts of methane hydrates, in which methane is trapped in ice in sediments along the continental margins, in quantities potentially larger than the total global amount of other fossil fuels. The methane in gas hydrates may become a source of energy, with exploratory drilling advanced furthest in the waters off Japan and India. Such exploration and use of gas hydrates might, however, lead to rapid global warming if drilling and use of methane hydrates inadvertently lead to uncontrolled destabilization. Destabilization of methane hydrates may have occurred in the past as a result of changes in sea level and/or changes in thermohaline circulation and subsequent changes in deep-ocean temperature[14]. Dissociation of gas hydrates and subsequent oxidation of methane in the atmosphere or oceans have been speculated to have played a role in the ending of ice ages, and in a major upheaval in the global carbon cycle and global warming. The influence on global climate and the global carbon cycle of methane hydrate reservoirs (with their inherent capacity to dissociate on timescales of a few thousand years at most) is as yet not well understood or documented.

Most methane hydrates are formed by bacterial action upon organic matter, and another unexpected discovery was that of the huge and previously unknown microbial biomass in seawater, also found buried deep in the sediments[15]. Fundamental issues such as the conditions that support and limit this biomass are still not understood and neither are their linkage to the remainder of the oceanic biosphere and the role of chemosynthesis and chemosymbiosis in the deep oceanic food supply and the global carbon cycle.

In contrast to these unexpected discoveries, paleoceanographers of a few decades ago could have predicted at least in part our increased knowledge of aspects of climate change such as the patterns of change in sea level at various timescales. The suddenness and common occurrence of rapid climate change events in Earth history, however, was unpredicted. On timescales of millions of years, the Earth's climate was warm globally during most of the Cretaceous and the early part of the Cenozoic (65-35 Ma), and the Earth had no large polar ice caps reaching sea level. Drilling in the Arctic Ocean, for instance, established that average summer surface water temperatures might have reached up to 18℃. The use of climate models has assisted in understanding such a warm world, but the models still cannot fully reproduce the necessary efficient heat transport to high latitudes at extremely low latitudinal temperature gradients.

Paleoceanographic research has provided considerable information on the major biogeochemical cycles over time. Geochemical models of the carbon cycle rely on carbon isotope data on bulk carbonates and on planktonic and benthic foraminifera in order to evaluate transfer of carbon from one reservoir (e.g., organic matter, including fossil fuel; methane hydrates) to another (e.g., limestone, the atmosphere, and dissolved carbon in the oceans)[16]. Information on pelagic carbonates and their microfossil content as well as stable isotope composition has assisted in delineating the rapidity and extent of the extinction at the end of the Cretaceous in the marine realm. Sedimentological data at many locations have documented the large-scale failure of the western margin of the Atlantic Ocean, with large slumps covering up to half of the basin floor in the North Atlantic as the result of the asteroid impact on the Yucatan Peninsula.

One of the major successes of paleoceanographic research has been the establishment of the nature and timing of polar glaciation during the Cenozoic cooling[17]. After a prolonged period of polar cooling in the middle to late Eocene, the East Antarctic Ice Sheet became established during a period of rapid ice volume growth (<100 ky) in the earliest Oligocene, 33.5 Ma. This establishment was followed by times of expansion and contraction of the ice sheet, and the West Antarctic Ice Sheet may have started to grow at 14 Ma. Until recently it was argued that the Northern Hemispheric Ice Sheets formed much later: these ice sheets increased in size around 3 Ma (in the Pliocene), and ever since have contracted and expanded on orbital timescales. There is now considerable evidence that the polar ice sheets in the Northern Hemisphere also became established, at least in part, in the earliest Oligocene, that is, at a similar time as the Southern Hemisphere ice sheets.

The cause(s) of the long-term Cenozoic cooling are not fully known. Possible long-term drivers of climate include the opening and closing of oceanic gateways, which direct oceanic heat transport[18]. Such changes in gateway configuration include the opening of the Tasman Gateway and Drake Passage which made the Antarctic Circumpolar Current possible, and closing of the Isthmus of Panama which ended the flow of equatorial currents from the Atlantic into the Pacific. Evidence has accumulated, however, that changes in atmospheric CO_2 levels may have been more important than changes in gateway configuration[19]. For instance, long-term episodes of global warmth (in timescales of millions of years) may have been sustained by CO_2 emissions from large igneous provinces, and short-term global warming (on timescales of ten to one or two hundred thousands of years) could have been triggered by the release of greenhouse gases from methane hydrate dissociation or burning/oxidation or organic material (including peat). Decreasing atmospheric CO_2 levels due to decreasing volcanic activity and/or increased weathering intensity have been implicated in the long-term Cenozoic cooling, and high-resolution paleoceanographic records show that climate change driven by changes in atmospheric CO_2 levels may have been modulated by changes in insolation caused by changes in orbital configuration. Recognition of variability in climatic signals in sediments at orbital frequencies has led to major progress in the establishment of orbitally tuned timescales throughout the Cenozoic.

Paleoceanographic research has led to greatly increased understanding of the Plio-Pleistocene Ice Ages as being driven by changes in the Earth's orbital parameters, and the correlation of data from oceanic sediments to records from ice cores on land[20]. Orbital forcing is the "pacemaker of the ice ages", but it is not yet fully understood how feedback processes magnify the effects of small changes in insolation into the major climate swings of the Plio-Pleistocene. Ice-core data and carbon isotope data for marine sediments show that changes in atmospheric CO_2 levels play a role. We also do not yet understand why the amplitude of these orbitally driven climate swings increased at c. 0.9 Ma, and why the dominant periodicity of glaciation switched from 40 000 (obliquity) to 100 000 (eccentricity) years at that time, the "mid-Pleistocene revolution"[21]. Effects of glaciation at low latitudes, including changes in upwelling, productivity, and monsoonal activity, are only beginning to be documented.

Great interest has been generated by the information on climate change at shorter timescales than 20 ky, the duration of the shortest orbital cycle, precession. Such climate variability includes the millennial-scale changes that occurred during the glacial periods and the climate variations of lesser amplitudes that have occurred since the last deglaciation[22]. These research efforts are beginning to provide information on timescales that are close to the human timescale, on such topics as abrupt climate change, and the fluctuations in intensity and occurrence of the El Nino-Southern Oscillation and the North Atlantic Oscillation (NAO), during overall colder and warmer periods of the Earth history.

The Future of Paleoceanography

Past progress in paleoceanography has been linked to advances in technology since the invention of the piston corer in the 1940s[23]. In the early to mid-1980s, paleoceanographic studies appeared to reach a plateau, with several review volumes published on, for example, Plio-Pleistocene ice ages, the oceanic lithosphere, and the global carbon cycle, as well as a textbook Marine Geology by J. P. Kennett. Paleoceanographic research, however, has benefited from research programs in the present oceans (such as the Joint Global Ocean Flux Program, JGOFS), and new technology has spurred major research activity. The extensive use and improvement of the hydraulic piston corer by DSDP/ODP/IODP led to recovery of minimally disturbed soft sediment going back in age through the Cenozoic, making high-resolution studies possible, including paleoceanographic and stratigraphic studies of sediment composition using the X-ray fluorescence (XRF) scanner. Progress in computers led to strongly increased use of paleoceanographic data in climate modeling, and to increased possibilities of remote sensing in drill holes. Developments in mass spectrometry led to the possibility to measure isotopes and trace elements in very small samples, such as those recovered in deep-sea cores, and new proxies continue to be developed, including proxies (e. g., on levels of oxygenation and temperature) using organic geochemical methods.

The IODP started in 2003 and is scheduled to start drilling in 2008 with three different platforms. Drilling activity has become integrated with that of the French research vessel Marion Dufresne, which has recovered many long piston cores (several tens of meters) for studies covering the last few hundred thousand years of the Earth history, in the International Marine Global Change Study (IMAGES) program. Examples of drilling by alternative platforms include the drilling in the Arctic Ocean, one of the frontiers in ocean science, and drilling in shallower regions than accessible to the drilling vessel Joides Resolution, such as coral drilling in Tahiti for studies of Holocene climate and rates of sea level rise. If these ambitious programs are carried out, we can expect to learn much about the working of the Earth system of lithosphere-ocean-atmosphere-biosphere, specifically about the sensitivity of the climate system, about the controls on the long-term evolution of this sensitivity, and about the complex interaction of the biospheric, lithospheric, oceanic, and atmospheric components of the Earth system at various timescales.

Vocabulary

paleoceanography: (n) 古海洋学

[paleo-]: 古代的

centennium: (n) 一百年

pl: centennia 几百年

adj: centennial 百年尺度的

millennium: (n) 一千年

pl: millennia 几千年

adj: millennial 千年尺度的

continent：(n) 洲,大陆
gateway：(n) 通道
　海洋中通常指海峡通道
geochemistry：(n) 地球化学
proxy：(n) 代理,指标
sea surface temperature：(n) 海表温度,通常缩写为 SST
thermohaline circulation：(n) 热盐环流
　[thermo-]：热的
primary productivity：(n) 初级生产力
planktic：(adj) 浮游生物的
benthic：(adj) 底栖生物的
microfossil：(n) 微体化石
　[micro-]：微小的
biota：(n) 生物区(系)
　[bio-]：生物的
evolutionary：(adj) 进化的,演变的
timescale：(n) 时间尺度
　[-scale]：标尺
ice age：(n) 冰河时期
uniformitarianism：(n) 均变论
ice cap：(n) 冰帽
paleoclimate：(n) 古气候,地质气候,古气候学
　adj：paleoclimatic
carbonate：(n) 碳酸盐
　[-ate]：(n) 盐类
microorganism：(n) 微生物
micropaleontology：(n) 微体古生物
calcareous：(adj) 钙质的,石灰质的
planktonic：(adj) 浮游生物的
siliceous：(adj) 硅质的
radiolarian：(n) 放射虫;(adj) 放射虫类的
heterotroph：(n) 异养生物
protist：(n) 原生生物
cyst：(n) 孢囊
autotroph：(n) 自养生物
dinoflagellate：(n) 腰鞭毛虫
diatom：(n) 硅藻
coccolithophore：(n) 颗石藻
metazoan：(n) 后生动物
ostracod：(n) 介形虫,介形亚纲动物
crustacea：(n) 甲壳纲动物;(adj) 甲壳纲的
biostratigraphic：(adj) 生物地层的
stable isotope：稳定同位素
Cenozoic：(n) 新生代;(adj) 新生界的
carbon isotope：碳同位素
organic geochemical：有机地球化学
biogeochemical：(adj) 生物地球化学的
radionuclide：(n) 放射性核素
geomagnetic：(adj) 地磁的
magnetic susceptibility：磁化率
My：一百万年
Ma：距今百万年前
　是 megaannum 的缩写
remote sensing：遥感
drill hole：钻孔,海洋里钻取岩芯的位置
orbital chronology：轨道年代
Pleistocene：(n/adj) 更新世(的)
methane hydrate：甲烷水合物
sea level：海平面
oxidation：(n) 氧化作用
bacterial action：细菌作用
microbial biomass：微生物生物量
chemosynthesis：(n) 化能合成作用
　[chemo-]：化学的
chemosymbiosis：(n) 化学共生
rapid climate change：快速气候变化
limestone：(n) 石灰岩

dissolved carbon：溶解碳
asteroid：（n）小行星；（adj）星状的
glaciation：（n）冰川作用
Eocene：（n/adj）始新世（的）
ice sheet：冰盖
ice volume：冰量
通常指冰川、冰盖、冰架等冰体的总体积
Pliocene：（n/adj）上新世（的）
Oligocene：（n/adj）渐新世（的）
Antarctic Circumpolar Current：绕南极流（自西向东横贯太平洋、大西洋和印度洋的全球性环流）
large igneous province：大火成岩省
peat：（n）泥煤，泥炭块，泥炭色
volcanic：（n）火山岩；（adj）火山的，火山引发的
insolation：（n）日晒，日射，日射量
~solar radiation：太阳辐射
orbital frequency：（n）轨道频率
古气候研究中通常指的是地球运行的轨道参数
Plio-Pleistocene Ice Age：上新世—更新世以来的冰期
ice core：冰芯，从冰川、冰盖等中钻取的柱状冰体
obliquity：（n）斜率
地球轨道斜率是地轴的倾斜角度
eccentricity：（n）偏心率
地球轨道偏心率是地球绕太阳运转的椭圆形轨道的离心率
mid-Pleistocene revolution：中更新世转型
orbital cycle：轨道周期
precession：（n）岁差，进动
millennial-scale：（n）千年尺度
deglaciation：（n）冰消期
piston corer：活塞取样方法
X-ray：（n）X 射线
fluorescence：（n）荧光，荧光性

Notes

[1] 此句主要解释了古海洋学研究对象涉及的地质时间尺度。它既包括了几十年、千百年（centennia and millennia）这些相对较短的时间段，也涉及百万年甚至几十亿年的长时间尺度。

[2] 古海洋学是一门交叉学科，研究内容包括大陆和海峡通道的位置变化，过去地球生态系统的演化，以及运用地球化学（geochemistry）方法研究沉积物特性并追溯海洋水文特征的演变历史。

[3] 由于古海洋学研究过去海洋环境，而这些海洋环境特征通常都处于不断变化的状态，不能保留至今，也就不能被直接观测，所以研究古海洋需要用到多种替代性指标来恢复过去的海洋环境。proxy 这里指古海洋研究中使用到的替代性指标，如下文提到需要结合多种替代性指标才能重建过去海表温度、大洋环流、海洋初级生产力等海洋特征。

[4] ice age 指地质历史时期中的大规模冰川、冰盖覆盖的时间段。与此相对应的是温暖

的时期。部分寒冷或者温暖的时期与现在的气候大不相同,气温变化幅度远超现代的气温范围。

[5] uniformitarianism 指均变论,意指当前发生的地质变化在地质历史时期中以类似的频率、强度和方法发生过。这句话主要解释了古海洋学研究中不同时间尺度的地质事件演变过程对于检测均变论的局限性具有重要意义。

[6] 这句话主要说明的是古海洋学通过研究过去气候、生态系统等的变迁,为地球未来的气候和生态系统发展提供信息。如同将今论古,古海洋研究为以古论未来提供了可能性。

[7] carbonate shell 是碳酸盐成分为主的生物壳体。大洋中的浮游和底栖碳酸盐骨骼的生物体在其生长过程中记录了丰富的海水温度、盐度等信息,因此常被用来重建过去洋流的温盐特征和洋流信息。

[8] micropaleontology 指微体古生物学。古海洋学研究最开始从微体古生物学角度出发,主要研究钙质的(calcareous)、硅质的(siliceous)等生物体的种群组合特征。

[9] biostratigraphic correlation 意指生物地层对比。生物的演化具有全球可对比性,生物地层对比是粗略建立地层年代的重要方法。地层年代的确立为生物演化进程研究及其与气候变化的联系提供了有价值的信息。

[10] 古海洋研究中常用到地球化学指标,如稳定同位素(stable isotope)。其中氧同位素和碳同位素是最常用到的古海洋环境特征的替代性指标。

[11] U_{37}^{K}和 TEX_{86}都是古海水温度指标,与氧同位素不同的是,它们是有机质中提取的古海水温度信息。

[12] 地层年代的定年方法除了生物地层年代方法外,放射性同位素(radionuclide)、地磁极性(geomagnetic polarity)转化等方法可以给出相对准确的数值定年。另外,还有如沉积物颜色反射率和磁化率变化的高分辨率研究也可为定年提供更多的信息。这些年代信息和地球轨道参数变化引起的气候变化相联系,促进了有关太阳日射量对不同纬度气候变化影响机制的研究。

[13] 古海洋研究的重要发现之一是气候变化不是渐变的,而是以快速变化或者阶段性变化为主。如南极冰帽的形成主要发生在几千年以内而不是在百万年的时间里逐渐形成,而更新世冰期也是在几十年内而不是几千年内快速地结束。

[14] 天然气水合物具有不稳定性,海平面(sea level)或者大洋环流的变化及其引发的深海温度变化均可以导致天然气水合物释放(dissociation)。下文也指出天然气水合物的释放及其氧化在全球气候变暖和碳循环中起着重要作用。

[15] 天然气水合物主要来自有机质与细菌作用的产物,这也是此前未知的存在于海水或者埋藏在海底的巨大海洋微生物量。

[16] 碳循环是生物地球化学循环中的重要环节。不同碳储库间(reservoir)的碳转移估算通常依靠计算碳酸盐中的碳同位素值变化幅度。

[17] Cenozoic cooling 指地球气候变化史中新生代以来气候总体呈现变冷的趋势。新生代以来,大约在 33 百万年前,南极大陆冰盖开始增长,而后大概在 3 百万年前,北极的冰盖逐渐形成。

[18] opening and closing of oceanic gateways 指连接大洋的通道打开或者关闭。海道的变化引起大洋环流和热量传输(oceanic heat transport)的改变,从而触发气候变化。下文提到的德雷克海峡的打开,引起的环南极流的形成,是新生代气候变冷机理的重要假说之一。

[19] 新生代变冷的原因除了海道变化外,另一种可能的影响因素是大气 CO_2 含量的变化。如下文提到的大火成岩省中释放的大量 CO_2 气体可能是百万年时间尺度上气候温暖的主要原因,而当火山活动减弱,风化作用增强时,大气 CO_2 含量减少也会导致新生代逐渐变冷。

[20] Plio-Pleistocene Ice Age 指上新世—更新世期间周期性出现的冰期现象,又称冰期旋回。一般认为这种冰期循环时受地球运行轨道参数周期性改变的影响。

[21] mid-Pleistocene revolution 指中更新世中气候转型,气候变化周期由此前的 4 万年的斜率周期变成 10 万年的偏心率周期。目前学界关于中更新世气候转型机制没有统一的认识。

[22] 气候变化周期除了斜率和偏心率周期外,另一个重要的地球轨道参数岁差(precession)周期也对冰期旋回中的更短时间尺度的气候变化产生重要影响,如千年尺度的气候变化甚至在冰消期出现小幅度的气候变化。

[23] 古海洋研究进展与海洋沉积物取样方法密切相关。其中活塞取样器(piston corer)的出现让岩芯取样深度大幅增加,极大地促进了古海洋研究的时间范围。

13. Milankovitch Cycles, Paleoclimatic Change, and Hominin Evolution

Text

Changes in Earth's orbit have helped pace climatic change for millennia. Scientists are now trying to understand whether and how these changes remodeled the landscapes our ancient ancestors inhabited.

The idea that critical junctures in human evolution and behavioral development may have been shaped by environmental factors has been around since Darwin. Although various hypotheses and models have been proposed, refined, and/or abandoned for at least a century, the concept of environmental determinism and hominin evolution is still a hot topic today. While it is ultimately local-level environmental processes acting upon individual populations that is one of the driving forces of evolutionary change, such shifts are often framed within the context of much larger regional or global climatic trends.

Long-Term Records of Paleoclimate

Direct measurements of climate components such as temperature and precipitation only exist for the last century or two. To reconstruct climate over longer time-scales, scientists indirectly measure these components by analyzing various proxies, or indicators, that are sensitive to climatic or environmental parameters and preserved in the geological record. Proxy records from marine sediment and ice cores provide the basis for much of our understanding of past climate. These long-term and relatively continuous natural archives are often used as references for comparison with local terrestrial-based paleoenvironmental reconstructions. For example, the record of oxygen and hydrogen isotope ratios preserved in glacial ice, and oxygen isotope ratios in the shells of marine organisms such as foraminifera and radiolaria[1], provide a record of past sea levels, ice volume, seawater temperature and global atmospheric temperature (Figure Ⅲ-10). Air bubbles trapped in ice cores also provide a direct record of the past chemical composition of the atmosphere, particularly CO_2. Carbon isotope ratios of shells in marine cores are equally valuable for estimates of water circulation and atmospheric CO_2 concentrations. Eolian dust preserved in both marine sediment and ice cores has been correlated with climate and environmental conditions in the dust's source region, specifically as a proxy for aridity. Continuous ice cores from Greenland record back to over 100,000 years ago, while those from Antarctica extend back to ~800,000 years ago. Thus, these records are relevant to the later members of the genus

Homo, such as H. erectus, H. heidelbergensis, H. neanderthalensis, and H. sapiens. Documenting a much longer timescale, marine sediment cores have been collected across the globe, and composite records have been compiled that extend beyond the Cenozoic[2], thus covering the entire duration of the Primate fossil record.

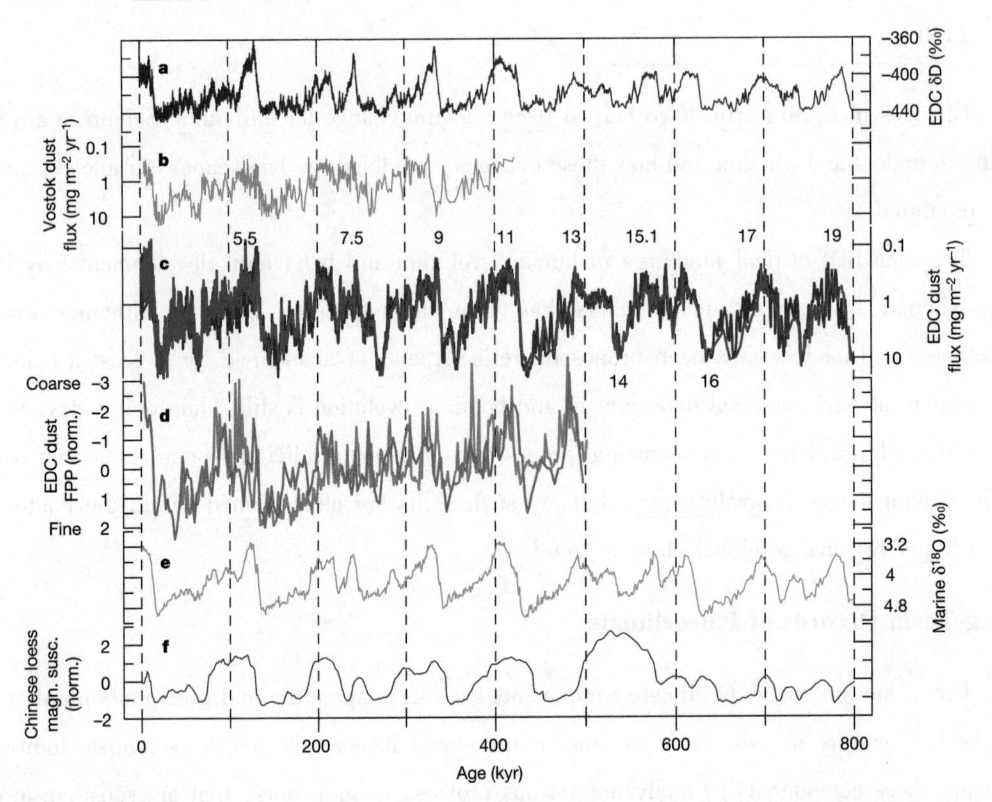

Figure Ⅲ－10: EPICA Dome C (EDC, Antarctica) data in comparison with other climatic indicators. (a) Stable isotope (δD) record from EDC. (b) Vostok dust flux record. (c) EDC dust flux records (numbers indicate Marine Isotope Stages). (d) EDC dust size data expressed as fine particle percentage. (e) Marine sediment δ18O stack (proxy for global ice volume). (f) Magnetic susceptibility stack record for Chinese loess. Peaks in most records depicted and odd MIS numbers indicate interglacial phases while troughs and even MIS numbers indicate glacial phases. [Source: Lambert et al., 2008]

There are a variety of other important high-resolution paleoclimate records relevant to hominin evolutionary history, but these are temporally or spatially restricted compared to marine cores. For example, the variation in thickness and grain size in Chinese loess deposits are related to extensive periods of cold, dry, winter Asian monsoon winds stretching back over the last 7 million years. Speleothems found in caves are also a rich archive of local paleoclimate information and, combined with uranium-thorium dating, can provide high-resolution records back to 500,000 years ago. Carbon and oxygen isotopic analysis as well as relative growth band thickness of speleothems have provided proxy data for local temperature, rainfall, aridity, and overlying vegetation (C3 vs. C4 plants)[3] at hominin sites in South Africa, Europe, the Levant, and Asia. Similar to the study of marine cores, an extensive arsenal of analytical methods have been applied to the study of lake cores, which serve as long, continuous archives of terrestrial climate change at annual to decadal scale for individual basins or watersheds. Existing lake cores in close proximity to paleoanthropological sites are typically restricted to the Holocene[4] but other cores in the Levant and Africa range from over 100 ka to 1 Ma. Additional scientific drilling initiatives are exploring thick lacustrine deposits directly associated with Plio-Pleistocene[5] paleoanthropological sites.

Astronomical Controls on Long-Term Climate Change

The pattern of incident solar radiation (insolation) received on the planet at a given place and time is an important factor in understanding both directional trends and variability observed in many paleoclimatic records, particularly those related to Quaternary[6] ice ages. Changes in insolation are, in turn, driven by Earth's natural orbital oscillations, termed Milankovitch[7] cycles. The three elements of Milankovitch cycles are eccentricity, obliquity, and precession. Eccentricity describes the degree of variation of the Earth's orbit around the Sun from circular to more elliptical. Eccentricity has two main periodicities, one cycle with an average of ~100,000 years and a longer cycle with a periodicity of ~413,000 years. Obliquity describes the tilt of the Earth's axis in relation to its orbital plane, which ranges from 22.1~24.5 degrees with a periodicity of ~41,000 years. Precession describes the motion of the Earth's axis of rotation, which does not point towards a fixed direction in the sky through time. Instead, the axis of rotation describes a clockwise circle in space, like the spinning of a wobbling top, with a periodicity of 19,000–23,000 years.

Solar radiation received at low-latitude is principally affected by variations in the cumulative effect of eccentricity and precession (eccentricity modulated precession), whereas higher latitudes are mainly affected by changes in obliquity. Since the Earth is tilted in its orbit, not all the Earth receives the same amount of energy, more energy being received at the equator than at the poles. Solar energy entering at a shallower angle at higher latitudes must travel further through the Earth's atmosphere compared to equatorial regions, reflecting some energy back to space. The same amount of solar energy also is spread over a larger area at higher latitudes. Increased tilt acts to amplify seasonal difference, while decreased tilt diminishes it. In its annual orbit, the Earth is currently closest to the sun (Perihelion) in early January, when the northern hemisphere is tilted away from the sun, and tilted towards the sun when the Earth is furthest from the sun (Aphelion) in early July. Thus, seasonality[8] is currently reduced in the northern hemisphere (but increased in the southern hemisphere) with the effect that northern hemisphere winters are not as cold as they could be, and summers are not as warm as they could be, a pattern that will be reversed in about 11,000 years. Although the interactions between orbital parameters are major external drivers of paleoclimatic changes, the internal dynamics of the climate system also exert important controls on temporal and spatial patterns of environmental change. Furthermore, both external and internal forcing[9] mechanisms can involve a complex series of feedbacks, and responses that may be linear or nonlinear, synchronous or delayed, or have a critical threshold ("tipping") point.

Paleoclimate and Hominin Evolution

One of the earliest examples that proposed a connection between climate-driven environmental change and hominin evolution was the "Savanna Hypothesis", which posited that the human lineage followed a simple trajectory from apelike to humanlike promoted by the challenges of an open savanna. While we now know that there is no single "magic bullet" that is responsible for the multitude of anatomical and behavioral changes documented in the hominin record, the concept that certain changes in the human lineage may have evolved in open habitat settings has persisted. With the establishment of the marine paleoclimatic framework, researchers began to evaluate hominin evolutionary processes and events in the context of global climatic oscillations, particularly the onset of Northern Hemispheric Glaciation (NHG) ~2.7 Ma. The "Turnover Pulse Hypothesis" championed by paleontologist Elisabeth Vrba proposed that a synchronous change in hominins, such as the origins of the genus *Homo*, and other African mammalian lineages, particularly speciation and extinction events in bovids[10], was caused by a shift from warm, moist conditions to cooler, drier, and more open habitats associated with a sharp transition in the marine oxygen isotope record associated with the onset of NHG. Other studies have since indicated that the record at specific East African hominin sites show either no faunal turnover at this time or that there were multiple pulses or prolonged periods of turnover set with a more gradual shift from forested to more open habitats.

A seminal study of terrigenous dust in marine cores off the coast of Africa by paleoceanographer Peter de Menocal suggested that subtropical African climate oscillated between markedly wetter and drier conditions, paced by Earth's orbital variations, with step-like increases in climate variability and aridity near 2.8, 1.7 and 1.0 Ma. These steps were coincident with changes in the dominant orbital cycles from precession to obliquity to eccentricity, and with the onset and intensification of high-latitude glacial cycles, respectively. Compared to the African fossil and geological record, these time periods also coincided with proposed diversification points in the hominin lineage (2.9 – 2.4 Ma), paleoenvironmental evidence for drier habitats and the expansion of *Homo* out of Africa (1.8–1.6 Ma), and the extinction of the *Paranthropus* lineage, the broadened range of *Homo erectus*, and the establishment of more modern savanna ecosystems (1.2–0.8 Ma). In addition to unidirectional shifts, de Menocal also highlighted the importance of "variability packets" of high- and low-amplitude paleoclimatic variability lasting 10,000 to 100,000 years in duration, paced by the orbital eccentricity modulation of precession. These alternating periods of relative paleoclimatic stability (low eccentricity) and instability (high eccentricity) as a mechanism for introducing genetic variance to natural selection are a key component of the "Variability Selection Hypothesis", which proposes that the wide variability in adaptive settings over time ultimately favored complex adaptations that were responsive to novel conditions (i.e., the evolution of adaptability).

Studies of East African lake records by geologist Martin Trauth and colleagues have also focused on critical intervals near 2.6, 1.8 and 1.0 Ma and documented the presence of large, but fluctuating lakes, indicating consistency in wetter and more seasonal conditions every 800,000 years. African monsoon intensity correlates with precession-paced insolation, and increased polar ice-volume acts to accentuate the pole-Equator thermal gradient, which leads to a north-south compression of the Intertropical Convergence Zone (ITCZ), the major control of monsoonal precipitation[11] patterns in Africa. Associated with major glacial events near 2.6, 1.8 and 1.0 Ma, Trauth and colleagues propose that global climate changes led to increased seasonality and regional climate sensitivity to insolation, which resulted in packages of precessionally forced alterations between episodes of large lakes and extreme aridity, possibly as rapid as every ~10,000 years during eccentricity maxima. They propose that these occurred during periods of eccentricity maxima every 800,000 years since 2.7 Ma (similar to de Menocal's variability packets). While some East African lake records provide strong evidence for this pattern, it may not be universal across space or time. Ultimately, this hypothesis proposes that periods of dramatic climatic oscillations between 2.7–2.5 Ma, 1.9–1.7 Ma, and 1.1–0.9 Ma led to rapid expansion then subsequent contraction/fragmentation of hominin habitats at precessional timescales with associated dispersal events and vicariance in the hominin lineage.

Discussion and Challenges

It seems intuitive that large-scale shifts and short-term variability in paleoclimate altered local to regional hominin habitats and resource availability that ultimately led to selection pressures on our fossil ancestors. However, climate systems are markedly complex and dynamic, and may change drastically over relatively short distances. It is important to maintain a critical perspective on the types, quality, and scale of empirical paleoenvironmental data, particularly when the volume and temporal resolution of proxy data far exceeds that of the hominin fossil record itself. For instance, error-bars on hominin first-appearance data (FADs) and last-appearance data

(LADs) that indicate the probability of true origination or extinction events are rarely reported. When accounting for influences such as sample size and geochronological uncertainties, the potential mismatch between a taxon's actual origination and its documented FAD in the fossil record (or extinction and LAD) is likely on the order of tens to hundreds of thousands of years. All hypotheses that propose causal links between paleoclimatic change and hominin evolution must ultimately reconcile global patterns with local responses, and extend far beyond a general temporal correlation between environmental change and an evolutionary event. Criteria for testing hypotheses of environmental forcing include a highly resolved time scale for the various records to validate cause-before-effect order, a robust correspondence between multiple lines of proxy evidence that shows similar patterns or trajectories, the ability to rule out alternative (non-environmental) hypotheses, and ultimately, a causal mechanism. Nevertheless, once the assumptions and limitations of utilizing global paleoclimatic data are appreciated, the almost dizzying array of natural archives of the past provide paleoanthropologists with a highly-resolved contextual framework within which they can develop research questions and test hypotheses.

Vocabulary

Milankovitch cycle:(n) 米兰科维奇旋回
Hominin:(n) 人及其直系原始祖先,人族
orbit:(n)(环绕地、日等运行的)轨道
  adj: orbital 轨道的
  ~orbital cycles:轨道周期
  pace:(vt) 为……设定节奏,调整……的节奏、步伐
millennia:(n) 千年
  adj: millennial 千年的
inhabit:(vt/vi) 居住于,栖居于,占据
juncture:(n) 接合点,连接
shape:(vt) 形成,影响,塑造
determinism:(n) 决定论,确定性
evolution:(n) 演化,进化(论),演变,发展
  v: evolve 逐步发展,演变
  adj: evolutionary 演变的,进化的,进化论的
frame:(n) 框架,边框;(v) 勾勒出
hydrogen:(n) 氢,氢气
  ~hydrogen isotope:氢同位素(如 δD)
eolian:(adj) 风的,风成的,风积的
  ~dust transport:风尘输送
Homo:(n) 人,人类,人属
H. erectus:直立人
H. heidelbergensis:海德堡人
H. neanderthalensis:尼安德特人
H. sapiens:智人
compile:(v) 汇编,收集
primate:(n) 灵长目动物(包括人,猴子等)
stack:(n) 堆,集群,堆栈
loess:(n) 黄土
monsoon:(n) 季风,季候风
  adj: monsoonal 季风的
  n: paleomonsoon 古季风
speleothem:(n) 洞穴堆积物,洞穴化学淀

积物,石笋,石钟乳
uranium-thorium dating:(n) 铀钍定年
Levant:(n) 黎凡特(西亚)
arsenal:(n) 可用于某目的的大量资源
watershed:(n) 流域,分水岭
proximity:(n) 靠近,近距,邻近
paleoanthropological:(adj) 古人类学的
ka: age in kiloanni:距今(公元 1950 年)千年
= kyr BP (age in kiloyears before 1950 CE)
= cal ka (BP)
initiative:(n) 措施,倡议
lacustrine:(adj) 湖的,湖成的
incident:(adj) 入射的
oscillation:(n) 振荡,振动,摆动
elliptical:(adj) 椭圆的
n: ellipse 椭圆(形)
periodicity:(n) 周期性,频率
rotation:(n) 旋转,转动,交替
wobble:(n/vt/vi) 摇摆,摇晃
principally:(adv) 主要地,大部分
cumulative:(adj) 积累的,渐增的
equator:(n) 赤道
adj: equatorial 赤道的,赤道上的
perihelion:(n) 近日点
aphelion:(n) 远日点
synchronous:(adj) 同步的,同时的
adj: synchronousd 在框架之内
threshold:(n) 门槛,阈,界
savanna hypothesis:(n) 稀树草原假说
lineage:(n) 血统,世系
trajectory:(n)(物体射向或抛向空中形成的)轨道,轨迹,(事业等的)发展轨迹,起落
apelike:(n) 类人猿,无尾猿
anatomical:(adj) 解剖的,结构上的
mammalian:(n) 哺乳类(的)
speciation:(n) 物种形成
n: species 物种,种
extinction:(n) 灭绝,消亡
adj: extinct 灭绝的,消亡的
seminal:(adj) 重大的,对未来有影响的,生殖的
glacial cycles:(n) 冰期旋回
n: glacier 冰川
n: glaciation 冰川作用
n: deglacial 冰消期
adj: glacial 冰的,冰期的
adj: interglacial 间冰期的
~ice-age transition 冰期转型
diversification:(n) 多元化,多样化
Paranthropus:(n) 傍人(属)
~*P. boisei*:鲍氏傍人
unidirectional:(adj) 单向的,单向性的,单方面的
alternating:(adj) 交替的,交互的
vicariance:(n) 地理分隔(同类动、植物因地壳变动产生的山脉和海洋阻隔所造成)
intuitive:(adj) 直觉的,易懂的
empirical:(adj) 经验主义的,以经验为依据的,经验的
first-appearance datum:(生物)首现面(地质记录中某个物种的首次出现)
last-appearance datum:(生物)末现面,指地质记录中某个物种的最后出现
geochronological:(adj) 地质年代的
taxon:(n) 分类单元,分类学;(adj) 分类的,分类学的

validate：（vt）批准，证实，使生效　　　　dizzying：（adj）令人昏乱的，极快的

Notes

[1] 有孔虫（foraminifera）是一种由碳酸钙构成外壳的大而多样的单细胞水生生物（主要是海洋生物）。而放射虫（radiolaria）是一种由二氧化硅构成外壳的单细胞海洋生物。

[2] 新生代，即从6500万年前到现在的地质时代。

[3] 光合作用中二氧化碳同化的不同途径。C3植物包括乔木、灌木和适寒冷气候的草（约占植物种类的95%）。C4植物包括温暖气候的草和谷物，在高热、高光和低二氧化碳水平的条件下是有利的。

[4] 全新世，即从大约11 700年前到现在的地质时代。

[5] 上新世和更新世的结合，从约530万年前持续到约11 700年前（全新世开始）。

[6] 第四纪包括更新世和全新世的地质时期，从约260万年前持续至今。在2009年之前，第四纪和更新世的开始时间被定为~180万年。

[7] Milutin Milankovitch（米卢廷·米兰科维奇，1879-1958），塞尔维亚数学家，他提出气候变化，特别是冰期是地球轨道参数变化的结果。

[8] 季节性指的是气候要素年内分布的时间、持续时间或强度的变化，但不包括年总量（如日晒、降水）的变化。

[9] external and internal forcings是指外部和内部强迫。外部强迫机制包括来自气候系统外部的作用因子（如米兰科维奇循环）。气候系统本身的内部机制起作用（如造山运动、板块构造、火山活动、海洋环流、大气成分）。

[10] 牛科的成员，包括羚羊、牛、山羊和绵羊。不分枝的角是由一层角蛋白围绕着骨核构成的，这是其决定性特征之一。

[11] precipitation是降水的总称，包含液态和固态的各种形式。其中降雨（rainfall）在大部分中低纬地区是最主要的形式，因此，有时不甚严格地将rainfall与precipitation混淆使用。

14. Plate Tectonics Theory

Text

The success of plate tectonics required an acceptance of continental drift, seafloor spreading, and thus a reinterpretation of the large-scale geological history of most of the Earth[1].

In the late 19th and early 20th centuries, geologists assumed that Earth's major features were fixed, and that most geologic features such as basin development and mountain ranges could be explained by vertical crustal movement, described in what is called the geosynclinal theory[2]. Generally, this was placed in the context of a contracting planet Earth due to heat loss in the course of a relatively short geological time.

It was observed as early as 1596 that the opposite coasts of the Atlantic Ocean—or, more precisely, the edges of the continental shelves—have similar shapes and seem to have once fitted together.

By 1915, after having published a first article in 1912, Alfred Wegener was making serious arguments for the idea of continental drift in the first edition of The Origin of Continents and Oceans. In that book (re-issued in four successive editions up to the final one in 1936), he noted how the east coast of South America and the west coast of Africa looked as if they were once attached. Wegener was not the first to note this (Abraham Ortelius, Antonio Snider-Pellegrini, Eduard Suess, Roberto Mantovani and Frank Bursley Taylor preceded him, just to mention a few), but he was the first to marshal significant fossil and paleo-topographical and climatological evidence to support this simple observation (and was supported in this by researchers such as Alex du Toit). Furthermore, when the rock strata of the margins of separate continents are very similar it suggests that these rocks were formed in the same way, implying that they were joined initially. For instance, parts of Scotland and Ireland contain rocks very similar to those found in Newfoundland and New Brunswick. Furthermore, the Caledonian Mountains of Europe and parts of the Appalachian Mountains of North America are very similar in structure and lithology.

Earlier theories by Alfred Wegener and Alexander du Toit of continental drift postulated that continents in motion "plowed" through the fixed and immovable seafloor. The idea that the seafloor itself moves and also carries the continents with it as it spreads from a central rift axis was proposed by Harold Hammond Hess from Princeton University and Robert Dietz of the U.S. Naval Electronics Laboratory in San Diego in the 1960s. The phenomenon is known today as plate tectonics. In locations where two plates move apart, at mid-ocean ridges, new seafloor is continually formed during seafloor spreading[3].

Seafloor spreading helps explain continental drift in the theory of plate tectonics. When oceanic plates diverge, tensional stress causes fractures to occur in the lithosphere. The motivating force for seafloor spreading ridges is tectonic plate slab pull at subduction zones, rather than magma pressure, although there is typically significant magma activity at spreading ridges. Plates that are not subducting are driven by gravity sliding off the elevated mid-ocean ridges a process called ridge push. At a spreading center, basaltic magma rises up the fractures and cools on the ocean floor to form new seabed. Hydrothermal vents are common at spreading centers. Older rocks will be found farther away from the spreading zone while younger rocks will be found nearer to the spreading zone.

In the 1960s, the past record of geomagnetic reversals of Earth's magnetic field was noticed by observing magnetic stripe "anomalies"[4] on the ocean floor. This results in broadly evident "stripes" from which the past magnetic field polarity can be inferred from data gathered with a magnetometer towed on the sea surface or from an aircraft. The stripes on one side of the mid-ocean ridge were the mirror image of those on the other side. By identifying a reversal with a known age and measuring the distance of that reversal from the spreading center, the spreading half-rate could be computed.

In some locations spreading rates have been found to be asymmetric; the half rates differ on each side of the ridge crest by about five percent. This is thought due to temperature gradients in the asthenosphere from mantle plumes near the spreading center.

By explaining both the zebra-like magnetic striping and the construction of the mid-ocean ridge system, the seafloor spreading hypothesis (SFS) quickly gained converts and represented another major advance in the development of the plate-tectonics theory. Furthermore, the oceanic crust now came to be appreciated as a natural "tape recording" of the history of the geomagnetic field reversals (GMFR) of Earth's magnetic field. Today, extensive studies are dedicated to the calibration of the normal-reversal patterns in the oceanic crust on one hand and known timescales derived from the dating of basalt layers in sedimentary sequences (magnetostratigraphy) on the other, to arrive at estimates of past spreading rates and plate reconstructions.

After all these considerations, Plate Tectonics (or, as it was initially called "New Global Tectonics") became quickly accepted in the scientific world.

Earth's lithosphere, which is the rigid outermost shell of the planet (the crust and upper mantle), is broken into seven or eight major plates (depending on how they are defined) and many minor plates. Where the plates meet, their relative motion determines the type of plate boundary: convergent, divergent, or transform. Earthquakes, volcanic activity, mountain-building, and oceanic trench formation occur along these plate boundaries (or faults).

The key principle of plate tectonics is that the lithosphere exists as separate and distinct tectonic plates, which ride on the fluid-like solid the asthenosphere. Plate motions range from 10 to 40 mm/year at the Mid-Atlantic Ridge, to about 160 mm/year for the Nazca Plate.

Tectonic lithosphere plates consist of lithospheric mantle overlain by one or two types of crustal material: oceanic crust (in older texts called sima[5] from silicon and magnesium) and continental crust (sial[6] from silicon and aluminum). The distinction between oceanic crust and continental crust is based on their modes of formation. Oceanic crust is formed at sea-floor spreading centers. Continental crust is formed through arc volcanism and accretion of terranes through plate tectonic processes. Oceanic crust is denser than continental crust because it has less silicon and more of the heavier elements than continental crust. As a result of this density difference, oceanic crust generally lies below sea level, while continental crust buoyantly projects above sea level.

Average oceanic lithosphere is typically 100 km thick. Its thickness is a function of its age. As time passes, it cools by conducting heat from below, and releasing it radiatively into space. The adjacent mantle below is cooled by this process and added to its base. Because it is formed at mid-ocean ridges and spreads outwards, its thickness is therefore a function of its distance from the mid-ocean ridge where it was formed. For a typical distance that oceanic lithosphere must travel before being subducted, the thickness varies from about 6 km thick at mid-ocean ridges to greater than 100 km at subduction zones. For shorter or longer distances, the subduction zone, and therefore also the mean, thickness becomes smaller or larger, respectively. Continental lithosphere is typically about 200 km thick, though this varies considerably between basins, mountain ranges, and stable cratonic interiors of continents[7].

The location where two plates meet is called a plate boundary. Plate boundaries are where geological events occur, such as earthquakes and the creation of topographic features such as mountains, volcanoes, mid-ocean ridges, and oceanic trenches. The vast majority of the world's active volcanoes occur along plate boundaries, with the Pacific Plate's Ring of Fire[8] being the most active and widely known. Some volcanoes occur in the interiors of plates, and these have been variously attributed to internal plate deformation and to mantle plumes.

Some pieces of oceanic crust, known as ophiolites[9], failed to be subducted under continental crust at destructive plate boundaries; instead these oceanic crustal fragments were pushed upward and they are now preserved within continental crust.

Vocabulary

plate tectonics: (n) 板块构造

continental drift: (n) 大陆漂移

basin: (n) 盆地

mountain range: (n) 山脉

crustal：（adj）地壳的
contract：（vt/vi）（使）收缩
　n：contraction 收缩
coast：（n）海岸
topographical：（adj）地形的，地形学的
climatological：（adj）气候学的
strata：（n）地层
structure：（n）结构，构造
axis：（n）轴线
tensional stress：（n）张应力
fracture：（n）破裂
elevate：（vt）使升高
　n：elevation 海拔
seabed：（n）海床
temperature gradient：（n）温度梯度
asthenosphere：（n）软流圈
sedimentary sequence：（n）沉积层序
magnetostratigraphy：（n）磁性地层学
rigid：（adj）刚性的，坚硬的
convergent：（adj）汇聚的，收敛的，聚合的
divergent：（adj）离散的
transform：（vt/n）（使）转换
　~ transform boundary：转换型边界
　~ transform fault：转换断层
mountain-building：（n）造山
solid：（n）固体；（adj）坚硬的
silicon：（n）硅
magnesium：（n）镁
aluminum：（n）铝
arc：（n）岛弧
terrane：（n）地体
buoyantly：（adv）漂浮的
　n：buoyancy 浮力
cratonic：（adj）有克拉通盆地特征的
　n：craton 克拉通

Notes

[1] 板块构造论是为了解释大陆漂移现象而发展出的一种地质学理论。该理论认为，地球的岩石圈是由板块拼合而成，现今全球可划分为六大板块，海洋和陆地的位置是不断变化的。根据这种理论，地球内部构造的最外层分为两部分：外层的岩石圈和内层的软流圈。这种理论基于两种独立的地质观测结果：海底扩张和大陆漂移。

[2] 地槽-地台说简称槽台说，其基本论点：地壳运动主要受垂直运动控制，地壳此升彼降造成振荡运动，而水平运动则是派生或次要的。驱动力主要是地球物质的重力分异作用。物质上升造成隆起，物质下降则造成凹陷。主要的构造单元有地槽和地台两类，地台是由地槽演化而来的。

[3] 海底扩张学说是在大陆漂移学说的基础上所发展出的进阶地球地质活动学说。在各大洋中有一带状分布的洋中脊，是下方地幔软流层的出口。熔岩自洋中脊不断涌出，冷却而成为刚性的洋壳。洋壳受到由洋中脊涌出的熔岩的推挤而不断向两旁移动，使海面积扩大，同时大陆地壳受到推挤而分离。

[4] 地磁条带异常（magnetic stripe anomalies）也可以称为条带状磁异常（striped magnetic anomalies）。新生成的洋壳中磁性矿物受地磁场影响，会记录其结晶时地球

的磁极状态。当地球磁极发生倒转时,新生成的洋壳所记录的磁极状态与先前形成的洋壳相反。随着海底扩张,地球反复的磁极倒转事件会在洋壳中记录为正磁性和反磁性交替出现的条带,从而形成磁异常条带。磁异常可以有不同的尺度,但尺度大小并不代表异常的大小。磁异常可造成空间环境的扰动和航天器异常。

[5] 硅镁层(sima)在地球化学中指由富镁硅酸盐矿物组成的地壳。硅镁层出露到地表的代表是玄武岩。洋壳主要是硅镁层。岩石学研究表明,相比硅铝层,硅镁层密度更大(2800~3300 kg/m^3),这是因为铁、镁含量更高,铝含量降低。硅镁层岩石也叫铁镁质。密度最大的硅镁层岩石是硅含量最少的超基性岩。

[6] 硅铝层(sial)在地球化学中指大陆地壳岩石圈,得名于其主要构成化学元素硅与铝。也称硅铝层岩石为长英质,因为其包含很高比例的硅铝长石矿物。

[7] 盆-岭构造(basin and range tectonics)通常是大陆伸展构造系统在浅表层次的构造表现,是指在伸展变形区域,由掀斜构造、阶梯状正断层、地堑、地垒等共同产出,形成由不对称的纵列单面山、山岭及其间的盆地组合而成的构造-地貌单元。北美科迪勒拉造山带盆岭省(Basinand Basin and Range Province),是建立盆-岭构造概念的经典地区。该区域从墨西哥北部经美国西部一直延伸到加拿大,全长约 3000 km,最宽处位于美国西部,宽达 1000 km,而在南、北两侧都明显变窄。盆岭省由一系列 NW—NNW 向延伸、相间分布的山脉和盆地组成。

[8] 环太平洋火山带(Ring of Fire)是一个围绕太平洋经常发生地震和火山爆发的地区,全长 40 000 km,呈马蹄形。环太平洋火山带上有一连串海沟、火山弧和火山带,和板块移动。它有 452 座火山,是占有世界上的活跃的和休眠火山的 75%以上。环太平洋火山带有时也被称为火环带。地球上 90%的地震以及 81%最强烈的地震都在该地带上发生,第二个最猛烈的地震带是从爪哇岛、苏门答腊岛伸延至喜马拉雅山脉、地中海以及大西洋的阿尔卑斯带,地球上 6%的地震和 17%最强烈的地震在这里发生。第三个最猛烈的地震带则是大西洋洋中脊。

[9] 蛇绿岩(ophiolite)是指一组由蛇纹石化超镁铁岩、基性侵入杂岩和基性熔岩以及海相沉积物构成的岩套,又称蛇绿岩套,代表性层序自下而上为橄榄岩、辉长岩、席状基性岩墙和基性熔岩以及海相沉积物。其可以形成于洋中脊、弧后盆地、弧前盆地、岛弧或活动大陆边缘等构造环境。蛇绿岩套代表了古大洋岩石圈的残留,是确定古板块边界的重要证据。

15. Seismology

Text

Geophysics is the study of the physics of the Earth and its environment in space. One emphasis is the exploration of the Earth's interior using physical properties measured at or above the Earth's surface, together with mathematical models to predict those properties[1]. Subdisciplines include seismology, the study of seismic waves; geomagnetism, the study of the magnetic field; and geodesy, the study of the gravitational field and the shape of the Earth's surface[2].

Every day there are about fifty earthquakes worldwide that are strong enough to be felt locally, and every few days an earthquake occurs that is capable of damaging structures. Each event radiates seismic waves that travel throughout Earth, and several earthquakes per day produce distant ground motions that, although too weak to be felt, are readily detected with modern instruments anywhere on the globe. Seismology is the science that studies these waves and what they tell us about the structure of Earth and the physics of earthquakes. It is the primary means by which scientists learn about Earth's deep interior, where direct observations are impossible, and has provided many of the most important discoveries regarding the nature of our planet. It is also directly concerned with understanding the physical processes that cause earthquakes and seeking ways to reduce their destructive impacts on humanity.

The natural sources of seismic waves—earthquakes—are of great practical and scientific importance, and their seismological fingerprints provide valuable insights into earthquake causes, mechanisms[3], and seismic hazards. Earthquakes usually occur by slip on faults, and the seismic signature allows the likely fault orientations to be constrained. Understanding is required of why earthquakes start suddenly—and stop—this is provided through descriptions of frictional behavior on fault surfaces. The surface consequences of earthquakes, notably ground motions and tsunamis, are direct hazards. Within the upper crust, other effects, such as changes in fluid pressures in porous rocks, are important as they may influence future earthquakes and also trigger mineral deposits. Volcano behavior often involves seismicity. Slow-slip earthquakes have been recently recognized as discontinuous, earthquake-like events that release energy over a period of hours to months, rather than the seconds to minutes, characteristic of a typical earthquake.

Earthquakes create rapid movements in the Earth's interior, which give rise to waves (oscillatory movements of material points), which, with sufficient amplitude, can traverse the entire planet. Elasticity theory is used to model such waves, which, like other types of waves, reflect and refract at boundaries and follow curved ray paths in heterogeneous media. When such waves arrive at the surface, the ground motions can be recorded and used to constrain the structure along the pathway. Different types of oscillatory movements give rise to waves that travel at different speeds. Primary (P) waves[4] involve oscillations parallel to the direction of travel and travel faster than secondary (S) waves[5], which involve oscillations transverse to the direction of travel. S waves cannot travel through liquids and provide evidence for the Earth's liquid outer core (Figure Ⅲ-11).

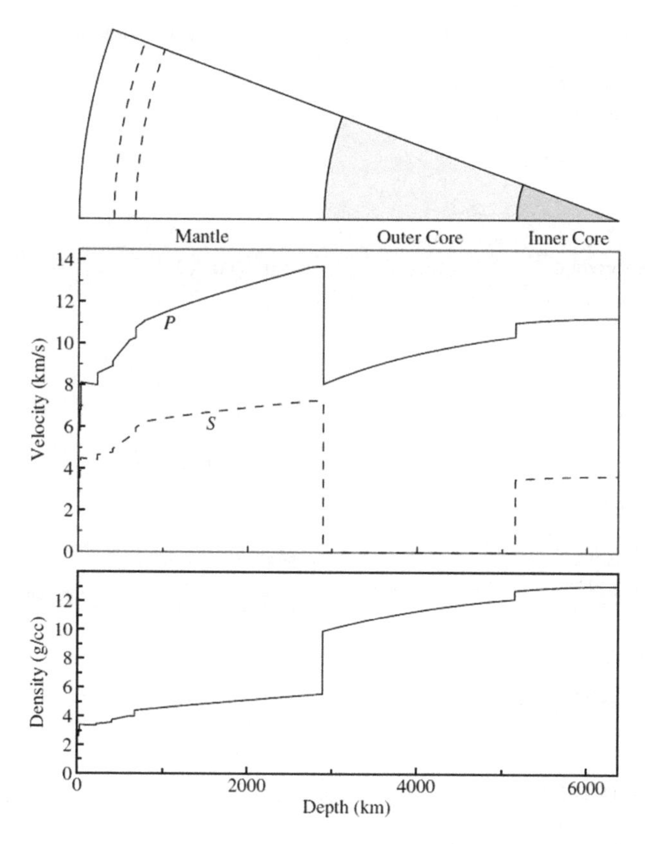

Figure Ⅲ-11: Earth's P velocity, S velocity, and density as a function of depth. [Source: Preliminary Reference Earth Model (PREM)]

Seismology occupies an interesting position within the more general fields of geophysics and Earth sciences. It presents fascinating theoretical problems involving analysis of elastic wave propagation in complex media, but it can also be applied simply as a tool to examine different areas of interest. Applications range from studies of Earth's core, thousands of kilometers below the surface, to detailed mapping of shallow crustal structure to help locate petroleum deposits[6]. Much of the underlying physics is no more advanced than Newton's second law ($F = ma$), but the complications introduced by realistic sources and structures have motivated sophisticated mathematical treatments and extensive use of powerful computers. Seismology is driven by observations, and improvements in instrumentation and data availability have often led to breakthroughs both in seismology theory and in our understanding of Earth structure.

The information that seismology provides has widely varying degrees of uncertainty. Some parameters, such as the average compressional wave travel time through the mantle, are known to a fraction of a percent, while others, such as the degree of damping of seismic energy within the inner core, are known only very approximately. The average radial seismic velocity structure of Earth has been known fairly well for over fifty years, and the locations and seismic radiation patterns of earthquakes are now routinely mapped, but many important aspects of the physics of earthquakes themselves remain a mystery.

Vocabulary

geophysics:(n) 地球物理学
exploration:(n) 探测,勘探
interior:(n) 内部
seismology:(n) 地震学
  adj: seismological
  adv: seismologically
geomagnetism:(n) 地磁学
geodesy:(n) 大地测量学
gravitational:(adj) 重力的,引力的
  radiate:(vt/vi) 辐射,放射;(adj) 辐射(状)的
detect:(vt) 发现,识别,调查
slip:(n) 滑移,错动;(vi) 滑行,滑动
orientation:(n) 方向,方位
constrain:(vt) 约束,限制
frictional:(adj) 摩擦的,摩擦产生的
consequence:(n) 结果,后果
porous:(adj) 多孔的,有孔的
trigger:(n/vt) 触发(器),引发
deposit:(n) 沉积物;(vt) 放置,沉积
seismicity:(n) 地震活动性
slow-slip:(n) 慢滑移,缓慢蠕动
discontinuous:(adj) 不连续的
release:(v) 释放,排放,解放;(n) 放松,释放
period:(n) 时间,时期,周期
oscillatory:(adj) 振荡的,振动的
sufficient:(adj) 足够的,充足的
traverse:(v) 横越,穿过;(n) 越过,横档
elasticity:(n) 弹性,弹力
reflect:(vt) 反射,反映
refract:(vt) 折射
curved:(adj) 弧形的,弯曲的;(vt) 使弯曲,变弯
ray:(n) 射线,光线;(vt) 放射,呈辐射状伸出
heterogeneous:(adj) 非均匀的
parallel:(adj) 平行的,并行的;(vi) 与……相平行
transverse:(adj) 横向的,横断的
propagation:(n) 传播,蔓延
petroleum:(n) 石油,原油
sophisticated:(adj) 复杂的,精密的
extensive:(adj) 广泛的,大规模的
uncertainty:(n) 不确定性
compressional:(adj) 压缩性的

damp：(n) 阻尼

approximately：(adv) 大约,大概

radiation：(n) 放射状的

~radiation pattern：能量辐射花样

Notes

[1] 此句介绍了地球物理学 (geophysics) 的定义及主要研究内容。physics 指地球的物理特性,physical properties 指物理属性参数(如密度),mathematical models 指针对地球建立的数字参数化模型。

[2] 此句介绍了地球物理学的诸多研究分支,包括地震学 (seismology)、地磁学 (geomagnetism)、大地测量学 (geodesy) 等。地震学是研究地震、地震波及其在地球内部传播等与地震有关的科学。地磁学是研究地球和大气圈之磁性的科学,主要研究有磁性的现象、来源、磁场等方面。大地测量学是研究关于地球重力的科学,研究范围包括地球上的重力现象、重力分布、重力场及其他相关性质的研究。

[3] mechanism 意指地震的产生、破裂机制,地震学研究的重要内容之一,是认清地震是如何产生的,即地震的震源机制。目前主流观点认为,天然大地震主要由活动断层的错动产生,大部分地震事件,发生在板块交界处或构造运动较活跃的区域。地震震源机制的研究,也为板块运动学说提供了有力支撑。

[4] primary wave,简称 P 波,即地震纵波,又称胀缩波,是地震时从震源传出的一种弹性波,传播它的介质质点振动方向和波的传播方向一致。纵波传播时,介质的密度会加密和变疏,体积的大小发生变化,但形态不改变,在未固定形状的介质中也能通过,即地震纵波在地球内部的各部分都能传播。纵波的传播速度比横波快,因此地震纵波总是最先到达观测点,故又称初至波 (primary wave),P 波即选用首字母缩写而成。

[5] secondary wave,简称 S 波,即地震横波,又称剪切波 (shearing wave),是地震时从震源传出的一种弹性波,传播它的介质质点振动方向与波的前进方向垂直,故称地震横波。横波经过时,介质的体积不变,但形状要改变,产生切变方式的变形。在地壳中横波传播的速度较慢。到达地面时人感觉摇晃,物体为摆动,对地面破坏很大。

[6] 地震学不仅研究天然地震,也研究某些人为的或自然因素所造成的(如地下爆炸、岩浆冲击、岩洞塌陷等)地震动,应用领域很广,既可以用于探测地球内部的结构与构造(天然地震学),又可用于寻找油气资源和能源(勘探地震学),此时所用震源是人工控制的,对地下结构的成像可以达到很高的精度,地震方法是石油勘探中必不可少的重要手段。

16. Subduction Zones

Text

Subduction zones are descending limbs of mantle convection cells and are the dominant physical and chemical system of Earth's interior (Figure Ⅲ-12). The sinking of lithosphere in subduction zones provides most of the force needed to drive the plates and cause mid-ocean ridges to spread, with the result that plate tectonics and subduction zones are surficial and interior expressions of Earth's dominant tectonic mode.

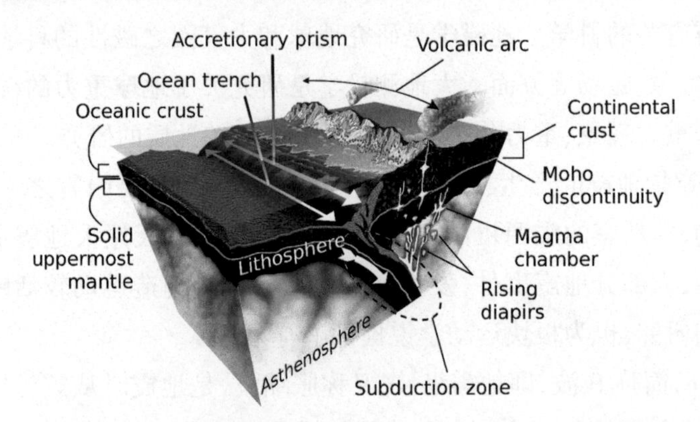

Figure Ⅲ-12: Diagram of the geological process of subduction.

Subduction zones are also our planet's largest recycling system. They deliver raw materials to the subduction factory, where oceanic lithosphere, sediments, and seawater re-equilibrate with ambient mantle, triggering melting and incidentally creating continental crust. What is not recycled in the upper few hundred kilometers of a subduction zone sinks to the core-mantle boundary[1], where this residue[2] may be reheated for a billion years or so until it is resurrected as a mantle plume. There is a continuum from subduction, involving normal oceanic lithosphere, to collision, involving continental lithosphere or oceanic lithosphere with thickened crust. When continental crust is caught in the maw of a subduction zone, the smooth operation of the subduction zone is spectacularly disrupted, and a mountain range such as the Himalayas may result. Nevertheless, continental crust has been carried down to depths of 100 km or more before resurfacing.

Among the terrestrial planets[3], only Earth appears to have subduction zones and plate tectonics. Mercury and Earth's moon are tectonically and magmatically dead[4], while Venus[5] is dominated by thick lithosphere and mantle plumes. Linear magnetic anomalies and the presence of rocks similar to andesites in the ancient southern highlands of Mars suggest that plate tectonics and subduction zones may have existed on that planet in the ancient past but no more. Mars has become a "single-plate" planet dominated by the immense volcanoes of the Tharsis region[6], the solar system's largest "hot spot" volcanoes.

Subduction zones burrow deeply but are imperfectly camouflaged, and we can use geophysics and geochemistry to watch them. Not surprisingly, the shallowest portions are known best. Subduction zones are strongly asymmetric for the first several hundred kilometers. Their dimensions are defined by deep trenches, lines of volcanoes parallel to the trenches, and inclined planar arrays of deep earthquakes that dip away from the trench beneath the volcanoes and extend down to the 660 km discontinuity[7]. Earthquakes in subduction zones occur at enormously greater depths than elsewhere on Earth, where seismicity is limited to the uppermost 20 km. This requires either that brittle conditions extend to much greater depths or that there are unusual causes of seismicity in subduction zones. Seismic tomography provides additional details. This technique uses seismic velocity information from many ray paths, crisscrossing Earth between various points near Earth's surface and reaching different depths in its interior, to produce a three-dimensional (3–D) model of relative velocity. Regions of anomalously fast mantle corresponding to the earthquake plane define the down-going lithospheric slab, and regions of anomalously slow mantle lie above this. Subducted lithosphere sometimes can be traced this way past the 660 km discontinuity into the lower mantle.

The term subduction zone is sometimes used interchangeably with convergent (destructive) plate margin or island arc, and while all three terms are intimately related, they are not synonymous. Subduction zones are the three-dimensional manifestation of convective downwelling, convergent plate margins are the surficial manifestations of downwelling, and arcs (better referred to as arc-trench complexes) are surficial and crustal manifestations of a subduction zone that is operating beneath it.

Subduction zones are formally distinct from convergent plate margins because the latter are required by plate tectonics, that is, the motions of rigid spherical shells around the Euler pole of rotation, whereas none of the geometric rules of plate tectonics applies to the behavior of the lithosphere once it descends below the surface. Subduction zones are defined by the inclined array of earthquakes known as the "Wadati-Benioff Zone" after the two scientists who first identified it[8]. Subduction zones are also distinct from arc-trench complexes. Arcs are consequences of subduction zone processes preserved in the overlying crust, geological evidence of deformation and chemical recycling caused by subduction. Not all arcs are truly island arcs, but the description fits because it captures the sense of a curved array of discrete volcanoes engendered by subduction. Although they are relatively minor components of subduction zones in terms of mass, arcs are important because they provide accessible samples of the products of interacting mantle and subducted materials and testify to the operation of ancient subduction zones. Arcs are also nurseries for continental crust, where thickened welts of low-density crust are generated and processed further to await accretion to other tracts of buoyant crust.

The relationship between subduction zones and convergent plate margins is better appreciated when their typical representations are considered together. Convergent plate margins appear in map view, whereas subduction zones are typically shown as cross sections. Combining these two perspectives more realistically presents the globe-encircling and mantle-permeating nature of the subduction zone system.

The cumulative length of convergent plate margins is >55,000 km, almost equal to that of mid-ocean ridges (60,000 km). Lithosphere consumed at subduction zones may pond at the base of the transition zone, 660 km beneath the surface, or may plunge to the core-mantle boundary, 2900 km beneath the surface. The subduction zone system is clearly the most extensive and pervasive feature of our planet. Most geodynamicists agree that the plates move because of the sinking of lithosphere in subduction zones. The scale and role of subduction zones indicate that they are the most important tectonic feature of our planet.

The mantle wedge is that part of the mantle that lies above the subduction zone and where subducted inputs are mixed with convecting mantle to generate magmas, fluids, and ultimately continental crust. Convecting asthenosphere is the important part, because this interacts with slab-derived fluids and melts to generate arc magmas. The dynamics of this magmagenetic system are fundamentally different from Earth's two other magma-producing regimes, hot spots (mantle plume) and mid-ocean ridges. Hot spot and ridge melts are associated with upwelling mantle, but the mantle wedge melts where the mantle descends and is associated with the coldest thermal regime found on our planet, two features otherwise expected to ensure that melting does not occur. It is the fluxing effect of dense aqueous solutions, squeezed from subducted materials and rising to interact with asthenospheric mantle, that is ultimately responsible for convergent margin magmas. It is this interaction between sinking slab, rising fluids, and moving mantle that defines the deep-Earth reprocessing system sometimes referred to as the "subduction factory."

Vocabulary

mantle convection：(n) 地幔对流
tectonics：(n) 构造，构造学
  adv：tectonically 构造上，构造地
  ~plate tectonics：(n) 板块构造(理论)
subduction factory：(n) 俯冲工厂
re-equilibrate：(vt) 重新平衡
continental crust：(n) 大陆地壳
Himalayas：(n) 喜马拉雅山脉
magnet：(n) 磁体，磁铁
  adj：magnetic 有磁性的，磁的

n: magnetism 磁性,磁力
andesite: (n) 安山岩
Mars: (n) 火星
camouflage: (v) 隐蔽,隐藏
asymmetric: (adj) 不对称的
discontinuity: (n) 不连续面
tomography: (n) 层析成像(技术)
ray path: (n) 射线路径
slab: (n) 板片
lower mantle: (n) 下地幔
synonymous: (adj) 同义的
convective: (adj) 对流的
downwelling: (n) 沉降流
surficial: (adj) 地表的
arc-trench complex: (n) 沟-弧复合体
Euler pole: (n) 欧拉极
Wadati-Benioff Zone: (n) 贝尼奥夫带
overlying crust: (n) 上覆地壳
discrete: (adj) 离散的
accretion: (n) 堆积,吸积
cross section: (n) 横截面
geodynamicist: (n) 地球动力学家
~geodynamic: (adj) 地球动力学的
mantle wedge: (n) 地幔楔
dynamics: (n) 动力学
magmagenetic: (adj) 岩浆产生的

Notes

[1] core-mantle boundary 意为核幔边界,指地核与下地幔的边界。核幔边界存在一些波速异常体,目前的一些研究认为这些波速异常体与古老的俯冲物质有关。

[2] residue 指俯冲板片 (subduction slab) 下沉到核幔边界的残留物质。

[3] terrestrial planets 指类地行星。太阳系中的行星分为以大气占体积主体的巨行星,以及以硅酸盐岩石或金属为主的类地行星。

[4] 这句话指现在的水星 (Mercury) 和月球没有板块构造和岩浆作用。

[5] 金星 (Venus) 是一个与地球在大小、质量、组成和离太阳的距离等方面都非常相似的行星,但是现今金星不存在类似地球上的板块构造运动,也没有内生磁场。根据金星的重力和地形的研究显示金星的岩石圈比较厚。金星上存在约 10 个类似地球上夏威夷下方的地幔柱,最新的金星快车探测资料显示其中几个地幔柱存在近期的火山活动。

[6] Tharsis region 即塔尔西斯地区,是火星西半球赤道附近一处辽阔的火山高原,该地区为太阳系中最大火山所在地。

[7] 660 km discontinuity 即 660 km 不连续面,是上地幔和下地幔之间的界面。

[8] 贝尼奥夫带(或译为毕乌夫带)是位于海沟处平行于海沟的地震带。日本地震学家和达清夫(K. Wadati)在 20 世纪 30 年代首先发现了这一震源带,50 年代美国地震学家贝尼奥夫(H. Benioff)进一步研究予以确定。该带连续分布着浅源、中源和深源地震,海沟附近为浅源地震,趋向岛弧、大陆依次为中源、深源地震。

17. Submarine Geodynamics

Text

Submarine geodynamics refers to the study of the geological processes and dynamics that occur beneath the seafloor, specifically in the oceanic crust and the underlying mantle. It encompasses the understanding of tectonic activities, volcanic processes, plate movements, and the formation and evolution of submarine features such as seamounts, mid-ocean ridges, and oceanic trenches[1] (Figure Ⅲ-13).

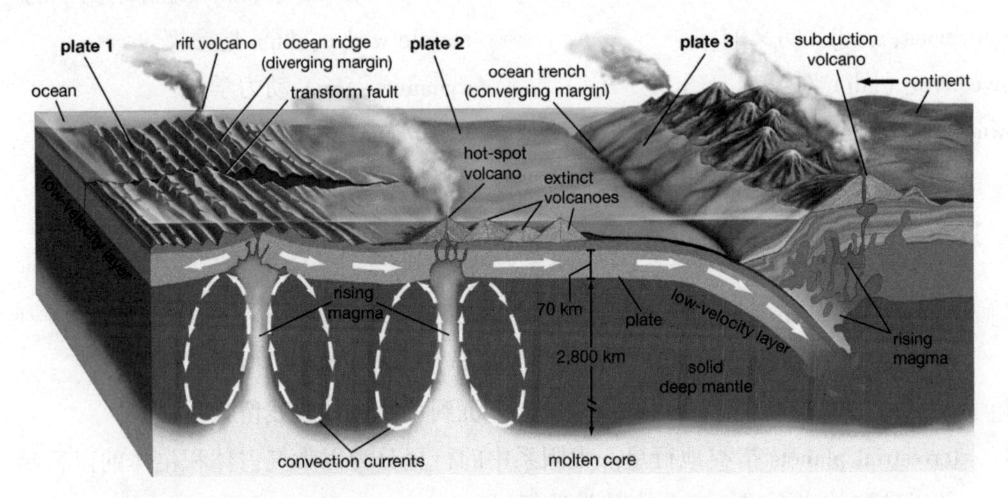

Figure Ⅲ-13: Schematic for submarine geodynamics. [Source: Encyclopedia Britannica]

Here are some key aspects and processes involved in submarine geodynamics:

Plate Tectonics: Submarine geodynamics is closely linked to plate tectonics, which describes the movement and interaction of Earth's lithospheric plates. The seafloor is divided into several tectonic plates that are in constant motion. Submarine geodynamics investigates the processes associated with plate boundaries, such as divergent boundaries (where new crust is formed), convergent boundaries (where crust is destroyed), and transform boundaries (where plates slide past each other).

Mid-Ocean Ridges: Mid-ocean ridges are underwater mountain ranges formed by the upwelling of molten material from the Earth's mantle at diverging margin. Submarine geodynamics studies the formation, spreading, and evolution of mid-ocean ridges, which play a crucial role in seafloor spreading, the creation of rift volcano and new oceanic crust and the formation of transform fault.

Subduction Zones: Subduction zones are areas where one tectonic plate is forced beneath another at ocean trench (converging margin). These zones are often associated with deep-sea trenches and volcanic activity (subduction volcano). Submarine geodynamics investigates the processes involved in subduction, including the sinking of oceanic lithosphere into the mantle, the generation of volcanic arcs, and the occurrence of earthquakes.

Seamounts and Volcanic Activity: Submarine geodynamics examines the formation and dynamics of seamounts, which are underwater volcanic mountains (hot-spot volcanoes and extinct volcanoes[2]). It involves the study of volcanic eruptions, magma generation and ascent, and the effects of volcanic activity on the seafloor and surrounding environment.

Hydrothermal Vent Systems: Hydrothermal vents are submarine hot springs that occur near mid-ocean ridges and volcanic regions. Submarine geodynamics investigates the formation and functioning of hydrothermal vent systems, including the circulation of seawater through the seafloor, the heat and chemical exchange between water and rocks, and the biological communities[3] that thrive in these extreme environments.

Understanding submarine geodynamics is crucial for unraveling Earth's geological history, studying the dynamics of plate tectonics, and exploring the processes that shape the seafloor and its associated ecosystems. It involves a combination of seafloor mapping[4], geophysical surveys[5], geochemical analyses[6], and numerical modeling[7] to gain insights into the complex interactions occurring beneath the ocean's surface.

Vocabulary

submarine: (adj) 海面下的，海底的

tectonic activity: (n) 构造活动

motion: (n) 运动

molten: (adj) 熔融的

volcanic arc: (n) 火山弧

volcanic activity: (n) 火山活动

hot spring: (n) 热泉

extreme: (adj) 极端的

crucial: (adj) 关键的

numerical: (adj) 数值的

Notes

[1] oceanic trench 意为海沟，是位于海洋中的两壁较陡、狭长的、水深大于 5000 m 的沟槽，多分布于汇聚板块边缘，其成因与板块俯冲相关。

[2] extinct volcano 意为死火山，是指在相当长的时间内没有喷发过，且不太可能再次喷发的火山，通常由于地质活动的变化或岩浆供应的停止而导致火山活动的终结。

[3] biological community 意为生物群落，是指在特定区域内，由不同种类的生物相互作用和共存所形成的群体，包括植物、动物和微生物等多种生物组成的生态系统。文中

这里特指热液喷口生物群落。

[4] seafloor mapping 意为海底地形测绘,是指使用各种技术和工具对海洋底部进行测绘,以获取关于海底地形、地质特征和水文条件的详细信息。

[5] geophysical survey 意为地球物理勘测,是通过测量、记录和解释地球物理现象(如重力、磁场、地震波等)的方法,用于研究地球内部结构、地壳运动、资源勘探等地球科学领域的调查和探测技术。

[6] geochemical analysis 意为地球化学分析,是通过对地球物质样本(如岩石、土壤、水样等)中元素、同位素和化学性质的测量和研究,以揭示地球化学过程、环境变化和地球系统的演化。

[7] numerical modeling 意为数值模拟,是使用计算机和数学模型来模拟和预测复杂系统的行为和现象,以便研究和理解自然和物理过程,如天气、气候、地球动力学等。

18. Submarine Mineral Resources

Text

Deep-Sea Mineral Resources and Their Formation

Hidden deep within the world's oceans are extensive, untapped reservoirs of mineral wealth, which encompass polymetallic nodules, cobalt-rich crusts, and hydrothermal sulfides. These remarkable formations are the result of intricate metallogenic systems that have shaped the seafloor over geological epochs. One of the most fascinating processes contributing to their formation is the interaction of hydrothermal fluids with the surrounding rocks. This hydrothermal activity occurs primarily along the mid-ocean ridges, where the Earth's tectonic plates diverge. Fast-spreading ridges, slow-spreading ridges, and ultraslow-spreading ridges are all hotspots for the creation of these mineral-rich deposits.

The deep-sea environment is home to a wide array of mineral resources, each processing unique characteristics. Seamounts, underwater mountains rising from the abyssal plains, often harbor seafloor placers, marine sands, and rare earth elements. In contrast, back-arc basins, which form in the aftermath of subduction zones, have the potential to host a variety of critical and rare metals[1]. These resources, along with dispersed and precious metals, constitute essential components of modern technology and industry, making their exploration and extraction increasingly vital in our resource-dependent world.

The Challenge of Exploring Deep-Sea Mineral Resources

Exploring these hidden treasures is an enormous undertaking, fraught with challenges. The abyssal basin's extreme depths, crushing pressures, and frigid temperatures demand advanced technologies and expertise. The quest for deep-sea mineral resources often begins with prospecting methods that employ remotely operated vehicles (ROVs) and autonomous underwater vehicles (AUVs). These robotic marvels descend into the inky depths, collecting samples and mapping the seafloor to assess the resource potential.

Resource evaluation is a meticulous process that involves assessing the quantity, quality, and distribution of mineral resources. Geologists specializing in process mineralogy examine rock and sediment samples to determine ore-forming processes and ore-controlling factors. This rigorous assessment is a crucial step in determining the feasibility of deep-sea mining operations, considering both technical and economic factors.

Responsible Development and Environmental Considerations

As the demand for mineral resources continues to rise, the prospect of developing deep-sea mining operations becomes increasingly attractive. However, the delicate ecosystems of the deep-sea are highly susceptible to disruption. Hydrothermal activities around black chimneys[2], for instance, create unique environments inhabited by extraordinary forms of life. Thus, assessing environmental impact is paramount when contemplating utilizing deep-sea mineral resources.

Phosphatization, authigenic processes, and biological mineralization all contribute to the delicate balance of the deep-sea ecosystem. The oxygen minimum zone, an area characterized by low oxygen levels, plays a pivotal role in the occurrence of these processes. Responsible development of deep-sea mineral resources necessitates a holistic approach that prioritizes the preservation of these unique habitats. Environmental impact assessments, regulatory frameworks, and sustainable metallurgy techniques are essential components of this endeavor.

In conclusion, the appeal of deep-sea mineral resources is undeniable, as they provide solutions to the world's resource requirements. However, their exploration, development, and utilization must be conducted with the utmost care and responsibility to protect the delicate ecosystems and unique geological formations. Balancing the pursuit of these treasures with the preservation of the world's most remote and enigmatic environments is a challenge that requires international cooperation and sustainable practices.

Vocabulary

polymetallic：(adj) 多金属的
hydrothermal：(adj) 热液的
sulfide：(n) 硫化物
metallogenic：(adj) 成矿的
placer：(n) 砂矿
rare earth element：(n) 稀土元素
back-arc basin：(n) 弧后盆地
precious metal：(n) 关键金属
abyssal：(adj) 深海的
prospecting method：(n) 找矿方法
resource potential：(n) 资源潜力
evaluation：(n) 评价,评估
process mineralogy：(n) 工艺矿物学
ore-controlling factor：(n) 控矿因素
mining：(n) 采矿
phosphatization：(n) 磷酸盐化
authigenic process：(n) 自生作用
oxygen minimum zone：(n) 最低含氧带
occurrence：(n) 赋存状态
sustainable：(adj) 可持续的
metallurgy：(n) 冶金,冶金术
utilization：(n) 利用

Notes

[1] rare metal 即稀有金属,指在地壳中含量较少、分布稀散的金属元素,具有特殊的物理和化学性质,常用于高科技产业和特殊材料制备。

[2] black chimney 即黑烟囱。“黑烟囱”是指海底富含硫化物的高温热液活动区,因热液喷出时形似“黑烟”而得名。喷溢海底热泉的出口,由于物理和化学条件的改变,含有多种金属元素的矿物在海底沉淀下来,尤其是喷溢口的周围连续沉淀,不断加高,形成了一种烟囱状的地貌,叫作黑烟囱。

Text Sources

Best, 2013. Igneous and Metamorphic petrology. Hoboken:John Wiley & Sons.

Campisano, 2012. Milankovitch Cycles, Paleoclimatic Change, and Hominin Evolution. Nature Education Knowledge, 4(3):5.

Coffin, Whittaker, 2015. Intraplate Magmatism// Harff, Meschede, Petersen, Thiede (eds). Encyclopedia of Marine Geosciences. New York:Springer.

Gill, 2010. Igneous Rocks and Processes: A Practical Guide. Hoboken:John Wiley & Sons.

Molnar, 1988. Continental Tectonics in the Aftermath of Plate Tectonics. Nature, 335(6186): 131-137.

Peppe, Deino, 2013. Dating Rocks and Fossils Using Geologic Methods. Nature Education Knowledge, 4(10):1.

Pinet, 2019. Invitation to Oceanography. Burlington:Jones & Bartlett Learning.

Plank, Manning, 2019. Subducting Carbon. Nature, (574): 343-352.

Rossi, Steel, 2016. The Role of Tidal, Wave and River Currents in the Evolution of Mixed-Energy Deltas: Example from the Lajas Formation (Argentina). Sedimentology, (63): 824-864.

Shearer, 2019. Introduction to Seismology. London:Cambridge University Press.

Stern, 2002. Subduction Zones. Reviews of Geophysics 40(4): 3-1~3-38.

Sverdrup, Johnson, Fleming, 1942. The Oceans: Their Physics, Chemistry, and General Biology. Upper Saddle River:Prentice-Hall.

Thomas, 2011. Paleoceanography// Steele, Thorpe, Turekian (eds.). Encyclopedia of Ocean Sciences. Cambridge:Academic Press.

USGS. https://www.usgs.gov/science/science-explorer/ocean.

Wilson, 1989. Igneous Petrogenesis. New York:Springer.

Winter, 2014. Principles of Igneous and Metamorphic Petrology. Upper Saddle River:Pearson Education.

Woods Hole Coastal and Marine Science Center, USGS. https://www. usgs. gov/centers/whcmsc/science.

PART IV

OCEAN TECHNOLOGY

1.Marine/ocean Acoustics

Text

Marine/ocean acoustics[1] is the study of sound and its behavior in the sea (Figure Ⅳ-1). When underwater objects vibrate, they create sound-pressure waves that alternately compress and decompress the water molecules as the sound wave travels through the sea. Sound waves radiate in all directions away from the source like ripples on the surface of a pond. The compressions and decompressions associated with sound waves are detected as changes in pressure by the structures in our ears and most man-made sound receptors such as a hydrophone[2], or underwater microphone. The basic components of a sound wave are frequency[3], wavelength[4] and amplitude[5].

Frequency is the number of pressure waves that pass by a reference point per unit time and is measured in Hertz (Hz) or cycles per second. To the human ear, an increase in frequency is perceived as a higher pitched sound, while a decrease in frequency is perceived as a lower pitched sound. Humans generally hear sound waves whose frequencies are between 20 and 20,000 Hz. Below 20 Hz, sounds are referred to as infrasonic, and above 20,000 Hz as ultrasonic[6].

Wavelength is the distance between two peaks of a sound wave. It is related to frequency because the lower the frequency of the wave, the longer the wavelength.

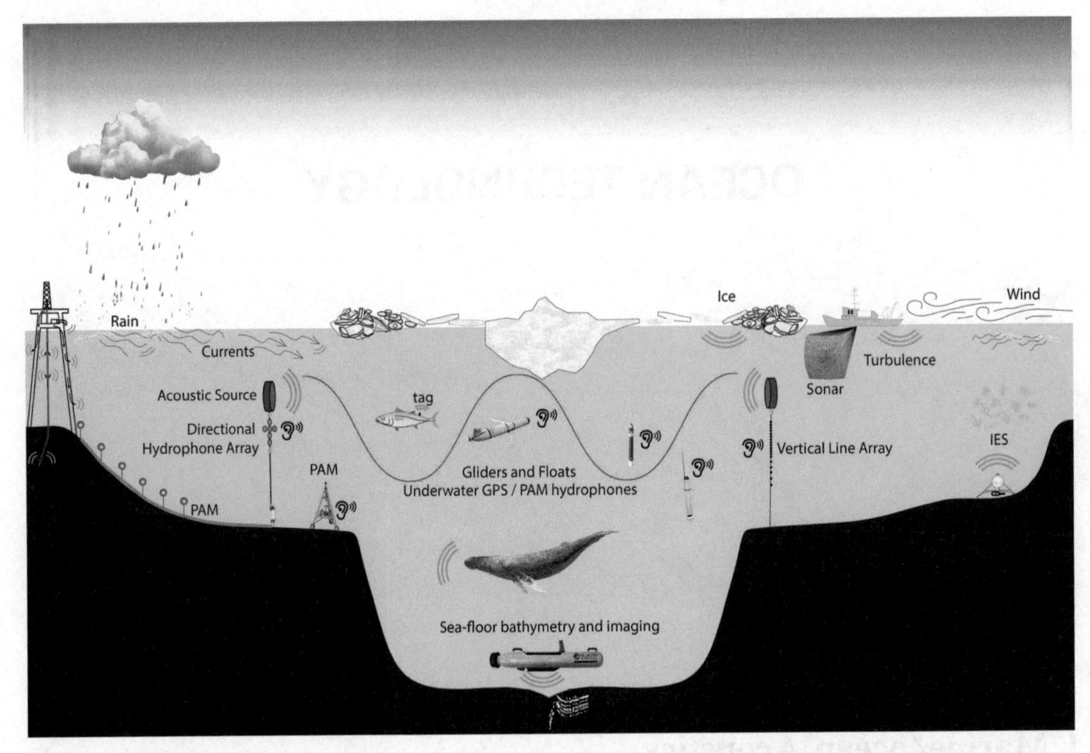

Figure Ⅳ-1: Acoustic monitoring as a marine conservation tool. [Source: imbrsea.eu]

Amplitude describes the height of the sound pressure wave or the "loudness" of a sound and is often measured using the decibel (dB) scale[7]. Small variations in amplitude ("short" pressure waves) produce weak or quiet sounds, while large variations ("tall" pressure waves) produce strong or loud sounds.

The decibel scale is a logarithmic scale used to measure the amplitude of a sound. If the amplitude of a sound is increased in a series of equal steps, the loudness of the sound will increase in steps which are perceived as successively smaller. A decibel doesn't really represent a unit of measure like a yard or meter, but instead a pressure value in decibels expresses a ratio between the measured pressure and a reference pressure. On the decibel scale, everything refers to power, which is amplitude squared. It is worth noting that the reference pressure in air differs from that in water. Therefore a 150 dB sound in water is not the same as a 150 dB sound in air. So when you are describing sound waves and how they behave it is very important to know whether you are describing sound in the sea or in air.

The speed of a wave is the rate at which vibrations move through the medium. Sound moves at a faster speed in water (1500 meters/sec) than in air (about 340 meters/sec) because the mechanical properties of water differ from air. Temperature also affects the speed of sound (e.g. sound travels faster in warm water than in cold water) and is very influential in some parts of the ocean. Remember that wavelength and frequency are related because the lower the frequency, the longer the wavelength. More specifically, the wavelength of a sound equals the speed of sound in either air or water divided by the frequency of the wave. Therefore, a 20 Hz sound wave is 75 m long in the water (1500/20 = 75), but only 17 m long (340/20 = 17) in air.

Sound in the sea[8] can often be "trapped" and effectively carried very long distances by the "deep sound channel" that exists in the ocean. This SOFAR[9] or SOund Fixing And Ranging channel is so named because it was discovered that there was a "channel" in the deep ocean within which the acoustic energy from a small explosive charge (deployed in the water by a downed aviator) could travel over long distances. An array of hydrophones could be used to roughly locate the source of the charge thereby allowing rescue of downed pilots far out to sea. Sound, and especially low-frequency sound, can travel thousands of meters with very little loss of signal.

The field of ocean acoustics provides scientists with the tools needed to quantitatively describe sound in the sea. By measuring the frequency, amplitude, location and seasonality of sounds in the sea, a great deal can be learned about our oceanic environment and its inhabitants. Hydroacoustic monitoring (listening to underwater sounds) has allowed scientists to measure global warming, listen to earthquakes and the movement of magma through the sea floor during major volcanic eruptions, and record low-frequency calls of large whales the world over. As our oceans become noisier each year, the field of ocean acoustics will grow and only become more essential.

Vocabulary

acoustics: (n) 声学
behavior: (n) 行为,性能,表现
vibrate: (vi/vt)(使)振动
  n: vibration 振动
alternately: (adv) 交替地,轮流地
  compress: (vt) 压缩
  ant: decompress 解压缩
molecule: (n) 分子
  adj: molecular 分子的
  ~atom: (n) 原子
  ~ion: (n) 离子
ripple: (n) 波纹;(vi) 扩散,涌起
pond: (n) 池塘
hydrophone: (n) 水听器,水中地震检波器
frequency: (n) 频率
wavelength: (n) 波长
amplitude: (n)(声音、无线电波等的)振幅
Hertz: (n) 赫,赫兹(声波频率单位)
  pl: Hertz
higher pitched sound: 更高的音调
infrasonic: (adj) 次声的,亚(低于)音频的
ultrasonic: (adj) 超声的;(n) 超声波
peak: (n) 高峰,尖端
loudness: (n) 响度,音量
decibel: (n) 分贝(声音强度的单位)
logarithmic: (adj) 对数(性,式)的

successively:(adv) 连续,先后,渐渐
  adj:successive
yard:(n) 码(长度单位,1 yard 等于 3 ft 或 0.9144 m)
square:(vt) 使成二次幂;(n) 正方形,平方
medium:(n) 介质
mechanical:(adj) 机械学的,力学的
trap:(vt) 卡住,捕获
explosive:(adj) 易爆炸的,突增的
aviator:(n) 飞行员,飞机驾驶员
rescue:(n) 援救,营救
quantitatively:(adv) 定量化地
seasonality:(n) 季节性
inhabitant:(n) 居民,栖息动物
hydroacoustic monitoring:水声监测
magma:(n) 岩浆,熔岩
volcanic eruption:火山爆发

Notes

[1] marine/ocean acoustic 即海洋声学,是一门专注于研究声波在海洋中的传播规律以及利用声波探测海洋的科学,具有极高的应用价值。除了军事领域的应用外,水下声技术已成为探测和开发海洋资源的重要手段。海洋声学的基本内容涵盖三个方面:一是探究声波在海洋中的传播规律以及海洋条件对声传播的影响,这包括不同水文条件和海底底质条件下声波的传播特征,海底对声波传播的作用,海水对声的吸收特性,声波的起伏现象,散射效应和海洋噪声等问题;二是利用声波进行海洋探测,这不仅可以测量海洋的深度,还能发现隐藏在海底的石油资源等;三是海洋声学技术和仪器的发展,其中,各种类型的声呐设备是海洋声学技术中的佼佼者。

[2] hydrophone 即水听器,也称为接收换能器,是一种能够将水中的声信号转换成电信号的装置。它主要用于接收水中的声信号。水听器广泛用于水下通信、探测、目标定位与跟踪等领域,是声呐系统中不可或缺的关键组件。无论是进行水下的探测与识别、通信交流,还是开展海洋环境监测与海洋资源开发,都离不开水听器的技术支持。

[3] frequency 即频率,是指单位时间内完成周期性变化的次数,是衡量周期运动频繁程度的物理量,单位为每秒。为了纪念德国物理学家赫兹的杰出贡献,人们把频率的单位命名为赫兹,简称“赫”,并用符号 Hz 来表示。每个物体都拥有一个由其本身性质决定的、与振幅无关的频率,这被称为物体的固有频率。频率的概念不仅在力学和声学领域有着广泛的应用,在电磁学、光学以及无线电技术等领域也经常被使用。

[4] wavelength 即波长,是指波动在一个完整的振动周期内所传播的距离。具体而言,它是沿着波的传播方向,相邻两个振动相位相差 2π 的点之间的距离。波长可通过波速与周期的乘积来计算得出。

[5] amplitude 即振幅,是指振动过程中物理量所能达到的最大偏离值,用于描述振动的

幅度和强度。在声振动中,振幅具体指的是声压(由声波引起的动态压强)相对于静止压强的最大变化量。声波振幅的大小决定了声音的强度,即振幅越大,听起来越响亮;反之,振幅越小,听起来越安静。

[6] 声音是一种机械振动,能够在各种物态的介质中传播。我们日常听到的声音是具有一定频率的声波。人耳的听觉频率范围大致为 20~20 000 Hz,超出这一频率范围的声音人耳无法察觉。频率低于 20 Hz 的声波被称为次声波,而频率高于 20 kHz 的声波被称为超声波。声音的频率越高,我们感知到的音调就越高;相应地,声音的频率越低,我们感知到的音调越低。

[7] decibel 即分贝,是声波振幅相对大小的单位,用于量度两个相同单位数量之间的比例关系。

[8] 海水中声速指的是声波在海水这一介质中的传播速度。影响声波在海水中传播速度的主要因素包括温度、盐度和压强(或深度)等,其中温度的变化对声速的影响最为显著。由于海水中温度、盐度的分布不均,会导致声速分布也不均匀,这种现象称为海洋中的声速梯度,声波在海水中的传播速度(约为 1531 m/s)比在空气中的传播速度快 3.5 倍。当已知海水的温度、盐度和深度时,可以计算出对应的声速值。用于直接测量海水中声速的仪器被称为声速仪,其工作原理是通过测量声脉冲在一段已知距离内传播的时间间隔,进而计算出声速。

[9] SOFAR 通道 (SOFAR channel),也称深海低频声学通道 (deep sound channel),是海洋中一种特殊的低频声音传播路径。在这个通道里,低频声波能够以最极小的能量损失传播至非常远的距离。SOFAR 通道之所以具有如此卓越的性能,是因为海洋中声速剖面的特殊分布产生了"通道折射"效应,这种效应使得声波能够避免与海面或海底的直接接触,从而减小了能量的损耗。

2. Sound Absorption

Text

In acoustics, absorption[1] refers to the process by which a material, structure, or object takes in sound energy when sound waves are encountered, as opposed to reflecting[2] the energy (Figure Ⅳ-2). Part of the absorbed energy is transformed into heat and part is transmitted through the absorbing body. The energy transformed into heat is said to have been "lost".

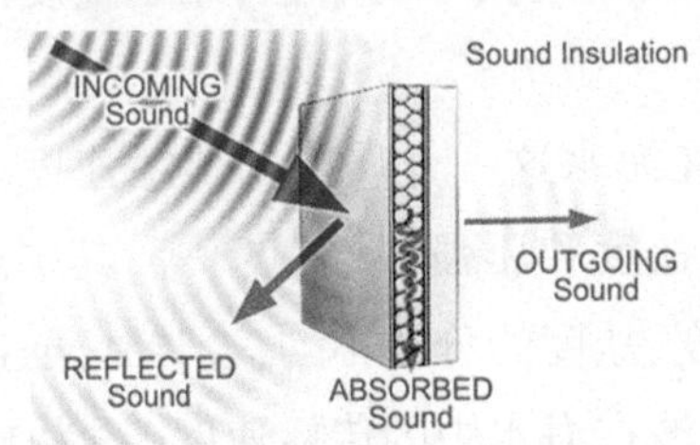

Figure Ⅳ-2: Absorption of sound. [Source: hushcitysp.com]

When sound from a loudspeaker collides with the walls of a room, part of the sound's energy is reflected back into the room, part is transmitted through the walls, and part is absorbed into the walls. Just as the acoustic energy was transmitted through the air as pressure differentials (or deformations), the acoustic energy travels through the material which makes up the wall in the same manner. Deformation causes mechanical losses via conversion of part of the sound energy into heat, resulting in acoustic attenuation, mostly due to the wall's viscosity. Similar attenuation mechanisms apply to the air and any other medium through which sound travels.

The fraction of sound absorbed is governed by the acoustic impedances[3] of both media and is a function of frequency and the incident angle. Size and shape can influence the sound wave's behavior if they interact with its wavelength, giving rise to wave phenomena such as standing waves and diffraction.

Acoustic absorption is of particular interest in soundproofing[4]. Soundproofing aims to absorb as much sound energy (often in particular frequencies) as possible converting it into heat or transmitting it away from a certain location. In general, soft, pliable, or porous materials (like cloths) serve as good acoustic insulators—absorbing most sound, whereas dense, hard, impenetrable materials (such as metals) reflect most.

The absorption of sound in seawater is the frequency-dependent reduction in sound intensity[5] due to the energy loss (conversion of acoustic energy into heat) through viscous and structural relaxation effects and through molecular relaxation processes associated with dissolved electrolytes in the seawater. Relaxation is the process where the medium, here the water, returns to its former state after having been exposed to the pressure variation in a sound wave. Absorption is measured in terms of decibel loss in sound pressure amplitude at a specific frequency over a characteristic length or the number of times the sound pressure amplitude has decreased by 1/e [1 Neper (Np)][6] of its original value over a characteristic length (i.e., typically, dB/km or Np/km). Neper is a dimensionless unit named after the inventor of the logarithms, John Napier[7]. The attenuation of sound in seawater includes absorption and other mechanisms that affect the amplitude of the observed sound wave, such as scattering from particles in suspension, fish, other biological entities, and other fluid inhomogeneities in the water column, as well as changes due to surface and bottom scattering[8]. Also spherical or cylindrical spreading of sound waves leads to attenuation of acoustic signals.

The absorption of sound in seawater is dependent upon factors, such as salinity, temperature, pH, and pressure[9]. Water is an associated polar liquid and in freshwater, i.e., water free of contaminates, the absorption of sound is due to dissipation of acoustic energy, i.e., the transfer of acoustic energy to thermal energy through the action of shear and bulk viscosity, where the bulk viscosity effect includes the effect of structural relaxation. On the passage of a sound wave through water some molecules break their structural bonds and move from the normal to the close-pack arrangement. This process will involve a time lag and hence acoustic absorption takes place. In seawater other loss mechanisms tend to dominate. In the past several decades, the research on the absorption of sound in seawater has focused on understanding the molecular chemical relaxation processes associated with electrolytes in seawater where ionic dissociation is alternately activated and deactivated by the sound wave's condensation and rarefaction phases, and the influence of acoustic energy on electrolyte association and dissociation rates.

Vocabulary

absorption:(n) 吸收
take in:吸收,领会,接待
reflect:(vt) 反射
transform:(vt) 使转化,使改变
transmit:(vt) 使通过,传(热、声等)
collide:(vi/vt)(使)碰撞
deformation:(n) 变形
conversion:(n) 转换
attenuation:(n) 衰减,稀释,变薄
viscosity:(n) 黏度,黏性
fraction:(n) 分数
impedance:(n) 阻抗
incident angle:入射角
standing wave:驻波
diffraction:(n) 衍射
soundproofing:(n) 隔音
pliable:(adj) 柔韧的,可塑的
porous:(adj) 多孔的,透气的,透水的

insulator：(n) 隔热(或绝缘、隔音等的)材料(或装置)
impenetrable：(adj) 不可逾越的，穿不过的
reduction：(n) 减少
electrolyte：(n) 电解质，电解液
dimensionless：(adj) 无量纲的
suspension：(n) 暂停，悬浮物
biological entity：生物体
fluid inhomogeneity：流体不均匀性
salinity：(n) 盐分，含盐量
polar liquid：极化液体
contaminate：(vt) 污染，使不纯
dissipation：(n) 损耗，耗散
thermal：(adj) 热(量)的
shear and bulk viscosity：剪切黏度和体积黏度
structural relaxation：结构弛豫
ionic dissociation：电离离解
condensation and rarefaction phases：凝聚和稀疏相

Notes

[1] sound absorption 即声吸收，指声波通过介质或射到介质表面时，其声能减少的过程。这一过程主要归因于介质的黏滞性、热传导性和分子弛豫过程，这些因素导致有序的声运动能量不可逆地转变为无序的热运动能量。

[2] acoustic reflection 即声反射，指当声波从一种介质传播到声学特性不同的另一种介质时，在两种介质的界面处将发生反射现象，使得入射声波的一部分能量被反射回第一种介质中。

[3] acoustic impedance 即声波阻抗，又称为声阻抗或音阻，是介质所具有的一种物理性质，它反映了介质对声波传播所产生的阻碍作用。

[4] soundproofing 即隔音，是指采取多种措施以降低声源对人体产生的声压影响。减少声音的方法有多种，比如增大声源与接收点之间的距离、利用隔音屏障或隔音板来反射和吸收声波，以及运用主动降噪技术来消除噪声。

[5] 海水中声波的吸收程度与声波的频率有关。

[6] Neper 是用于以自然对数的形式来表示两个场量之间比值的一种单位或度量方式。

[7] John Napier (1550–1617)，苏格兰数学家，对数的发明者。

[8] 海水中声波的衰减主要包括声吸收和声散射等现象。

[9] 海水中声波的吸收与多个因素有关，包括海水的温度、盐度、压强和 pH 值等。

3.Sound Scattering

Text

Seawater has bubbles, suspended particles[1], organisms[2], and many other things that affect sound as it travels through seawater it. The use of a flashlight can be used as an example to illustrate what happens. Most of the time, the flashlight creates a circle of light on objects at which it is pointed (Figure Ⅳ-3).

However, what happens when it is foggy out? You may have noticed that you can now see the beam of the flashlight. What is different? When it is foggy out, the air contains many water molecules. When you shine the flashlight in fog, the light is scattered in all directions off the water molecules, making the beam of the flashlight visible. When the light is scattered, it does not travel as far. Therefore on foggy nights the headlights of a car do not project as far as on clear nights.

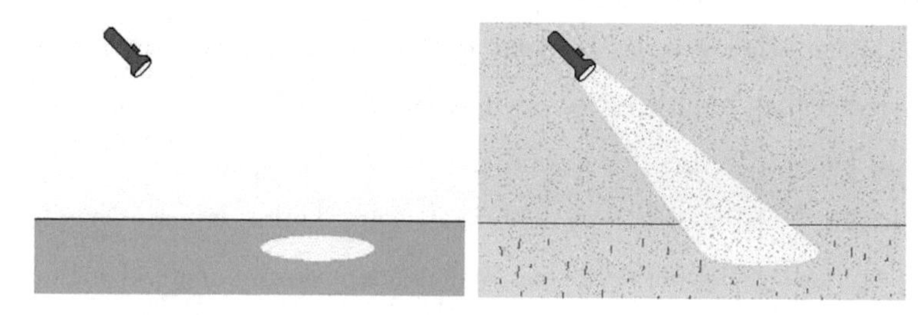

Figure Ⅳ-3: Diagrams of light scattering. [Source: dosits.org]

Sound is a wave just like light. The same thing happens in the ocean to sound waves that happens to the flashlight beam. The amount of scattering is affected by the size of the object (the scatterer) and the wavelength of the sound[3]. An object will be a significant scatterer if its size is comparable to or bigger than the wavelength of the sound. You can calculate the wavelength of sound in seawater by dividing the speed of sound (approximately 1500 meters per second) by the frequency of the sound. For example, a 200 Hz sound will have a wavelength of 7.5 meters (1500 meters per second divided by 200 cycles per second = 7.5 meters). If the object is much smaller than one wavelength of the sound, the sound will tend to travel around the object in its path and not be significantly affected. That is why scientists use high frequency sound to look for small objects, such as fish, in the ocean.

Scattering happens when the path of a sound wave is broken up by objects (volume scattering[4]) or by the sea floor or sea surface (boundary scattering[5]).

Suspended particles and inhomogeneities of the ocean water cause scattering. Scatterers can be small, like suspended sediment, bubbles, or plankton, or larger, like fishes or whales. In most ocean environments, marine life is the primary source of volume scattering. The degree of scattering depends on the wavelength of the sound relative to the size of the particle, the particle density, and particle shape. Inhomogeneities are fluctuations in salinity, density, or temperature that cause scattering. These can be ocean fronts[6], eddies[7], or stratified layers of the ocean with a density gradient at the boundary on large or small scales.

Volume scattering usually decreases with increasing ocean depth, because at greater depth fewer particles exist and density gradients and other inhomogeneities are reduced. One exception is the Deep Scattering Layer (DSL)[8], a concentrated layer of marine organisms that creates strong scattering that can sometimes resemble surface scatter.

A rough sea surface can be a highly effective but also highly variable scatterer[9]. This roughness can have scales ranging from millimeter ripples to large storm waves. Scattering at the sea surface is complex due to the variability of wind and waves and the presence of bubbles. This scattering depends on the angle at which the sound approaches the sea surface, the wind speed, the wavelength, and the presence of bubbles. Greater scattering generally occurs at smaller wavelengths and higher wind speeds. In polar regions, sea ice may cause scattering levels as much as 40 dB greater than in an ice-free environment[10]. The roughness of the underside of the ice is a significant factor affecting the scattering.

Bottom scattering[11] is another type of surface scattering but is more complicated to predict than sea surface scattering. Bottom scattering is dependent on the angle of incidence, the sound wavelength, the surface roughness, seafloor type (rock or sediment, and composition), sediment layering, grain size, and grain size distribution. The relationship between grain size and scattering is unclear in many cases; surface roughness may be a better indicator of scattering strength. When the grain size of the bottom is representative of sand (0.0625 to 2 millimeters), individual grains may act as scatterers along with the surface as a whole at some frequencies. Understanding bottom scattering may also be complicated by additional factors, such as porosity and sound transmission into the seafloor and scattering from subsurface sediment layers.

Vocabulary

bubble:(n)气泡

suspended particle:悬浮颗粒

organism:(n)生物,生物体,有机体

flashlight:(n)手电筒

beam:(n)光束

scatter:(vi)分散

visible:(adj)可见的

scattering:(n)散射

comparable:(adj) 可比较的,相当的
volume scattering:体积散射
boundary scattering:边界散射
suspended sediment:悬沙
plankton:(n) 浮游生物
fluctuation:(n) 波动,起伏,涨落
front:(n) 锋面,前面
eddy:(n) 旋涡,涡流
stratified layer:分层
density:(n) 密度
gradient:(n) 梯度
concentrate:(vi) 使……集中,聚集
resemble:(vt) 像,类似于
rough:(adj) 粗糙的
millimeter:(n) 毫米
polar region:极地
complicated:(adj) 复杂的
seafloor:(n) 海底
grain size distribution:晶粒度分布
porosity:(n) 孔隙度,疏松

Notes

[1] suspended particle 即悬浮颗粒,指能够在海水中保持悬浮状态一段时间的固体颗粒。

[2] organism 即生物,又称生物体、有机体、机体,是指任何以“单一实体”形式运作的有机生命系统,或在自然界中能够展现出生命现象(如代谢、生长、发育、感应、运动、生殖)的个体。

[3] 声散射的强弱与散射体的大小以及入射声波的波长有关。

[4] volume scattering 即体积声散射,指由于海水中存在的各种不均匀体(如鱼类等)所导致的声散射现象。

[5] boundary scattering 即边界声散射,指发生在海洋边界(包括海面和海底)上的声散射现象。

[6] ocean front 即海洋锋面,指特性上存在显著差异的两种或多种水体之间形成的狭窄过渡带。这些特性可以用温度、盐度、密度、速度、颜色和叶绿素等要素的水平梯度,或者它们的更高阶导数来描述;也就是说,一个锋面的具体位置可以通过一个或多个上述要素的特征强度来界定。

[7] ocean eddy 即海洋涡旋,也称中尺度涡旋,指海洋中直径为 100~300 km、寿命为 2~10 个月的涡旋。相较于常见的用肉眼可见的涡旋,中尺度涡旋具有更大的直径和更长的寿命;然而,与全年存在的海洋环流相比,其规模又小得多,故称为中尺度涡旋。中尺度涡旋通常分为两类:气旋式涡旋(在北半球表现为逆时针旋转)和反气旋式涡旋(在南半球表现为逆时针旋转)。

[8] deep scattering layer (DSL) 即深水散射层,指深海(极地除外)中厚度达数百米能强烈散射声波的水层。这种散射现象主要由鱼类、小虾和乌贼等小型海洋动物的群集活动所引起。由于不同的海洋生物群体分布在不同的深度,因此深水散射层通常呈现出多层结构。同时,这些生物具有趋光性,深海散射层会表现出昼夜迁移的特性,

即白天下降、夜间上升,且夜晚的散射强度通常比白天更为显著。深水散射层的存在对主动声呐系统造成了混响干扰,但同时也为潜艇的隐蔽行动提供了有利条件。

[9] 海表散射是指由于海面不平整以及海面存在气泡薄层等因素,导致声波在传播过程中发生散射的现象。由于海表面的多变性和复杂性,理论计算的结果和实际观测到的数据之间往往存在一定的差异。

[10] 在极地海区,海冰对声波产生的散射作用更为强烈。

[11] 海底散射是指由于海底的不平整性和粗糙表面特性,导致声波在传播过程中与之相互作用而发生散射的现象。

4. Ocean Optics

Text

Ocean optics[1] is the study of how light interacts with water and the materials in water. Although research often focuses on the sea, the field broadly includes rivers, lakes, inland waters, coastal waters, and large ocean basins. How light behaves in water is critical to how ecosystems function underwater. Knowledge of ocean optics is needed in aquatic remote sensing research in order to understand what information can be extracted from the color of the water as it appears from satellite sensors in space. The color of the water as seen by satellites is known as ocean color[2]. While ocean color is a key theme of ocean optics, optics is a broader term that also includes the development of underwater sensors using optical methods to study much more than just color, including ocean chemistry, particle size, imaging of microscopic plants and animals, and more (Figure Ⅳ-4).

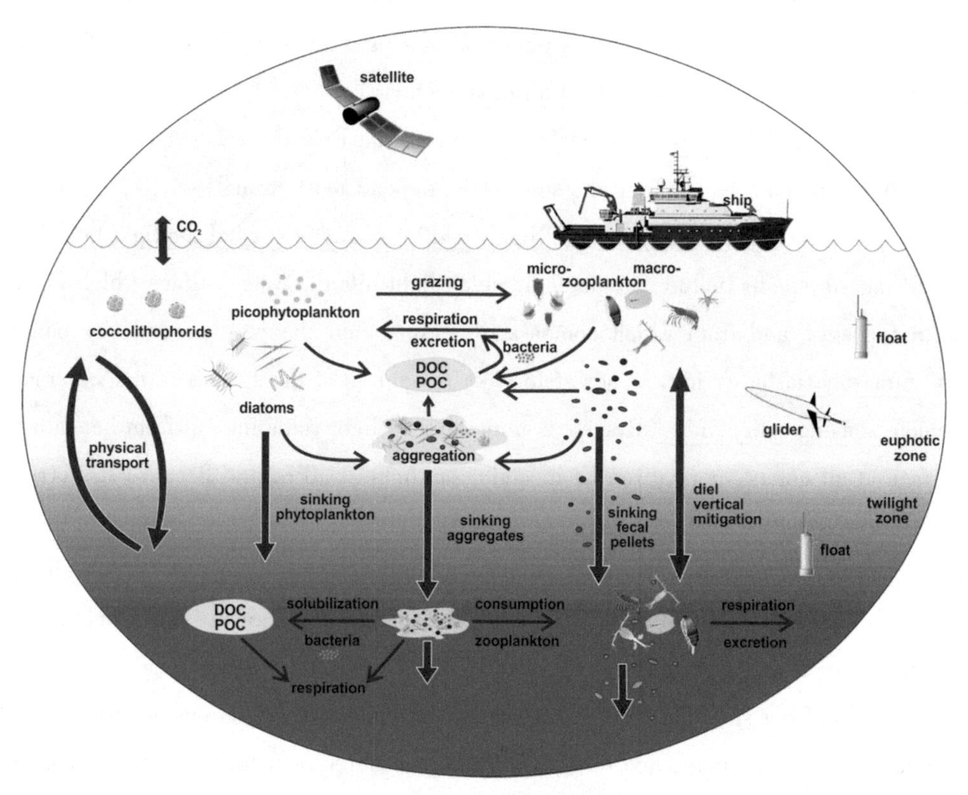

Figure Ⅳ-4: Schematic of processes that need to be measured to fully understand ocean productivity and carbon sequestration. Many of these topics involve optical measurements. [Source: Siegel et al., 2016]

Inherent optical properties (IOPs)[3] depend on what is in the water only. These properties stay the same no matter what the incoming light is doing (daytime or nighttime, low sun angle or high sun angle).

Water with large amounts of dissolved substances, such as lakes with large amounts of colored dissolved organic matter (CDOM)[4], experiences high light absorption. Besides, phytoplankton and other particles also absorb light.

Areas with sea ice, estuaries with large amounts of suspended sediments, and lakes with large amounts of glacial flour are examples of water bodies with high light scattering. All particles scatter light to some extent, including plankton, minerals, and detritus[5]. Particle size affects how much scattering happens at different colors; for example, very small particles scatter light exponentially more in the blue colors (wavelengths) than other colors, which is why the ocean and the sky are generally blue (called Rayleigh scattering[6]). Without scattering, light would not "go" anywhere (outside of a direct beam from the sun or other source) and we would not be able to see the world around us.

Attenuation in water, also called beam attenuation or the beam attenuation coefficient, is the sum of all absorption and scattering. Attenuation of a light beam in one specific direction can be measured with an instrument called a transmissometer[7].

Apparent optical properties (AOPs)[8] depend on what is in the water (IOPs) and what is going on with the incoming light from the sun. AOPs depend most strongly on IOPs and only depend somewhat on incoming light, aka the "light field". Characteristics of the light field that can affect AOP measurements include the angle at which light hits the water surface (high in the sky vs. low in the sky, and from which compass direction) and the weather and sky conditions (clouds, atmospheric haze, fog, or sea state, aka roughness of the surface of the water).

Remote sensing reflectance (Rrs)[9] is a measure of light radiating out from beneath the ocean surface at all colors, normalized by incoming sunlight at all colors. Because Rrs is a ratio, it is slightly less sensitive to what is going on with the light field (such as the angle of the sun or atmospheric haziness). Rrs is measured using two paired spectroradiometers[10] that simultaneously measure light coming in from the sky and light coming out from the water below at many wavelengths. Since it is a measurement of a light-to-light ratio, the energy units cancel out, and Rrs has the units of per steradian due to the angular nature of the measurement (upwelling light is measured at a specific angle, and incoming light is measured on a flat plane from a half-hemispherical area above the water surface).

Kd is the diffuse (or downwelling) coefficient of light attenuation (Kd)[11], also called simply light attenuation, the vertical extinction coefficient, or the extinction coefficient. Kd describes the rate of decrease of light with depth in water, in units of per meter. The "d" stands for downwelling light, which is light coming from above the sensor in a half-hemispherical shape (aka half of a basketball). Scientists sometimes use Kd to describe the decrease in the total visible light available for plants in terms of photosynthetically active radiation (PAR)[12]—called "Kd(PAR)." In other cases, Kd can describe the decrease in light with depth over a spectrum of colors or wavelengths, usually written as "Kd(λ)." At one color (one wavelength) Kd can describe the decrease in light with depth of one color, such as the decrease in blue light at the wavelength 490 nm, written as "Kd(490)." In general, Kd is calculated using Beer's Law[13] and a series of light measurements collected from just under the water surface down through the water at many depths.

"Closure" refers to how optical oceanographers measure the consistency of models and measurements. Models refer to anything that is not explicitly measured in the water, including satellite-derived variables that are estimated using empirical relationships (for example, satellite-derived chlorophyll-a concentration is estimated from the ratios between green and blue remote sensing reflectance using an empirical relationship). Closure includes measurement closure, model closure, model-data closure, and scale closure. In model-data closure experiments, misalignment between data and models may occur, due to measurement error, issues with the model, both, or some other external factor.

Ocean optics has been applied to study topics like primary production, phytoplankton, zooplankton, shallow-water habitats like seagrass beds and coral reefs, marine biogeochemistry, heating of the upper ocean, and carbon export to deep waters by way of the ocean biological pump. The portion of the electromagnetic spectrum usually involved in ocean optics is from ultraviolet through infrared, covering wavelengths of about 300 nm to less than 2000 nm wavelengths.

Vocabulary

ocean optics: (n) 海洋光学
basin: (n) 盆地
aquatic: (adj) 水生的,水的
ocean color: (n) 海洋水色
theme: (n) 主题,主旋律
microscopic: (adj) 微小的
inherent optical properties: 固有光学特性
sun angle: 太阳角
dissolved: (adj) 可溶的
substance: (n) 物质
colored dissolved organic matter:有色可溶性有机物,黄色物质
phytoplankton: (n) 浮游植物
estuary: (n)(江河入海的)河口
glacial flour: 冰川粉粒
mineral: (n) 矿物质
detritus: (n) 碎屑
Rayleigh scattering: 瑞利散射
transmissometer: (n) 透射计,能见度计
apparent optical properties: 表观光学特性

light field：光场
characteristic：（n）特征，特点
atmospheric haze：大气霾雾
remote sensing：（n）遥感
reflectance：（n）反射比，反射率
spectroradiometer：（n）光谱仪，分光辐射度计
simultaneously：（adv）同时地
steradian：（n）立体角
diffuse/downwelling coefficient of light attenuation：漫衰减系数
extinction：（n）灭绝，消失
photosynthetically active radiation：光合有效辐射
Beer's Law：比尔定律
explicitly：（adv）明确地，明白地
empirical relationship：经验关系
chlorophyll-a：（n）叶绿素 *a*
concentration：（n）浓度
misalignment：（n）未对准，偏移
primary production：初级生产力
zooplankton：（n）浮游动物
seagrass bed：海草床
coral reef：珊瑚礁
marine biogeochemistry：海洋生物地球化学
carbon export：碳输出
biological pump：生物泵
electromagnetic：（adj）电磁的
ultraviolet：（adj）紫外线的
infrared：（adj）红外线的

Notes

[1] ocean optics 即海洋光学，是一门专注于探究海洋的光学性质、光在海洋中的传播规律及运用光学技术探测海洋的学科。它既是海洋物理学的分支学科，也属于光学的分支学科。光电子学方法是海洋光学测量的主要手段，基础研究中包括实验和理论两个方面。

[2] ocean color 即海洋水色，指海洋水体所呈现的颜色，主要由海水自身的光学性质，具体包括其对光的吸收和散射特性所决定。为了最大程度上削弱海面反射光（即自然白光）的干扰，我们通常从海面的垂直上方进行观察，此时所看到的颜色即为海洋水色，它反映了海水内在的物理和化学特性。常用的海洋水色观测方法主要包括两类：一类是基于传统手段的水色计比色法；另一类是利用现代技术的水色遥感观测法。

[3] inherent optical properties（IOPs）即固有光学特性，指水体自身的吸收和散射性质的各种量度，这些特性不依赖周围环境光场的条件，是水体光学性质的内在表现。

[4] colored dissolved organic matter（CDOM）即有色可溶性有机物，也称黄色物质，是溶解于水体中的一类可通过光学手段进行测量的有机成分。CDOM 主要源自腐烂物质释放的单宁酸等化合物。在光谱的短波区域，尤其是蓝色至紫外波段，CDOM 展现出强烈的吸收特性。

[5] detritus 即海洋碎屑，是海水悬浮颗粒物的重要组成部分，主要以颗粒有机碳的形式

存在。其来源广泛,既包括通过河流径流或大气传输进入海洋的陆源物质,也涵盖海洋环境中自生的物质。海洋碎屑中含有浮游植物细胞、浮游动物的外骨骼碎片及其排泄的粪粒等。在近岸海域,碎屑中还可能包含海绵骨针、海草残骸以及人类活动产生的污染物等。在近岸,海洋碎屑的含量呈现出显著的垂直变化特征,在真光层以下,其含量随着深度的增加而逐渐减小,且减少的速率相对恒定。此外,随着深度的增加,碎屑颗粒的尺寸也呈现出逐渐减小的趋势。

[6] Rayleigh scattering 即瑞利散射,又称“分子散射”,指当散射粒子的尺寸远小于入射光波长(小于波长的 1/10)时,散射光在各个方向上的强度分布不均的一种现象。此现象中,散射光的强度与入射光波长的四次方成反比关系。由于这一特性,太阳光谱中波长较短的蓝紫光相较于波长较长的红色光,会受到更为显著的散射作用。而在短波成分中,蓝光又具有最大的能量。因此,在雨过天晴或秋高气爽时(空气中较粗的微粒相对较少,主要以分子散射为主),大气分子对蓝色光有着强烈的散射作用,使得蓝色光弥漫至整个天空,从而呈现出蔚蓝色的景象。

[7] transmissometer 即透射计,用于测量海水的消光系数。消光系数(extinction coefficient)反映了被测溶液对光的吸收程度,被测溶液的浓度较高时,溶液显色后的颜色相对较深,对光的吸收作用也更为显著,从而导致光线的透射率低;反之,若溶液浓度较低,光线的透射率则会较高。

[8] apparent optical properties (AOPs) 即表观光学特性,是指水体光学性质的直接观测结果,他们强烈依赖于 IOPs,同时也受环境光场的显著影响。

[9] remote sensing reflectance (Rrs) 即遥感反射率,指海面离水辐亮度与入射到海面的辐照度之比。

[10] spectroradiometer 即分光辐射度计,用来测定光源辐射度(或任何其他的辐射量)随波长的具体分布情况。

[11] diffuse (or downwelling) coefficient of light attenuation (Kd) 即漫衰减系数,描述水体中光强随水深增加而减弱的程度。

[12] photosynthetically active radiation (PAR) 即光合有效辐射,指能够对植物光合作用产生促进效应的太阳辐射成分,其波长区间主要覆盖 400~700 nm,与可见光光谱范围大致相符。随太阳高度角的提升,PAR 在太阳直接辐射中所占的比例会随之增加,最高可达 45%。相比之下,在散射辐射中,PAR 的比例更为显著,可达 60%~70%,这意味着在多云天气条件下, PAR 占总辐射的比例反而会有所提升。总体而言,PAR 平均约占太阳总辐射的 50%左右。

[13] Beer's Law 即比尔定律,是分光光度分析领域的基本定律,它阐述的是物质对特定波长光的吸收强度与该物质的浓度及其所在溶液层厚度之间的关系。

5. The Solar Spectrum

Text

At the mean distance of the Earth from the Sun, the solar irradiance[1] from photons of all wavelengths, *Es*, is

$$Es = 1368 \pm 5 \text{ W} \cdot \text{m}^{-2}$$

Although *Es* is historically called the solar constant[2], its value varies by a fraction of a percent on time scales of minutes to decades. A better term is therefore the total solar irradiance. Moreover, the total solar irradiance received by the Earth varies from about 1322 to 1413 $\text{W} \cdot \text{m}^{-2}$ over the course of a year, owing to the ellipticity of the Earth's orbit about the sun.

The energy of a photon is inversely proportional to its wavelength. Furthermore, the number of solar photons per wavelength interval is not uniform over the electromagnetic spectrum. Figure Ⅳ-5 shows the measured wavelength dependence of the solar spectral irradiance $Es(\lambda)$. The sharp dips in the $Es(\lambda)$ curve are Fraunhofer lines[3], which are due to selective absorption of solar radiation by elements in the sun's outer atmosphere. These lines are typically less than 0.1 nm wide, and they are much "deeper" (the irradiances within the lines are much less) than indicated in Figure Ⅳ-5, which gives $Es(\lambda)$ values averaged over much wider bands. The resolution of Figure Ⅳ-5 is 1 nm below 630 nm, 2 nm between 630 and 2,500 nm, and 20 nm beyond 2,500 nm. For example, the prominent line centered at $\lambda = 486.13$ nm decreases to 0.2 of the $Es(\lambda)$ values just outside the line at $\lambda = 486.05$ nm and 486.25 nm; the line depth is thus said to be 0.2.

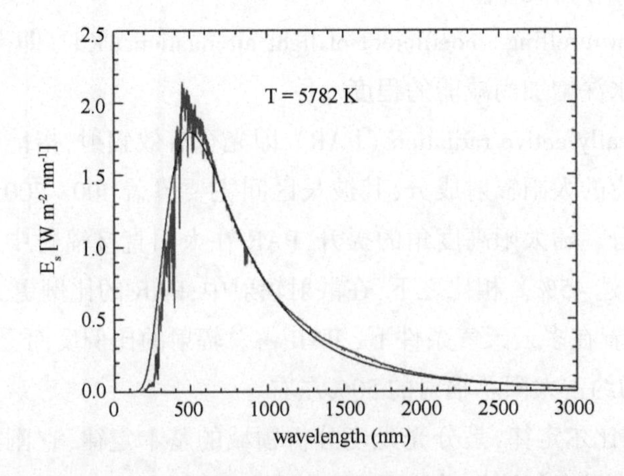

Figure Ⅳ-5: The solar spectrum. [Source: oceanopticsbook.info]

The blue curve in Figure Ⅳ-5 shows the blackbody[4] irradiance for a temperature of 5,782 K. This is the temperature for which a perfect absorber and emitter, or blackbody, emits the same total irradiance as the total solar irradiance of 1368 Wm^{-2}. The blackbody spectrum is a reasonably good approximation for the sun's spectral irradiance at infrared wavelengths, where the solar spectrum is never more than 25% different from the blackbody curve. However, the solar spectrum differs greatly from the black body curve at ultraviolet and visible wavelengths. Solids and liquids emit radiation that is well approximated by the blackbody curve at the appropriate temperature. Gases, however, show selective absorption and emission over very narrow wavelength ranges, as seen in the Fraunhofer lines. The gases in the solar atmosphere thus do not absorb and emit like a blackbody.

It is seldom necessary for optical oceanographers to concern themselves with the detailed wavelength dependence of $Es(\lambda)$. It is usually sufficient to deal with $Es(\lambda)$ values averaged over bandwidths of size $\Delta\lambda \approx 5$ to 20 nm, which correspond to the bandwidths of the optical instruments routinely used in underwater measurements and remote sensing.

What's relevant to oceanography is not the solar irradiance at the top of the atmosphere, but rather the sunlight that actually reaches the sea surface. The magnitude and spectral dependence of the solar radiation reaching the Earth's surface are highly variable functions of the solar zenith angle (i.e., of the time of day, date, and latitude) and of atmospheric conditions (cloud cover, humidity, aerosols, ozone concentration, etc.)[5]. For example, ozone (which was 300 Dobson units in these computations) greatly reduces the irradiance reaching the surface at near 300 nm. The prominent dip near 765 nm is due to oxygen, while the broad dip starting near 930 nm is due to water vapor.

Vocabulary

irradiance: (n) 辐照度
photon: (n) 光子,光量子
solar constant: 太阳常数
ellipticity: (n) 椭圆率
inversely: (adv) 相反地,成反比地
furthermore: (adv) 此外
uniform: (adj) 一致的,统一的
sharp: (adj) 急剧的(变化)
dip: (n)(暂时性的)下降,衰退
Fraunhofer lines: 夫琅禾费线
selective: (adj) 选择性的
typically: (adv) 典型地,通常,一般
prominent: (adj) 显著的,突出的
blackbody: (n) 黑体
approximation: (n) 近似
gas: (n) 气体
seldom: (adv) 很少,难得,不常
oceanographer: (n) 海洋学家
sufficient: (adj) 足够的
solar zenith angle: (n) 太阳天顶角

zenith：(n) 天顶
ant：nadir (n) 天底
humidity：(n) 湿度
aerosol：(n) 气溶胶
ozone：(n) 臭氧
water vapor：水蒸气

Notes

[1] irradiance 即辐照度，是指电磁辐射入射于某一曲面时，单位面积上所接受到的功率大小。在国际单位制中，辐照度的标准单位为瓦特每平方米（W/m^2）。辐照度综合反映了各频率辐射能量的总和。为更深入地探究辐射的特性，物理学家常常会针对辐射频谱的每一个单独频率进行独立分析。当对入射于曲面的辐射进行此类频率分解时，所得到的辐射量被称为光谱辐照度（spectral irradiance）。在国际单位制下，其单位为 $W/(nm \cdot m^2)$。

[2] solar constant 即太阳常数，是一个表征太阳电磁辐射通量的重要参数，它指的是在日地平均距离处，于地球大气层顶界上，垂直于太阳光线的单位面积在每秒钟内所接受到的太阳辐射能量。太阳常数需在地球大气层之外、垂直于入射光的平面上测量。值得注意的是，太阳常数涵盖了所有形式的太阳辐射，不仅仅局限于可见光部分，还包括紫外线、红外线等其他波段的辐射能量。

[3] Fraunhofer lines 即夫琅禾费线，是以德国物理学家约瑟夫·夫琅禾费（1787-1826）的名字命名的一系列光谱线。这些谱线最初是在太阳光谱中被观测到的暗特征线。

[4] blackbody 即黑体，一般指绝对黑体，是一个理想化的参考模型，它能够完全吸收所有入射的电磁辐射，而不会发生任何反射或透射。黑体会以电磁波的形式向外辐射能量，这种辐射称作黑体辐射。

[5] 地球表面接收到的太阳辐射强度受到太阳天顶角以及大气状况的综合影响。

6. Inherent and Apparent Optical Properties

Text

The inherent optical properties (IOPs) are properties of a medium that do not depend on the ambient light field in the medium. The IOPs are various measures of the absorption and scattering properties of a water body. The apparent optical properties (AOPs) are properties of a medium that (1) depend on both on the medium (the IOPs) and on the geometric (directional) structure of the ambient light field, and (2) display enough regular features and stability to be useful descriptors of the water body. Commonly used AOPs are various reflectances, average cosines, and diffuse attenuation coefficients. The IOPs and AOPs are mutually exclusive descriptors of optical properties, that is, something cannot be both an IOP and an AOP. However, there are other measures of the optical state of a water body that are neither an IOP nor an AOP. In particular, the various irradiances give information about the optical properties of a water body, but they are neither IOPs nor AOPs—they are radiometric variables. The radiative transfer equation, which connects the optical properties of the water body and the light within the water, is introduced next.

Figure Ⅳ-6 shows the most commonly used IOPs, AOPs, radiometric variables, and their relationships. This figure shows that the absorption coefficient[1] and the volume scattering function (VSF)[2] are at the "top" of the IOP family. All other IOPs, such as the scattering[3] and beam attenuation coefficients[4], can be derived from the absorption coefficient and the VSF. The IOPs are inputs to the radiative transfer equation (RTE), along with boundary conditions that specify the radiance incident onto a water body and how radiance is transmitted through the sea surface and reflected by the surface and bottom. The IOPs along with the boundary conditions are needed to solve the RTE for the radiance distribution. If the radiance distribution is known, then all other radiometric variables (various irradiances) can be computed, as can the AOPs.

Natural waters, both fresh and saline, are a witch's brew of dissolved and particulate matter. These solutes and particles are both optically significant and highly variable in kind and concentration. Consequently, the optical properties of natural waters show large temporal and spatial variations and seldom resemble those of pure water.

The great variability in the optical properties of natural waters is the bane of those who desire precise and easily tabulated data or simple models. However, the coupling between constituent properties and optical properties implies that optical measurements can be used to deduce information about aquatic ecosystems. Indeed, it is the connections between the optical properties

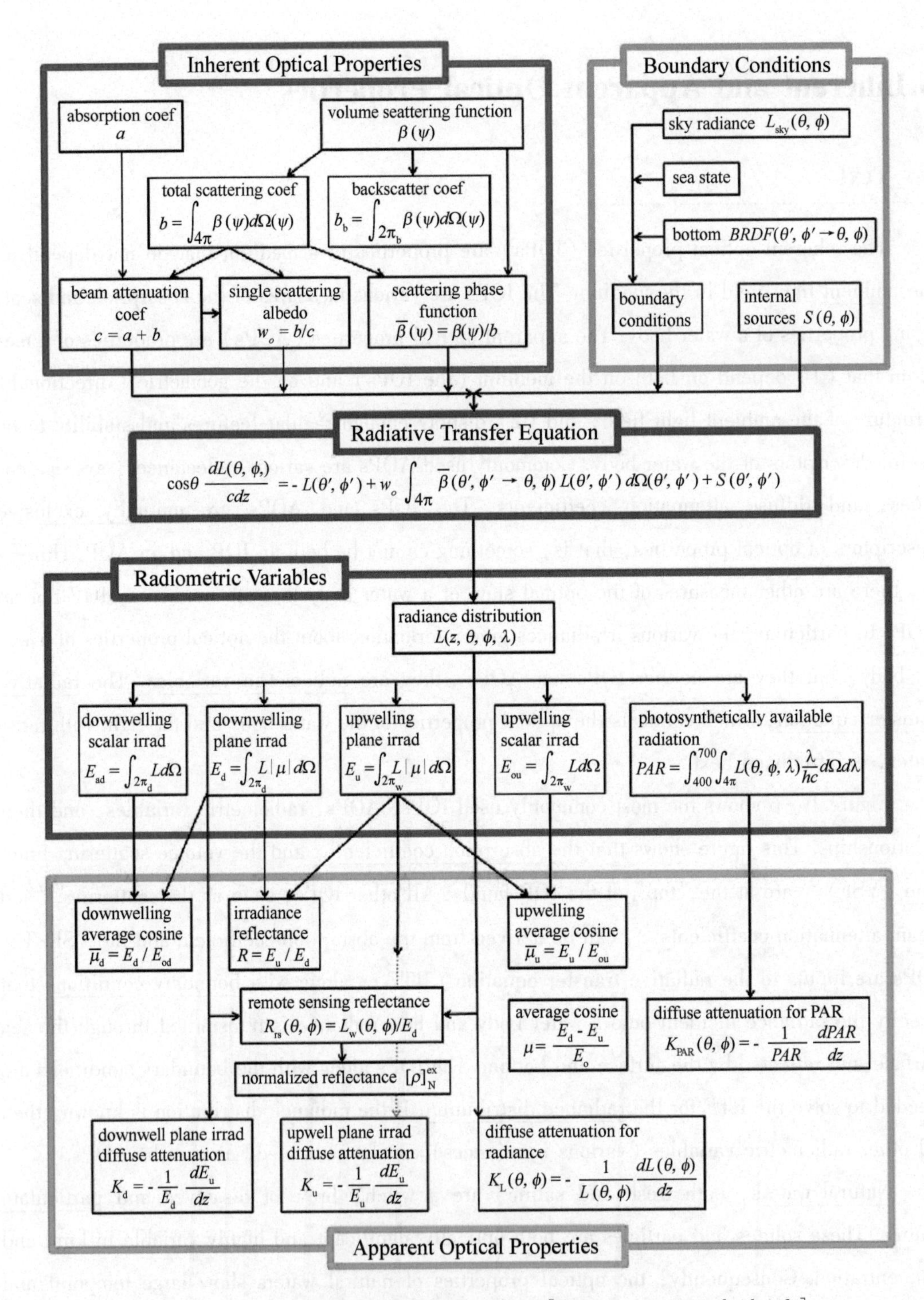

Figure Ⅳ-6: An ocean-optics organization chart. [Source: oceanopticsbook.info]

and the biological, chemical and geological constituents of natural waters that define the critical role of optics in aquatic research. For just as optical oceanography utilizes results from the biological, chemical, geological and physical subdisciplines of limnology and oceanography, so do those subdisciplines incorporate optics. This synergism is seen in such areas as biological oceanography, marine photochemistry, mixed-layer thermodynamics, lidar bathymetry, underwater visibility, and "ocean color" remote sensing of biological productivity, sediment load, pollutants, or bathymetry and bottom type.

IOPs

When light interacts with matter, one of two things can happen. The light can disappear, with its energy being converted to another form, such as heat or the energy contained in a chemical bond. This process is called absorption. The light can also change its direction and/or wavelength. Either of these processes is called scattering.

The absorption and scattering properties of a medium such as sea water are described by its inherent optical properties (IOPs). IOPs are properties of the medium and do not depend on the ambient light field. That is, a volume of water has well-defined absorption and scattering properties whether or not there is any light there to be absorbed or scattered. This means that IOPs can be measured in the laboratory on a water sample, as well as in situ in the ocean.

The absorption coefficient is the fundamental IOP that describes how a medium absorbs light. The volume scattering function likewise describes how the medium scatters light. If you know these two IOPs, then you know everything there is to know about the medium interacts with unpolarized light. Other IOPs are sometimes convenient, and they can be defined in terms of the absorption coefficient and the volume scattering function. The inherent optical properties usually employed in optical oceanography are the absorption and scattering coefficients, which are respectively the absorptance and scatterance per unit distance in the medium.

The IOPs depend on the composition, morphology, and concentration of the particulate and dissolved substances in the ocean[5]. Composition refers to what materials make up the particle or dissolved substance, in particular to the index of refraction of that material relative to that of the surrounding water. Morphology refers to the sizes and shapes of particles. Concentration refers to the number of particles in a given volume of water, which is described by the particle size distribution, or to the amount of a dissolved substance in the water. Different materials absorb much differently as a function of wavelength. Particles with different shapes scatter light differently, even if the particles have the same volume. Particles with different volumes scatter light differently even if they have the same shape. Consider, for example, atmospheric visibility through rain or fog. Even if the total water content per cubic meter of air is the same, you can "see through" a few large raindrops, whereas many small fog droplets greatly reduce visibility.

Because the physical characteristics of the dissolved and particulate substances in the ocean vary by orders of magnitude, so do the IOPs. For example, both the absorption and scattering coefficients of pure water are less that 0.01 m^{-1} at 440 nm. However, in turbid coastal waters with high concentrations of phytoplankton, mineral particles, and dissolved organic matter, the absorption and scattering coefficients can be four orders of magnitude larger. The volume scattering function (at a given scattering angle and wavelength) can also vary by orders of magnitude between open ocean and coastal waters. Understanding how variability in IOPs is determined by the various constituents of the ocean is a fundamental problem of optical oceanography[6].

AOPs

One of the primary goals of optical oceanography is to learn something about a water body, e.g., its chlorophyll or sediment concentration, from optical measurements. Ideally, one would measure the absorption coefficient and the volume scattering function, which tell us everything there is to know about the bulk optical properties of a water body, and those IOPs do indeed tell us a lot about the types and concentration of the water constituents. However, in the early days of optical oceanography, it was difficult to measure in situ IOPs other than the beam attenuation coefficient. On the other hand, it was relatively easy to measure radiometric variables such as the upwelling and downwelling plane irradiances. This led to the use of apparent optical properties (AOPs) rather than IOPs to describe the bulk optical properties of water bodies. A "good" AOP will give useful information about a water body, e.g., the types and concentrations of the water constituents, from easily made measurements of the light field.

Vocabulary

ambient: (adj) 周围环境的
geometric: (adj) 几何(学)的
commonly: (adv) 通常
cosine: (n) 余弦
mutually exclusive: 互相排斥的
radiometric: (adj) 辐射度测量的
variable: (n) 变量
radiative transfer equation: 辐射传输方程
absorption coefficient: 吸收系数
function: (n) 函数
boundary condition: 边界条件
incident: (adj) 入射的
radiance distribution: 辐亮度分布
natural water: 天然水
fresh: (n) 淡水;(adj) 淡的,新鲜的
saline: (n) 盐水;(adj) 咸的
particulate matter: 颗粒物
solute: (n) 溶解物,溶质
pure water: 纯水
bane: (n) 祸根,造成困扰的事物
tabulate: (vt) 列成表格,列表显示
constituent: (adj) 组成的,构成的;(n) 成分,构成要素
subdiscipline: (n) 分支学科
limnology: (n) 湖沼学
synergism: (n) 协同作用

thermodynamics：（n）热力学
bathymetry：（n）水深测量
sediment load：输沙量
pollutant：（n）污染物，污染物质
disappear：（vi）消失
ambient light field：环境光场
fundamental：（adj）基础的，基本的
likewise：（adv）同样地，类似地
unpolarized：（adj）非极化的
convenient：（adj）方便的，便利的
composition：（n）成分，构成
morphology：（n）形态学
refraction：（n）折射（角）
droplet：（n）液滴
order：（n）量级
turbid：（adj）浑浊的，污浊不清的

Notes

[1] absorption coefficient 即吸收系数，是衡量物质吸收光能力的一个重要参数，在比尔定律（Beer's law）中被表示为一个常数。吸收系数可通过光度法进行测量。光度法是基于物质对光吸收的特征及吸收的程度而进行定性、定量分析的一类方法。根据测定时所用的光源不同，分光光度法可分为可见分光光度法、紫外分光光度法及红外光谱法等。分光光度法灵敏度高，特别适用于微量组分的测定。

[2] volume scattering function（VSF）即体散射函数，指在某一特定方向上，单位散射体积内、单位立体角内散射辐射的强度与入射辐射照度之间的比值。

[3] scattering coefficient 即散射系数，是一个用来量化物质对光散射能力的参数。当光穿透物质时，部分光会被物质散射，而余下的光可能被吸收或反射。

[4] beam attenuation coefficient 即光束衰减系数，指沿着光束传输方向上，单位传输距离内由于水体散射和吸收而损失的辐射通量与入射到该介质的初始辐射通量之间的比值。

[5] IOPs 与海水中的颗粒物质以及溶解物质的组成、形态和浓度密切有关。

[6] 理解 IOPs 随海水成分如何变化是光学遥感的根本问题。

7. Absorption of Light

Text

When light interacts with matter one of three things can occur: (1) Absorption[1]: The light can disappear, with its radiant energy being converted into other forms, such as the energy of a chemical bond or heat. (2) Elastic Scattering: The light can change direction, without a change of wavelength. (3) Inelastic Scattering: The light can undergo a change of wavelength, usually to a longer wavelength, and usually in a different direction (Figure Ⅳ-7).

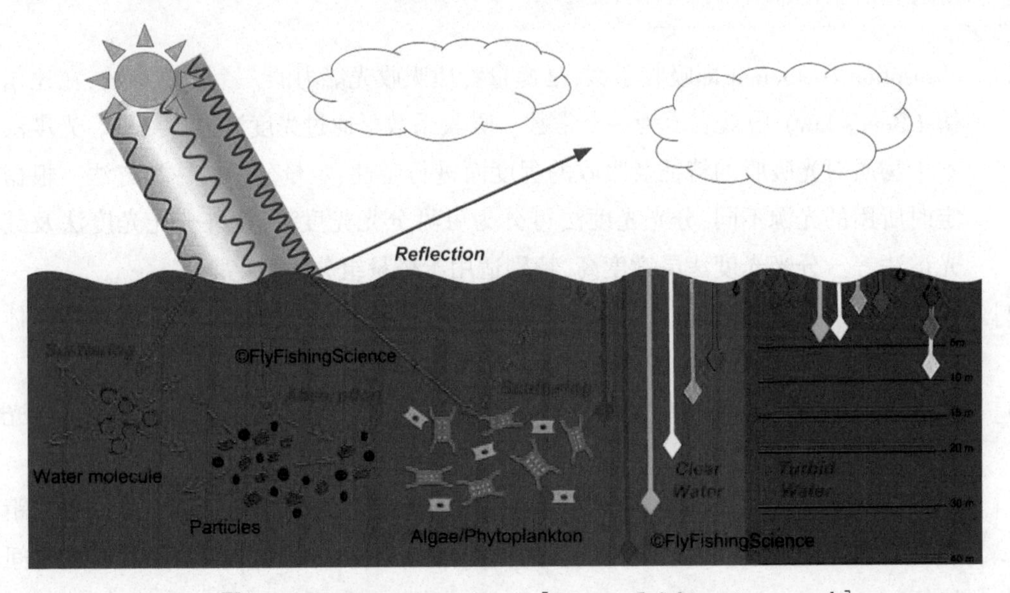

Figure Ⅳ-7: Light attenuation in water. [Source: flyfishingscience.co.uk]

If you are interested in phytoplankton physiology[2], absorbed light is what matters. Light absorbed by phytoplankton provides the energy that drives photosynthesis[3]; light elastically scattered by phytoplankton has no effect on photosynthesis and is of less interest to a plankton biologist. Therefore, many more papers have been published on absorption by phytoplankton than on scattering by phytoplankton. Similarly, it is only the light that is absorbed by water molecules that heats the water.

Phytoplankton absorption spectra are determined by the pigments in phytoplankton, hence measurement of phytoplankton absorption spectra gives information about what pigments are present and in what amounts. Thus, absorption is fundamental to understanding phytoplankton physiology. Similarly, measurement of total absorption can be used to extract the amounts of phytoplankton, colored dissolved organic matter (CDOM), non-algal particles, or pollutants (if present) in the water column.

Unlike scattering, which is weakly spectrally dependent, absorption is highly spectrally dependent[4]. The magnitude and spectral features depend upon the concentration and composition of the particulate and dissolved constituents and water itself. The inherent optical properties (IOPs) are conservative properties and therefore the magnitude of the absorption coefficient varies linearly with the concentration of the absorbing material. Theoretically, the absorption coefficient can be expressed as the sum of the absorption coefficients of each component.

Practically, however, it is not possible to measure the absorption properties of each individual absorbing component and thus the individual components can be grouped into similarly absorbing constituents based upon similarity in their optical properties and/or analytically-based groupings:

$$a(\lambda) = a_w(\lambda) + a_{phyt}(\lambda) + a_{NAP}(\lambda) + a_{CDOM}(\lambda)$$

where subscripts w, phyt, NAP and CDOM indicate water, phytoplankton, non-algal particles, and colored dissolved organic matter, respectively.

Absorption by water is weak in the blue and strong in the red, and it varies with temperature and salinity. Absorption by particles is separated into phytoplankton and non-algal particles (NAP) using spectrophotometry and an extractive technique. Phytoplankton absorption demonstrates the most spectral variations of any of the components due to the individual pigment absorption spectra but in general exhibits peaks in the blue and red regions of the spectrum due to the ubiquitous presence of chlorophyll-a. Non-algal particle absorption is strongest in the blue, decreasing approximately exponentially to the red. This component, operationally-defined, includes living zooplankton and bacteria, as well as the non-pigmented parts of phytoplankton (cell walls, membranes etc.), detrital material as well as inorganic particles. For oligotrophic environments with very low concentrations of suspended and dissolved material, the absorption coefficient is dominated by water and the wavelength of minimum absorption is in the blue, hence the blue color of the seawater. For eutrophic[5] and/or coastal environments with high concentrations of suspended and dissolved material, the absorption coefficient is dominated by that material and the wavelength of minimal absorption shifts to the green, lending green color to that environment.

Vocabulary

elastic scattering: 弹性散射

inelastic scattering: 非弹性散射

physiology: (n) 生理学

photosynthesis: (n) 光合作用

biologist: (n) 生物学家

pigment: (n) 色素

non-algal: (adj) 非藻类的

conservative: (adj) 保守的

theoretically: (adv) 理论上

practically: (adv) 实际上

individual：(adj) 单独的，个别的
similarity：(n) 相似性
analytically-based：(adj) 基于分析的
subscript：(n) 下标
  ant：superscript (n) 上标
spectrophotometry：(n) 分光光度法
ubiquitous：(adj) 无处不在的
exponentially：(adv) 以指数方式地
bacterium：(n) 细菌
  pl：bacteria
membrane：(n)(植物的)细胞膜
detrital material：碎屑物质
oligotrophic：(adj) 贫营养的，寡营养的
eutrophic：(adj) 富营养的

Notes

[1] absorption 即吸收。光吸收是光(电磁辐射)在海水中传播时，与其中的物质成分发生相互作用，导致电磁辐射的部分能量被转化为其他能量形式的能量的物理现象。传播过程中，未被吸收的光能则会以反射、散射或透射的方式逸出，这些光学效应共同决定了我们所观察到的物体色彩表现。

[2] physiology 即生理学，是生物学的一个重要分支学科，专注于研究生物机体内部复杂多样的各种生命现象，尤其是生物机体各组成部分的功能特性，以及这些功能得以实现的内在生理机制与调控过程。

[3] photosynthesis 即光合作用，主要发生在绿色植物(包括藻类)体内，它描述的是这些生物体利用自身的光合色素捕获光能，进而驱动化学反应，将二氧化碳与水转化为富含能量的有机物质，并在此过程中释放出氧气的现象。光合作用对于实现自然界中的能量转换、维持大气的碳-氧平衡具有重要意义。

[4] 海水中光吸收的特性与光的波长密切相关。在可见光光谱的范围内，红光和橙光波长最长，而紫光和蓝光波长最短。当太阳光穿透海洋水体时，较长波长的红光和橙光更容易被海水中的物质吸收，而较短波长的紫光和蓝光更倾向于被海水中的物质反射和散射。然而，由于人类视觉系统对紫光的敏感度相对较低，因此我们往往对反射回来的紫光视而不见。这正是导致海水看起来是蓝色的原因。

[5] 水体富营养化 (eutrophication) 是指受人类活动的影响，大量氮、磷等生物必需的营养元素涌入湖泊、河流、海湾等流速缓慢的水域中，导致藻类及其他浮游生物迅速繁殖，进而引发水体溶解氧含量显著下降，水质严重恶化，最终造成鱼类及其他水生生物大量死亡的现象。在自然条件下，湖泊等水体也会经历从贫营养状态向富营养状态的缓慢过渡，但这一自然过程极其缓慢。相比之下，由人类排放含有丰富营养物质的工业废水和生活污水所引起的水体富营养化，则能在短时间内迅速出现。当水体发生富营养化时，浮游藻类会大量暴发，形成水华(淡水水体中藻类大量繁殖的一种自然生态现象)。由于占据优势的浮游藻类种类及其色素差异，水面往往会呈现绿色、红色、棕色、乳白色等多种颜色。

8.Scattering of Light

Text

If you are interested in phytoplankton physiology, absorbed light is all that matters. Light absorbed by phytoplankton provides the energy that drives photosynthesis; light elastically scattered by phytoplankton has no effect on photosynthesis and is not of interest to a plankton biologist. Therefore, many more papers have been published on absorption by phytoplankton than on scattering by phytoplankton. Similarly, it is only the light that is absorbed by water molecules that heats the water. Parameterizations of upper-ocean heating rates often use Gershun's Law to compute how much light is absorbed and turned into heat without explicit consideration of scattering.

However, both absorption and scattering are equally important to the prediction and understanding of light propagation.

Both absorption and scattering affect visibility[1] in ways that are illustrated in Figure Ⅳ-8. Absorption removes light from a beam and thus makes a distant image darker. Scattering through small angles blurs the edges of an image, and can cause other effects such as twinkling or mirages, which affect the appearance of distant objects.

Ocean-color remote sensing essentially measures light leaving the oceans to determine how absorption in the water column changed sunlight in what was measured. The measured light comes from scattering within the water column—from either backscattering at angles greater than 90 degrees or from multiple forward scattering at angles of a few tens of degrees or greater[2]. Without scattering, there would be no remote sensing of the oceans to understand how absorption has modified the in-water light field. To be more specific, without elastic scattering, downwelling sunlight would continue to head downward into the ocean depths, eventually to be totally absorbed[3]. The ocean would appear a dim red color due to chlorophyll fluorescence. Without inelastic scatter, the ocean would be black because there would be no Raman scattering[4] or fluorescence to create even a small amount of upwelling light.

Scattering also enhances the effect of absorption. If light is scattered many times, the distance traveled by a light ray in going from depth z1 to depth z2 is greater than just the straight-line geometric distance that would be traveled by the initial ray. This gives the light a greater chance of being absorbed before reaching z2. Hence the saying, "A little bit of absorption goes a long way if you have a lot of scattering."

Any interaction of light with matter can cause a change in direction or wavelength. It does

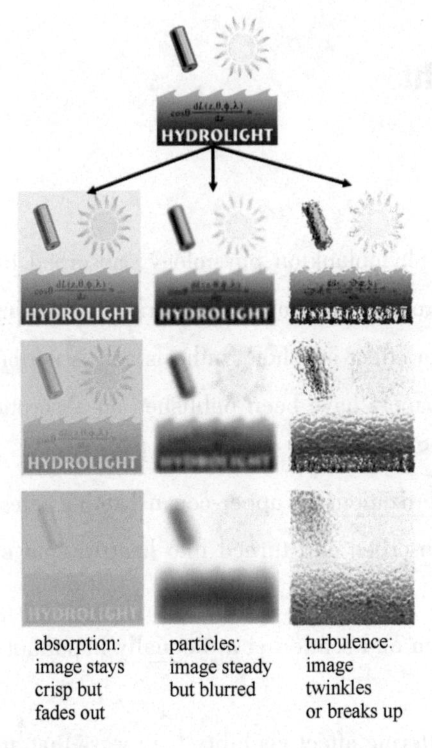

Figure Ⅳ - 8: Illustration of absorption and scattering effects on visibility. [Source: oceanopticsbook.info]

not matter whether the change in direction comes from reflection or refraction by a surface (such as sunlight reflected or transmitted by the sea surface), from reflection or refraction by a particle that is much larger than the wavelength of light, or from the light's propagating electric field causing electrons in an atom or molecule to oscillate and radiate light into all directions. Although you will sometimes see terms like "surface scattering" (for reflection by a surface) or "volume scattering" (for interactions within a volume of water), these processes are all scattering and are all caused by essentially the same underlying physics.

Vocabulary

parameterization: (n) 参数化

Gershun's Law: 格顺定律

consideration: (n) 考虑,顾及

equally: (adv) 同样地,相等地

propagation: (n) 传播

visibility: (n) 能见度,可见度

blur: (vt) (使)变得模糊不清

twinkling: 闪烁,闪耀

mirage: (n) 海市蜃楼,幻象

backscattering: (n) 后向散射

forward scattering: (n) 前向散射

dim: (adj) 暗淡的

fluorescence: (n) 荧光

Raman scattering: 拉曼散射

ray：(n)(热或其他能量的)射线，光线

electron：(n) 电子

atom：(n) 原子

oscillate：(n) 振荡

physics：(n) 物理学

Notes

[1] visibility 即能见度，又称可见度，是指视力正常的人能看清楚目标轮廓的最大水平距离。相较于大气环境，水中能见度显著降低。海水中的能见度通常为大气中的千分之一左右，这一显著差异主要归因于光在海水介质中遭遇的吸收和散射作用远比在大气中强烈。

[2] backscattering 即后向散射，在物理学领域中，指的是波、粒子或信号向其入射方向的反向传播现象。这一过程产生的反射效果不同于镜面反射的定向性，而更多地表现为散射导致的漫反射特性。后向散射在天文学、摄影和医学超声检查等多个领域中有着重要应用。

[3] elastic scattering 即弹性散射，在物理学范畴内，指的是粒子在碰撞过程中，仅涉及动能交换，而不改变粒子种类及其内部运动状态的现象。反之，若碰撞不仅导致动能交换，还引发了粒子内部状态的改变或粒子的种类变化，则此过程被称为非弹性散射 (inelastic scattering)。对光而言，弹性散射意味着光与物质相互作用后，光的频率、能量和波长均保持不变，仅光的传播方向发生偏转；而非弹性散射则是指光与物质相互作用导致光的频率、能量和波长发生变化的情形。

[4] Raman scattering 即拉曼散射，也称拉曼效应，是 1928 年由印度物理学家拉曼发现的一种物理现象。当光线投射到物质表面时会，会发生散射现象，其中，散射光不仅包含与入射光频率相同的弹性成分，即瑞利散射，还存在频率低于或高于入射光的成分，这一独特现象被称为拉曼效应。鉴于其在光的散射研究领域的杰出贡献，拉曼于 1930 年荣获诺贝尔物理学奖。

9. Radiative Transfer

Text

Radiative transfer[1] (also called radiation transport) is the physical phenomenon of energy transfer in the form of electromagnetic radiation. The propagation of radiation through a medium is affected by absorption, emission, and scattering processes. The equation of radiative transfer describes these interactions mathematically. Equations of radiative transfer have application in a wide variety of subjects including optics, astrophysics, atmospheric science, and remote sensing. Analytic solutions to the radiative transfer equation (RTE) exist for simple cases, but for more realistic media, with complex multiple scattering effects, numerical methods are required.

In order to formulate the RTE, it is convenient to imagine the total light field as many beams of electromagnetic radiation of various wavelengths coursing in all directions through each point of a water body. We then consider a single one of these beams. This beam and the processes affecting it are illustrated in Figure Ⅳ-9.

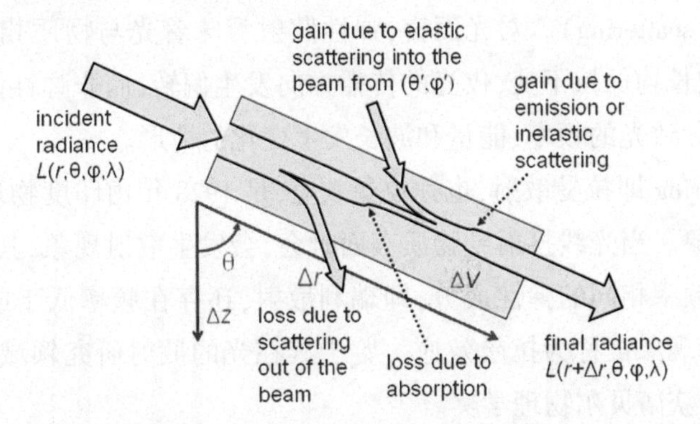

Figure Ⅳ-9: Illustration of a single beam of radiance and the processes that affect it as it propagates a distance. [Source: oceanopticsbook.info]

Now think of all the ways in which that beam's energy can be decreased or increased. Bearing in mind the preceding comments, the following six processes are both necessary and sufficient to write down an energy balance equation for a beam of light at the phenomenological level[2]:

Process 1: loss of energy from the beam through annihilation of the light and conversion of radiant energy to non-radiant energy (absorption)

Process 2: loss of energy from the beam through scattering to other directions without change in wavelength (elastic scattering)

Process 3: loss of energy from the beam through scattering (perhaps to other directions) with change in wavelength (inelastic scattering)

Process 4: gain of energy by the beam through scattering from other directions without change in wavelength (elastic scattering)

Process 5: gain of energy by the beam through scattering (perhaps from other directions) with a change in wavelength (inelastic scattering)

Process 6: gain of energy by the beam through creation of light by conversion of non-radiant energy into radiant energy (emission)

Vocabulary

radiative transfer: 辐射传输
emission: (n) 发射
mathematically: (adv) 数学上地
optics: (n) 光学
astrophysics: (n) 天体物理学
analytic solution: 解析解
realistic: (adj) 现实的,实际的
numerical method: 数值计算法
formulate: (vt) 用公式表示
phenomenological: (adj) 现象的,现象学的
annihilation: (n) 摧毁,毁灭

Notes

[1] radiative transfer 即辐射传输,是一种涉及电磁辐射能量在空间中传递的物理过程。在这一过程中,辐射能量通过介质传播时,会受到吸收、发射以及散射等多种物理过程的影响。辐射传输方程就是以数学方式描述这些交互作用,被广泛应用在光学、天文物理学和大气科学等多个领域。

[2] six processes in radiative transfer 包含了 3 个辐射能损失的过程,即吸收、弹性散射和非弹性散射;同时也涵盖了 3 个辐射能增加的过程,分别是发射、弹性散射和非弹性散射。

10. Remote Sensing in Oceanography

Text

Remote sensing in oceanography[1] is a widely used observational technique which enables researchers to acquire data of a location without physically measuring at that location. Remote sensing in oceanography mostly refers to measuring properties of the ocean surface with sensors on satellites or planes, which compose an image of captured electromagnetic radiation (Figure Ⅳ-10). A remote sensing instrument can either receive radiation from the earth's surface (passive), whether reflected from the sun or emitted, or send out radiation to the surface and catch the reflection (active)[2]. All remote sensing instruments carry a sensor to capture the intensity of the radiation at specific wavelength windows, to retrieve a spectral signature for every location. The physical and chemical state of the surface determines the emissivity and reflectance for all bands in the electromagnetic spectrum, linking the measurements to physical properties of the surface. Unlike passive instruments, active remote sensing instruments also measure the two-way travel time of the signal, which is used to calculate the distance between the sensor and the imaged surface. Remote sensing satellites often carry other instruments which keep track of their location and measure atmospheric conditions[3].

Remote sensing observations, in comparison to (most) physical observations, are consistent in time and have good spatial coverage. Since the ocean is fluid, it is constantly changing on different spatial and temporal scales. Capturing the spatial variation of the ocean with remote sensing is considered extremely valuable and is on the frontier of oceanographic research. The high variability of the ocean surface is also the deterministic factor in the differences between land and ocean remote sensing.

Characteristics

Remote sensing is actively used in various fields of natural sciences like geology, physical geography, ecology, archeology and meteorology but, remote sensing of the ocean is vastly different. Unlike most land processes the ocean, just like the atmosphere, is variable on way shorter time scales over its entire spatial scale; the ocean is always moving. The temporal variability in the object of study determines the usability of specific data and the applicable methods and is the reason why remote sensing methods differ materially between ocean and land surfaces[5]. A single wave on the surface of the ocean cannot be tracked by satellites of today. Ocean waves crash or disappear before a new observation is made, features with this time scale

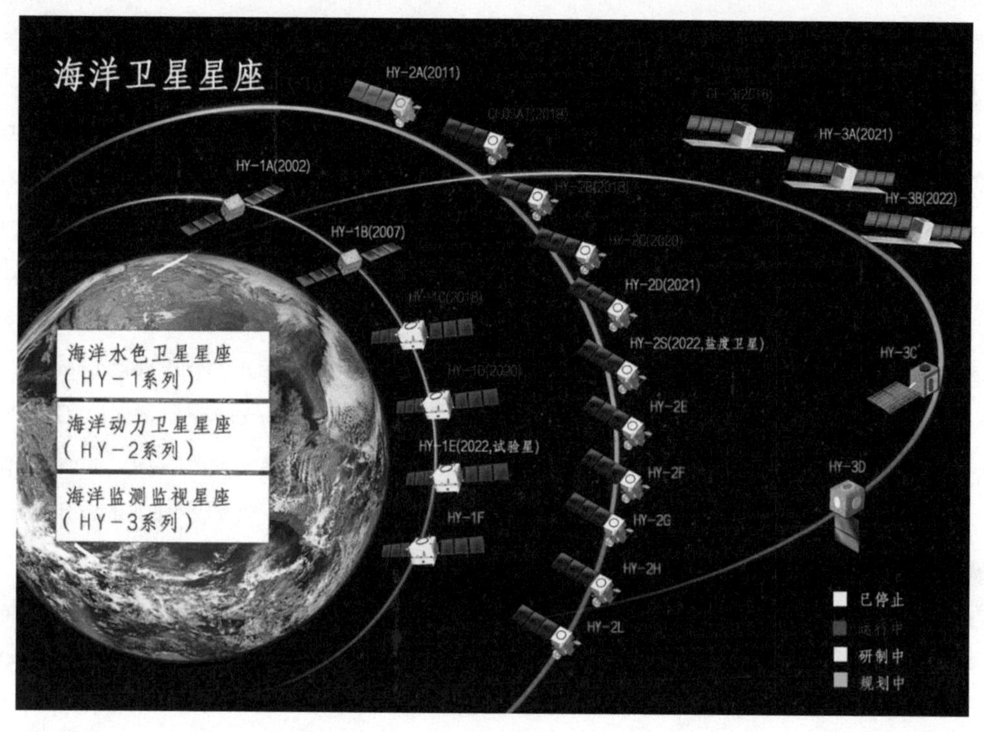

Figure Ⅳ-10: Remote sensing satellites developed by China (Haiyang series)[4]. [Source: kepuchina.cn]

are rarer on land. Unlike vegetation, snow and other land covers the ocean is opaque to most electromagnetic radiation (except for visible light) therefore the ocean surface is easy to monitor but it is a challenge to retrieve information of deeper layers. Remote sensing enables temporal analysis over vast spatial scale, since satellites have a constant revisit time, provide a wide image and are often operational for multiple consecutive years. This concept of constant data in time and space was a breakthrough in oceanography[6], which previously relied on measurements from drifters, coastal locations like tide gauges, ships and buoys. All in-situ measurements either have a small spatial footprint or are varying in location and time, so do not deliver constant and comparable data.

History

Remote sensing as we know it today started with the first earth orbiting satellite Landsat 1 in 1973[7]. Landsat 1 delivered the first multi-spectral images of features on land and coastal zones all over the world and already showed effectiveness in oceanography, although not specifically designed for it. In 1978, NASA made the next step in remote sensing for oceanography with the launch of the first orbiting satellite dedicated to ocean research, Seasat[8]. The satellite carried 5 different instruments: a Radar altimeter for retrieving sea surface height, a microwave scatterometer to retrieve wind speed and direction, a microwave radiometer to retrieve sea surface temperature (SST), an optical and infrared radiometer to check for clouds and surface characteristics and lastly the first Synthetic Aperture Radar (SAR) instrument. Seasat was only operational for a few months but, together with the Coastal Zone Color Scanner (CZCS)[9] on Nimbus-7, proved the feasibility of many techniques and instruments in ocean remote sensing. TOPEX/Poseidon, an altimeter launched in 1992, provided the first continuous global map of sea surface topography and continued on the possibilities explored by Seasat. The Jason-1, Jason-2 and Jason-3 missions continue the measurements from 1992 to today to form a complete time-series of the global sea surface height. Also, other techniques hosted on Seasat found continuation. The Advanced Very-High-Resolution Radiometer (AVHRR) is the sensor carried on all NOAA missions and made SST retrieval accessible with a continuous time-series since 1979. The European Space Agency (ESA) further developed SAR with the ERS-2, ENVISAT, and now Sentinel-1 missions by providing larger spatial footprints, lowering the resolution and flying twin missions to reduce the effective revisit time. Optical remote sensing of the ocean found continuation after the CZCS with polar orbiting missions ENVISAT, OrbView-2, MODIS and very recently with Sentinel-3, to form a continuous record since 1997. Sentinel-3 is now one of the best equipped missions to map the ocean hosting a SAR altimeter, multispectral spectrometer a radiometer and several other instruments on multiple satellites with alternating orbits providing exceptional temporal and spatial resolution.

Methods

The physical and chemical state of a surface or object have direct impacts on the emissivity, reflectance, and refractance of electromagnetic radiation. Sensors on remote sensing instruments capture radiation, which can be translated back to deduce the physio-chemical properties of the surface. Water content, temperature, roughness and color are characteristics often deduced from the spectral characteristics of the surface. A sensor on a satellite returns the composite signal for a certain area inside the footprint called a cell, the size of the unique cells is referred to as the spatial resolution[10]. The spatial resolution of a sensor is determined by the distance from earth and the available bandwidth for data transfer. A satellite passes over the same location consistently through time with the same interval called the revisit-time or temporal resolution[11]. Sensors cannot have both a very high temporal and spatial resolution, so a tradeoff has to be made specific for the goal of the mission. Sensors on satellites have measuring errors, caused by for example atmospheric interference, geolocation imprecision, and topographic distortion. Complete derived products from remote sensing often use simple calculations or algorithms to transform the spectral signature from a cell to a physical value. All methods of transferring spectral data have certain biases which can contribute to the measurement errors of the final result. Often surface characteristics can be deduced with very low error margins due to data corrections, using onboard data or models, and a physically correct translation of spectral characteristics to physio-chemical characteristics.

Although it is interesting to know the surface characteristics at a certain moment, research is more interested in documenting the change of a surface over time or the transport of characteristics through space. Change detection leverages the consistent temporal component of remote sensing data to analyze the change of surface properties in time. Change detection relies on having at least two observations taken at different times to analyze the difference between the two images visually or analytically. In land remote sensing, change detection is used, for example, to assess the impact of a volcano eruption, check the growth of plants over time, map deforestation, and measure ice sheet melt. In oceanography, the surface changes more quickly than the revisit time of a satellite, making it difficult to monitor certain processes. Change detection in oceanography requires the characteristic to change continuously, like sea level rise, or change spatial scale slower than the revisit time of the satellite, like algal blooms. Another way to infer change from only one acquisition is by computing the dynamical component and direction from a static image, which is leveraged in RADAR altimetry to deduce surface current velocity.

Vocabulary

intensity: (n) 强度
emissivity: (n) 发射率,放射率
spatial: (adj) 空间的
temporal: (adj) 时间的
scale: (n) 尺度
archeology: (n) 考古学
meteorology: (n) 气象学
opaque: (adj) 不透明的
revisit time: 重访时间
drifter: (n) 漂流浮标
tide gauge: 验潮仪,验潮站
buoy: (n) 浮标
in-situ: (adj) 原位的,现场的
multi-spectral: (adj) 多光谱的
altimeter: (n) 高度计
microwave scatterometer: 微波散射计
microwave radiometer: 微波辐射计
Synthetic Aperture Radar: 合成孔径雷达
feasibility: (n) 可行性
topography: (n) 地形
time-series: (n) 时间序列
polar orbiting mission: 极地轨道飞行任务
spectrometer: (n) 分光光度计
composite: (n) 复合的
tradeoff: (n) 权衡
interference: (n) 干扰
imprecision: (n) 不精确
distortion: (n) 畸变,失真
leverage: (vi) 利用;(n) 杠杆作用
deforestation: (n) 毁林
ice sheet: 冰盖
algal bloom: 藻华
velocity: (n) 速度

Notes

[1] remote sensing in oceanography 即海洋遥感，指以海洋及海岸带作为监测、研究对象的遥感，包括物理海洋学遥感、生物海洋学遥感以及化学海洋遥感等多个方面。海洋遥感利用传感器对海洋进行远距离非接触观测，以获取海洋景观和海洋要素的图像信息与数据资料。

[2] passive/active remote sensing 即被动/主动遥感。被动遥感，又称无源遥感系统，指自身不携带辐射源的遥感系统。在进行遥感探测时，该系统依赖探测仪器捕获并记录目标物体自身发射或反射自外界辐射源（如太阳）的电磁波信息。常用的被动遥感器涵盖航空摄影机、电视摄影机、红外及多光谱扫描仪、微波辐射计和光谱辐射计等多种类型。主动遥感，又称有源遥感，指从遥感平台上的人工辐射源向目标物体发射特定形式的电磁波，随后由传感器接收并记录其反射波的遥感系统。该系统的主要优势在于不依赖于太阳辐射，能够全天候工作，而且可以根据具体的探测需求，主动选择电磁波的波长和发射模式。主动遥感常用的电磁波包括微波波段和激光，大多采用脉冲信号，部分则使用连续波束。微波散射计、海面高度计、雷达、合成孔径雷达（SAR）、干涉 SAR 以及激光雷达等均属于主动遥感系统的范畴。

[3] 遥感卫星还需装备精密的定位系统以及专门用于大气测量的传感器组件，以支持后续的遥感数据反演处理过程。

[4] Haiyang 即海洋系列卫星，为中国自主研制和发射的海洋环境监测卫星系列，涵盖海洋一号、海洋二号和海洋三号三大类别。海洋一号（HY-1）卫星是海洋水色卫星，主要任务是探测海洋水色环境要素，诸如叶绿素浓度、悬浮泥沙含量、可溶性有机物、水温、污染物以及浅海水深和水下地形等。海洋二号（HY-2）卫星是海洋动力环境卫星，主要任务是探测海洋的海面风场、温度场、海面高度、浪场及流场等关键海洋动力环境数据。海洋三号（HY-3）卫星是海洋监视监测卫星，主要任务是探测海上目标，并对海洋环境进行实时监测，实现全天时、全天候海面目标与环境监测。

[5] 海洋与陆地最大的区别：海洋是流体，处在实时变化中。

[6] 海洋遥感的显著优势主要体现在两大方面：一是具有同步、大范围且实时获取资料的能力，加之高频次的观测特性，使得海洋遥感能够精确捕捉并记录大尺度的海洋现象，进而实现对这些现象的动态观测；二是具备全天时（昼夜）、全天候的工作能力，确保海洋遥感能持续稳定地提供高质量的观测数据。

[7] Landsat 1 即陆地卫星 1 号，是美国国家航空航天局（NASA）于 1972 年发射的一颗遥测卫星。它是 NASA 长期遥感卫星计划——陆地卫星计划的第一个成员。该人造卫星属于最早的地球资源卫星之一，对后来各国发射的一系列类似卫星有很大影响。

[8] Seasat 是世界上第一颗海洋卫星，搭载了合成孔径雷达（SAR），是人类历史上首颗

装备 SAR 的卫星。除 SAR 外,该星还搭载了微波散射计、雷达高度计以及红外辐射计等一系列传感器。该星的核心使命在于验证各类传感器在监视海洋现象方面的可行性和效能,并确定技术指标与性能要求。

[9] Coastal Zone Color Scanner (CZCS) 即海岸带水色扫描仪,是首个专门用来测量海洋水色的航天传感器。尽管在此之前已有若干航天传感器能够在一定程度上探测海洋水色,但它们的光谱波段、空间分辨率和观测范围等核心参数都是基于陆地或者气象观测的需求进行优化,在应用于海洋水色检测时存在显著的局限性和不足。

[10] 像元亦称像素点或像元点,是构成数字化影像的最小单元。在遥感数据采集过程中,如扫描成像时,像元是传感器对地面景物进行扫描采样的最小单元;在数字图像处理中,它是对模拟影像进行扫描数字化时的采样点。

[11] 卫星重访周期是指卫星再次经过地球表面同一地点的时间间隔,这个时间长度取决于卫星的轨道类型与运行高度。

11. Infrared Radiometer

Text

All objects with a temperature above absolute zero emit electromagnetic radiation. The wavelengths and intensity of radiation emitted are related to the temperature of the object. Terrestrial surfaces (e.g., soil, plant canopies, water, snow) emit radiation in the mid-infrared portion of the electromagnetic spectrum (approximately 4–50 μm)[1].

Infrared radiometers[2] are sensors that measure infrared radiation, which is used to determine surface temperature without touching the surface (when using sensors that must be in contact with the surface, it can be difficult to maintain thermal equilibrium without altering surface temperature). Infrared radiometers are often called infrared thermometers because temperature is the desired quantity, even though the sensors detect radiation.

Typical applications of infrared radiometers include plant canopy temperature measurement for use in plant water status estimation, road surface temperature measurement for determination of icing conditions, and terrestrial surface (soil, vegetation, water, snow) temperature measurement in energy balance studies.

The AVHRR[3] instrument is a space-borne sensor that measures the reflectance of the Earth in five spectral bands that are relatively wide by today's standards. AVHRR instruments are or have been carried by the National Oceanic and Atmospheric Administration (NOAA) family of polar orbiting platforms (POES) and European MetOp satellites. The instrument scans several channels; two are centered on the red (0.6 micrometers) and near-infrared (0.9 micrometers) regions, a third one is located around 3.5 micrometers, and another two the thermal radiation emitted by the planet, around 11 and 12 micrometers.

The first AVHRR instrument was a four-channel radiometer. The last version, AVHRR/3, first carried on NOAA-15 launched in May 1998, acquires data in six channels (Figure Ⅳ-11)[4]. The AVHRR has been succeeded by the Visible Infrared Imaging Radiometer Suite, carried on the Joint Polar Satellite System spacecraft.

NOAA has at least two polar-orbiting[5] meteorological satellites in orbit at all times, with one satellite crossing the equator in the early morning and early evening and the other crossing the equator in the afternoon and late evening. The primary sensor on board both satellites is the AVHRR instrument. Morning-satellite data are most commonly used for land studies, while data from both satellites are used for atmosphere and ocean studies. Together, they provide twice-daily global coverage, and ensure that data for any region of the Earth are no more than six hours

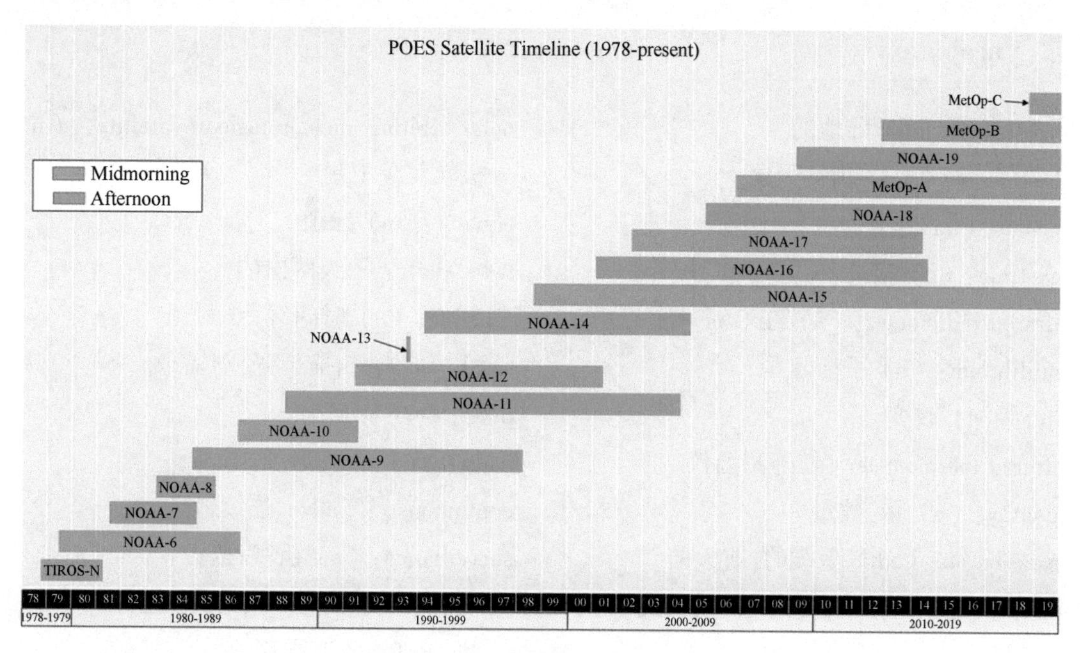

Figure Ⅳ-11: A timeline of NOAA and MetOp missions that carried the AVHRR.

[Source: Kalluri et al., 2021]

old. The swath width[6], the width of the area on the Earth's surface that the satellite can "see", is approximately 2,500 kilometers (~1,540 miles). The satellites orbit between 833 and 870 kilometers (±19 kilometers, 516–541 miles) above the surface of the Earth.

The highest ground resolution that can be obtained from the current AVHRR instruments is 1.1 kilometers (0.68 miles) per pixel at the nadir. AVHRR data have been collected continuously since 1981.

The primary purpose of these instruments is to monitor clouds and measure the thermal emission of the Earth. These sensors have proven useful for a number of other applications, however, including the surveillance of land surfaces, ocean state, aerosols, etc. AVHRR data are particularly relevant to study climate change and environmental degradation because of the comparatively long records of data already accumulated (over 40 years)[7]. The main difficulty associated with these investigations is to properly deal with the many limitations of these instruments, especially in the early period (sensor calibration, orbital drift, limited spectral and directional sampling, etc.).

The AVHRR instrument also flies on the MetOp series of satellites. The three planned MetOp satellites are part of the EUMETSAT Polar System (EPS) run by EUMETSAT. Operational experience with the MODIS sensor onboard NASA's Terra and Aqua led to the development of AVHRR's follow-on, VIIRS, which is currently operating on board the Suomi NPP and NOAA-20 satellites.

Vocabulary

terrestrial：（adj）陆地的
plant canopy：（n）植被冠层
portion：（n）部分
spectrum：（n）谱
infrared radiometer：（n）红外辐射计
equilibrium：（n）平衡
alter：（vt）改变
infrared thermometer：红外测温仪
quantity：（n）量，数量
space-borne：（adj）星载的，航天的
scan：（vt）（X射线、超声波、电磁波等）扫描
channel：（n）通道
near-infrared：（adj）近红外的
polar-orbiting meteorological satellite：（n）极轨气象卫星
equator：（n）赤道
twice-daily：一天两次的
timeline：（n）时间线
swath width：扫描带宽，刈幅
pixel：（n）像素
nadir：（n）星下点
continuously：（adv）连续地
surveillance：（n）监视，监控
degradation：（n）毁坏，恶化（过程）
orbital drift：轨道漂移
directional sampling：定向取样

Notes

[1] 海面辐射，又称海面发射辐射、海面热辐射或海面红外辐射，通常指的是海面向空中释放的长波辐射。其光谱特性近乎理想黑体辐射，在8～14 μm范围内的中红外波段具有显著的辐射峰值。海面辐射的强度主要取决于海面温度，可根据斯特藩-玻尔兹曼定律由实测温度来计算。相反地，也可根据遥感探测的海面辐射推算出海面温度。

[2] infrared radiometer即红外辐射计，能够被装载在航天器及航空观测平台上，通过测量海面红外辐射能量实现对海面温度的有效探测。

[3] AVHRR全称为Advanced Very High Resolution Radiometer，是美国国家海洋与大气管理局（NOAA）系列气象卫星上搭载的传感器。自1979年TIROS-N卫星成功发射以来，AVHRR传感器便持续不断地进行着对地观测任务。AVHRR是一种多光谱通道的扫描辐射仪，可监测云层和地球表面的辐射状况，能够获取多种参数信息，包括地表温度（LST）、海表温度（SST）和归一化植被指数（NDVI）等。

[4] AVHRR传感器的发展经历了三代革新。其中，AVHRR/1作为初代产品，配备了4个原始波段，但仅有一个热红外波段，被部署于TIROS-N、NOAA-6、NOAA-8和NOAA-10等卫星上。随后推出的AVHRR/2在继承前代的基础上，增加了一个热红外波段，这一改进促进了地表温度的精确反演。最新的AVHRR/3则搭载于AVHRR

KLM、NOAA18/19 和 Metop A/B 等卫星平台上,又额外增加了一个中红外波段。

[5] polar-orbiting meteorological satellite 即极轨气象卫星,也叫太阳同步轨道气象卫星,其运行轨道位于地球上空 650~1500 km 的范围内,沿地球南北两极方向进行飞行,完成一次绕地周期大约需要 115 min。中国的风云一号系列气象卫星便是此类卫星的典型代表。极轨气象卫星的显著优势在于器全球覆盖能力。

[6] swath width 即刈幅,指卫星扫描带宽,反映了卫星扫描或观测时所能覆盖的地表范围宽度。

[7] AVHRR 提供自 1981 年以来的海表温度数据,可用于研究全球变暖背景下海表温度的变化情况。

12.Microwave Radiometer

Text

A microwave radiometer (MWR)[1] is a radiometer that measures energy emitted at one millimeter-to-meter wavelengths (frequencies of 0.3-300 GHz), known as microwaves. Microwave radiometers are very sensitive receivers designed to measure thermally-emitted electromagnetic radiation. They are usually equipped with multiple receiving channels to derive the characteristic emission spectrum of planetary atmospheres, surfaces or extraterrestrial objects. Microwave radiometers are utilized in a variety of environmental and engineering applications, including remote sensing, weather forecasting, climate monitoring, radio astronomy, and radio propagation studies.

Using the microwave spectral range between 1 and 300 GHz provides complementary information to the visible and infrared spectral range. Most importantly, the atmosphere and also vegetation are semi-transparent in the microwave spectral range. This means components like dry gases, water vapor, or hydrometeors interact with microwave radiation, but overall, even the cloudy atmosphere is not completely opaque in this frequency range.

For weather and climate monitoring, microwave radiometers are operated from space as well as from the ground. As remote sensing instruments, they are designed to operate continuously and autonomously often in combination with other atmospheric remote sensors like for example cloud radars and lidars. They allow the derivation of important meteorological quantities such as vertical temperature and humidity profiles, columnar water vapor quantity, and columnar liquid water path with a high temporal resolution on the order of minutes to seconds under nearly all weather conditions[2] (Figure Ⅳ-12). Microwave radiometers are also used for remote sensing of Earth's ocean and land surfaces, to derive ocean temperature and wind speed, ice characteristics, and soil and vegetation properties.

Solids, liquids (e.g. the Earth's surface, ocean, sea ice, snow, vegetation) but also gases emit and absorb microwave radiation. Traditionally, the amount of radiation a microwave radiometer receives is expressed as the equivalent blackbody temperature also called brightness temperature[3]. In the microwave range several atmospheric gases exhibit rotational lines. They provide specific absorption features shown at a figure on the right which allow to derive information about their abundance and vertical structure. Examples for such absorption features are the oxygen absorption complex (caused by magnetic dipole transitions) around 60 GHz which is used to derive temperature profiles or the water vapor absorption line around 22.235 GHz (dipole rotational

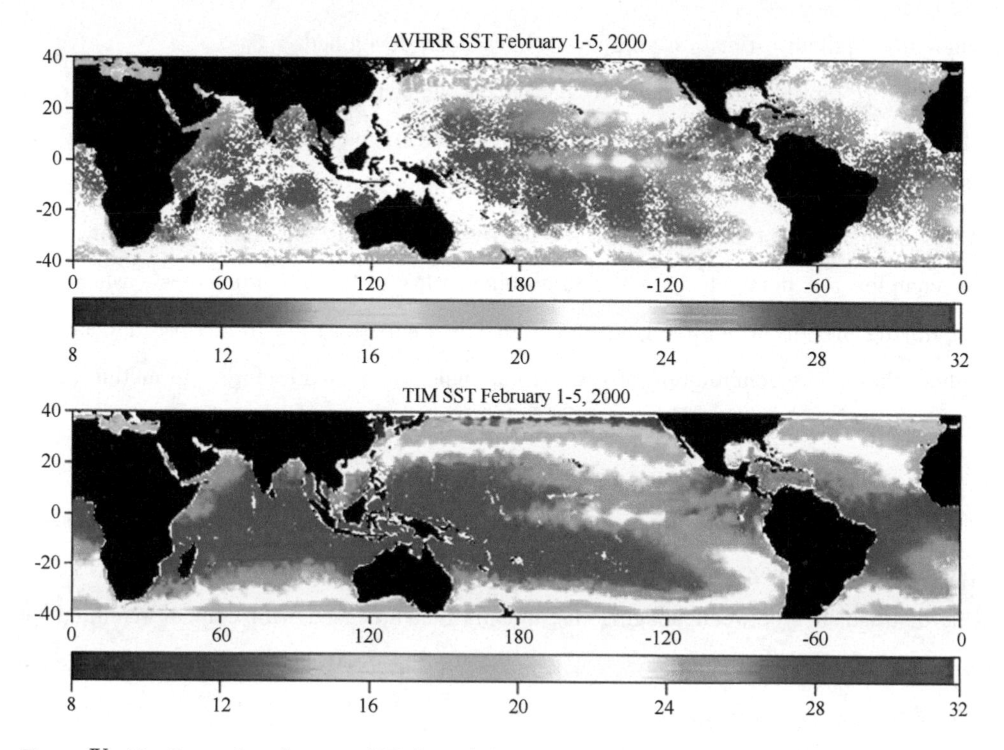

Figure Ⅳ-12: Comparison between SST derived from IR measurements (top image) and a microwave derived SST (bottom image). [Source: eumetrain.org]

transition) which is used to observe the vertical profile of humidity. Other significant absorption lines are found at 118.75 GHz (oxygen absorption) and at 183.31 GHz (water vapor absorption, used for water vapor profiling under dry conditions or from satellites). Weak absorption features due to ozone are also used for stratospheric ozone density and temperature profiling.

Besides the distinct absorption features of molecular transition lines, there are also non-resonant contributions by hydrometeors (liquid drops and frozen particles). Liquid water emission increases with frequency, hence, measuring at two frequencies, typically one close to the water absorption line (22.235 GHz) and one in the nearby window region (typically 31 GHz) dominated by liquid absorption provides information on both the columnar amount of water vapor and the columnar amount of liquid water separately (two-channel radiometer). The so-called "water vapor continuum" arises from the contribution of faraway water vapor lines.

Larger rain drops as well as larger frozen hydrometeors (snow, graupel, hail) also scatter microwave radiation especially at higher frequencies (>90 GHz). These scattering effects can be used to distinguish between rain and cloud water content exploiting polarized measurements but also to constrain the columnar amount of snow and ice particles from space and from the ground.

The retrieval of physical quantities using microwave radiometry (e.g. temperature or water vapor profiles) is not straightforward and comprehensive retrieval algorithms (using inversion techniques like optimal estimation approach) have been developed.

Temperature profiles are obtained by measuring along the oxygen absorption complex at 60 GHz. The emission at any altitude is proportional to the temperature and density of oxygen. As oxygen is homogeneously distributed within the atmosphere and around the globe, the brightness temperature signals can be used to derive the temperature profile. Signals at the center of the absorption complex are dominated by the atmosphere closest to the radiometer (when ground-based). Moving into the window region, the signal is a superposition from close and far regions of the atmosphere. The combination of several channels contains therefore information about the vertical temperature distribution. A similar approach is used to derive vertical profiles of water vapor utilizing its absorption line at 22.235 GHz.

Microwave instruments are flown on several polar orbiting satellites for Earth observation and operational meteorology as well as part of extraterrestrial missions.

One distinguishes between imaging instruments that are used with conical scanning for remote sensing of the Earth surface, e. g., AMSR[4], SSMI, WindSat[5], and sounding instruments that are operated in cross-track mode, e.g., AMSU/MHS. The first type uses lower frequencies (1 - 100 GHz) in atmospheric windows to observe sea-surface salinity, soil moisture, sea-surface temperature, wind speed over ocean, precipitation and snow. The second type is used to measure along absorption lines to retrieve temperature and humidity profile. Furthermore, limb sounders, e.g., MLS, are used to retrieve trace gas profiles in the upper atmosphere.

Other examples of microwave radiometers on meteorological satellites include the Special Sensor Microwave/Imager, Scanning Multichannel Microwave Radiometer, WindSat, Microwave Sounding Unit and Microwave Humidity Sounder. The Microwave Imaging Radiometer with Aperture Synthesis is an interferometer/imaging radiometer capable of resolving soil moisture and salinity over small regions of surface.

Vocabulary

sensitive:(adj) 敏感的,灵敏的

equip:(vt) 装备

n: equipment 装备

planetary:(adj) 行星的

extraterrestrial:(adj) 地球外的

astronomy:(n) 天文学

range:(n) 范围

complementary:(adj) 互补的

semi-transparent:(adj) 半透明的

hydrometeor:(n) 水汽凝结体

derivation:(n) 得到,导出

columnar:(adj) 柱状的

liquid:(n) 液体

traditionally:(adv) 传统来说

brightness temperature：亮温
rotational：（adj）旋转的
magnetic dipole transition：磁偶极子跃迁
stratospheric：（adj）平流层的
molecular：（adj）分子的
non-resonant：（adj）非共振的
continuum：（n）连续体
graupel：（n）霰，软雹
hail：（n）冰雹
distinguish：（vi）区分
straightforward：（adj）直截了当的，易懂的
optimal estimation：最优估计
proportional：（adj）正比于
homogeneously：（adv）均匀地
superposition：（n）叠加
conical：（adj）圆锥的
soil moisture：土壤水分，土壤湿度
precipitation：（n）沉淀，降水，降水量（包括雨、雪、冰等）
aperture：（n）光圈，相对孔径
interferometer：（n）干涉仪

Notes

[1] microwave radiometer 即微波辐射计，是一种用于测量毫米至米波长范围内电磁波（微波）的辐射计。微波辐射计通常配备有多个接收通道，用于推导行星大气层、地表或地外天体的特征发射光谱。微波辐射计广泛应用于各种环境和工程领域，包括遥感、天气预报、气候监测、射电天文学以及无线电传播研究。

[2] 相较于可见光和红外遥感器，微波辐射计具备全天候、全天时的工作能力。具体而言，可见光遥感器受限于光照条件，只能在白天进行有效观测；而红外遥感虽然能在夜晚工作，但其观测能力受到云雾遮挡的严重限制，无法穿透云层进行观测。相比之下，微波辐射计则不受这些因素的制约，能够持续稳定地进行测量。

[3] brightness temperature 即亮度温度，简称亮温，指在同一波长下，当实际物体与黑体的光谱辐射强度相等时，所对应的黑体温度。

[4] AMSR 全称为 Advanced Microwave Scanning Radiometer（高级微波扫描辐射计），是一种改进型的多频率、双极化被动微波辐射计。2001 年，该仪器被搭载于日本的 ADEOS-II对地观测卫星上并发射升空。

[5] WindSat 是全球首颗实现星载全极化微波辐射测量的仪器，于 2003 年搭乘美国 Coriolis 试验卫星发射升空。它能捕获全球海表微波辐射的 4 个 Stokes 参量，开创了全极化微波卫星遥感观测海面风场、风速和风向的先河。

13. Radar Altimeter (TOPEX/Poseidon)

Text

TOPEX/Poseidon[1] was a joint satellite altimeter mission between NASA, the U.S. space agency; and CNES, the French space agency, to map ocean surface topography[1] (Figure Ⅳ-13). Launched on August 10, 1992, it was the first major oceanographic research satellite. TOPEX/Poseidon helped revolutionize oceanography by providing data previously impossible to obtain. Oceanographer Walter Munk[3] described TOPEX/Poseidon as "the most successful ocean experiment of all time"[4]. A malfunction ended normal satellite operations in January 2006.

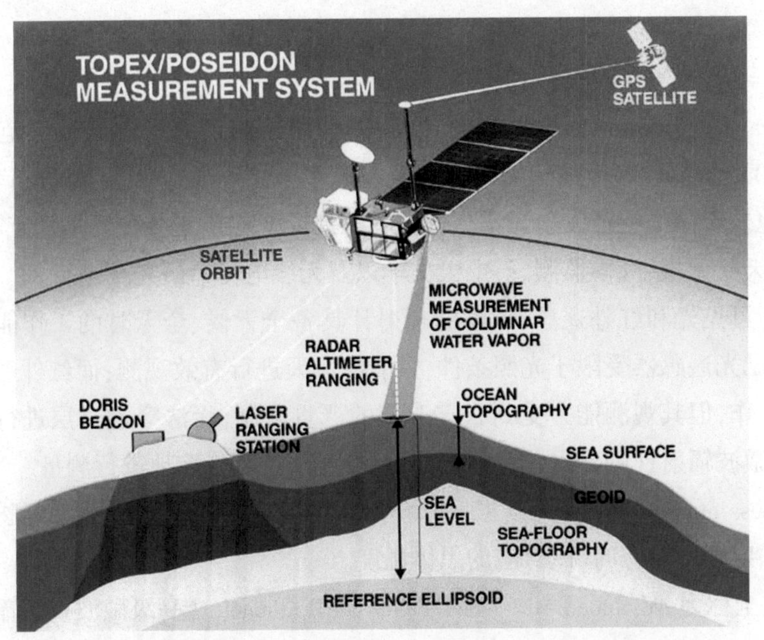

Figure Ⅳ-13: TOPEX/Poseidon measurement system. [Source: Lefauve (2014)]

Before TOPEX/Poseidon, scientists had only a brief glimpse of Earth's ocean as a whole from the pioneering but short-lived Seasat satellite. TOPEX/Poseidon's radar altimeter provided the first continuous global coverage of the surface topography of the oceans. From orbit 1,330 kilometers above the Earth, TOPEX/Poseidon provided measurements of the surface height of 95 percent of the ice-free ocean to an accuracy of 3.3 centimeters. The satellite's measurements of the hills and valleys of the sea surface led to a fundamentally new understanding of ocean circulation[5] and its effect on climate.

The mission's most important achievement was to determine the patterns of ocean circulation—how heat stored in the ocean moves from one place to another. Since the ocean holds most of the Earth's heat from the Sun, ocean circulation is a driving force of climate. TOPEX/Poseidon made it possible for the first time to compare computer models of ocean circulation with actual global observations and use the data to improve climate predictions.

While a three-year prime mission was planned, TOPEX/Poseidon delivered more than 10 years of data from orbit. In those years, the mission:

- Measured sea level with an unprecedented accuracy
- Mapped global tides for the first time
- Monitored the effects of currents on global climate change and produced the first global views of seasonal changes of currents
- Monitored large-scale ocean features like Rossby and Kelvin waves and studied such phenomena as El Niño[6], La Niña[7], and the Pacific Decadal Oscillation[8]
- Mapped basin-wide current variations and provided global data to validate models of ocean circulation
- Mapped year-to-year changes in heat stored in the upper ocean
- Improved our knowledge of Earth's gravity field
- Observed the temperature of the ocean and main seas for over a period of 10 years

TOPEX/Poseidon was launched using an Ariane 42P expendable launch vehicle, along with Korea Institute of Technology's Kitsat-1 satellite and France's S80/T satellite. Lift-off from Kourou in French Guiana took place on August 10, 1992. At lift-off, the mass of the satellite was 2,402 kilograms. The mission was named after the ocean TOPography EXperiment and the Greek god of the ocean Poseidon.

In October 2005, after more than 62,000 orbits, TOPEX/Poseidon stopped providing science data after a momentum wheel malfunctioned, and the satellite was turned off on January 18, 2006. TOPEX/Poseidon's follow-on mission, Jason-1, was launched in 2001 to continue the ongoing measurements of sea surface topography. The two satellites, TOPEX/Poseidon and Jason-1, flew in a tandem mission for three years providing twice the coverage of the sea surface and allowing scientists to study smaller features than could be seen by one satellite.

The record of global sea surface height begun by TOPEX/Poseidon and Jason-1 continues into the future with the Ocean Surface Topography Mission on the Jason-2 satellite, which launched in June 2008. The Jason-3 mission launched January 17, 2016[9].

TOPEX/Poseidon flew two onboard altimeters sharing the same antenna, but only one altimeter was operated at any time, with TOPEX given preference (on average 9 in 10 cycles during the first 10 years of the mission).

TOPEX: The NASA-built Nadir-pointing Radar Altimeter using C band (5.3 GHz) and Ku band (13.6 GHz) for measuring height above sea surface.

Poseidon: The CNES-built solid state Nadir-pointing Radar Altimeter using Ku band (13.65 GHz).

The satellite was also equipped with instruments to accurately pinpoint its location. Precise orbit determination is crucial because errors in locating the spacecraft would distort the sea level measurement calculated from the altimeter readings.

Three independent tracking systems determined the position of the spacecraft. The first, the NASA laser retroreflector array (LRA) reflected laser beams from a network of 10 to 15 ground-based laser ranging stations under clear skies. The second, for all-weather, global tracking, was provided by the CNES Doppler Orbitography and Radiopositioning Integrated by Satellite tracking system receiver (DORIS). This device uses microwave doppler techniques (changes in radio frequency corresponding to relative velocity) to track the spacecraft. DORIS consists of an on-board receiver and a global network of 40 to 50 ground-based transmitting stations.

The third system used an on-board experimental Global Positioning System (GPS) demonstration receiver to precisely determine the satellite's position continuously by analyzing the signals received from the U.S. Air Force's GPS constellation of Earth-orbiting satellites. TOPEX/Poseidon was the first mission to demonstrate that the Global Positioning System could be used to determine a spacecraft's exact location and track it in orbit. Knowing the satellite's precise position to within 2 centimeters (less than 1 inch) in altitude was a key component in making accurate ocean height measurements possible.

A number of satellites use exotic dual-band radar altimeters to measure height from a spacecraft. That measurement, coupled with orbital elements (possibly from GPS), enables determination of the topography. The two lengths of radio waves permit the altimeter to automatically correct for varying delays in the ionosphere.

Vocabulary

revolutionize:(vt) 革命

previously:(adv) 以前

malfunction:(n) 故障

normal:(adj) 正常的

ant: abnormal

glimpse:(n) 短暂的感受;(vt) 瞥见,看一眼

pioneering:(adj) 开拓性的,先驱性的,探索性的

hill:(n) 小山

valley (n) 山谷

achievement:(n) 成就

hold:(vt) 持有,保留

driving force: 驱动力

prime:(adj) 主要的,首要的

unprecedented:(adj) 空前的,前所未有的,

没有先例的

adv：unprecedented

tide：（n）潮汐

seasonal：（adj）季节的

phenomenon：（n）现象

pl：phenomena

El Niño：厄尔尼诺现象

La Niña：拉尼娜现象

Pacific Decadal Oscillation：太平洋年代际振荡，十年涛动

validate：（vi）验证

gravity：（n）重力

expendable：（adj）可消耗的

lift-off：（n）（航天器的）发射，起飞

momentum wheel：动量轮

turn off：关掉

follow-on：（adj）后继的

ongoing：（adj）持续存在的，不断发展的，正在进行的

tandem：（adj）（两匹马）前后纵列

antenna：（n）天线

preference：（n）偏爱，喜爱

pinpoint：（vi）明确指出

crucial：（adj）关键的，至关重要的

independent：（adj）独立的，不相关的

ant：dependent

laser retroreflector array：激光后向反射器阵列

constellation：（n）星座，一系列（相关的想法、事物）

exotic：（adj）奇异的

dual-band：（adj）双频带的

ionosphere：（n）电离层

Notes

[1] TOPEX/Poseidon 卫星由美国国家航空航天局喷气推进实验室（NASA/JPL）和法国国家空间研究中心（CNES）联合研制。NASA 提供卫星平台和 5 个科学仪器，并负责卫星的运营。CNES 则贡献两个科学仪器，并负责发射服务。该卫星专注于海面地形的精密测量，从而观测并解析海洋环流。

[2] ocean surface topography 即海面地形，是指海面相对于大地水准面（geoid）的垂直位移。

[3] Walter Munk（沃尔特·芒克，1917—2019），美国物理海洋学家，曾任加利福尼亚州拉霍亚斯克里普斯海洋研究所（Scripps Institution of Oceanography）的海洋学领域主席。Munk 被《纽约时报》誉为“海洋学界的爱因斯坦”，他在海洋环流、潮汐和深海混合、海啸和地震波，以及地球自转等方面做出了里程碑式的贡献。

[4] TOPEX/Poseidon 被 Walter Munk 高度赞誉，称其为“迄今为止最为成功的海洋科学实验”。

[5] ocean circulation 即海洋环流，其主要表现形式为地转流。海面地形的梯度能够揭示地转流的异常。

[6] El Niño 即厄尔尼诺，系西班牙语，指在太平洋的秘鲁和厄瓜多尔沿岸区域出现的一

种海温异常升高现象。这是一种大规模的海洋和大气相互作用的现象。

[7] La Niña 即拉尼娜,指赤道太平洋东部和中部海表温度大范围持续异常变冷的现象,也称为反厄尔尼诺。

[8] Pacific Decadal Oscillation 即太平洋年代际振荡,是北太平洋海表温度年代际尺度上最主要的变化,其周期通常为 20~30 a。PDO 的主要特征是,在太平洋 20°N 以北的区域,表层海水温度会出现异常偏高或偏低的现象。具体而言,在 PDO 的“暖相位”(或“正相位”)期间,西太平洋偏冷而东太平洋偏暖;相反,在“冷相位”(或“负相位”)期间,西太平洋偏暖而东太平洋偏冷。

[9] Altimeter Mission Series: TOPEX/Poseidon(1992—2006 年)、Jason-1(2001—2013 年)、OSTM/Jason-2(2008—2019 年)、Jason-3(2016 年至今)、Sentinel-6 Michael Freilich(2020 年至今)、SWOT(2022 年至今)、Sentinel-6B(计划中)。

14. Radar Scatterometer

Text

One of the most familiar variables in the Earth's atmosphere is the wind. Wind speed can range from light, almost imperceptible breezes to the violent, nearly unimaginable power of wind in the eyewall of a hurricane. Measuring wind on land requires only anemometers[1] and a sufficiently large number of weather stations equipped with them. The vast distances at sea, however, deter the efficient collection of wind speed data. For decades, global weather models and forecasts relied on infrequent wind speed data gathered by ships—and these ships were likely to avoid problem areas like storms and hurricanes.

The measurement of wind speed over the oceans is accomplished by a technique called scatterometry (Figure Ⅳ-14). Originally discovered as a source of noise in radar (it affected the early radar used in World War Ⅱ), the effect of winds on the ocean was first analyzed by a scatterometer flown on the Skylab missions in 1973 and 1974. The first orbital deployment of a scatterometer was the Seasat-A satellite scatterometer (SASS), which operated as part of the June-October 1978 Seasat mission. Despite the short lifetime of the mission, SASS demonstrated the ability to make accurate wind speed measurements over the oceans and enabled better location of storms at sea. More than a decade would pass until the ESA's European Remote Sensing Satellite-1 (ERS-1) carried the Active Microwave Instrument (AMI) into orbit in 1991. AMI functioned as both a scatterometer and a SAR, so it did not provide full-time scatterometry data. Despite that limitation, oceanographers welcomed this new source of wind data. ERS-2, which was launched in April 1995, also carried AMI. The next full-time scatterometer was the NASA scatterometer (NSCAT), was flown on the ill-fated Japanese Midori satellite. NSCAT functioned from August 1996 to June 1997, when Midori suffered a catastrophic power loss. NSCAT's sophisticated design and flawless performance fully validated the technique of scatterometry, allowing marked improvement in ocean weather forecasting and ocean modeling, and even demonstrated some land surface cover applications. Due to the demonstrated value of scatterometry data from this mission, NASA proposed, funded, built, and launched the QuikSCAT[2] mission, carrying the SeaWinds instrument, in an unprecedentedly short of time. QuikSCAT launched in June 1999, only 2 years after the demise of NSCAT, and produced excellent data.

Scatterometry is based on a simple principle. It is an example of an "active" microwave radar instrument, operating by transmitting pulses of microwaves at the ocean surface and then measuring the scattering of the return reflection by capillary waves, small ripples that form right

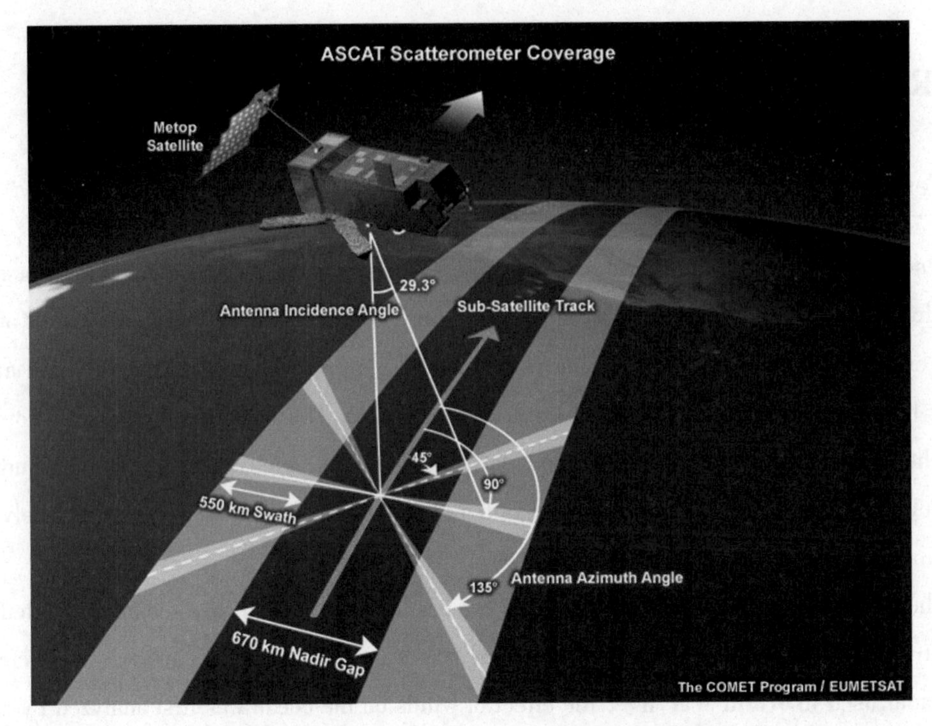

Figure Ⅳ-14: ASCAT Scatterometer. [Source: eumetrain.org]

on the surface in response to the wind. As the roughness[3] of the ocean surface increases, the reflection of scattered energy also increases. The instrument detects the return reflection strength and converts it to wind speed and direction. Seasat used a two-antenna system that provided two wind direction possibilities oriented 180 degrees from each other. NSCAT used a three-antenna system that allowed unambiguous wind-vector determination. SeaWinds employs a single rotating dish antenna system creating two spot beams that sweep the oceans in a circular pattern, which also provides data allowing precise wind vector determination. Both NSCAT and SeaWinds operate in the 13 GHz range; SeaWinds beams 110 W pulses with a 189 Hz repetition frequency. SeaWinds covers 90% of the global ocean every day with a swath that is 1800 km wide. Its wind speed measurements are accurate to approximately 2 m/sec for wind speeds between 3 and 20 m/sec.

Scatterometry data are applicable to numerous weather and climate phenomena. One of the most effective uses of the data is the ability to locate ocean storms. Scatterometry data can detect the circular rotation of winds around tropical lows much earlier than conventional forecasts, sometimes almost two days earlier, aiding detection of tropical storm and hurricane formation. Once such storms have formed, the data provide accurate wind speed and location information.

The exchange of heat, water vapor, and gases (such as CO_2) at the ocean surface is dependent on wind speed[4], and scatterometry data enable improved modeling of these processes on a global scale. Scatterometry data may also unlock one of the mysteries of the El Niño phenomenon: why the winds that normally blow unceasingly over the south Pacific Ocean suddenly falter and diminish, allowing waves of warm water to migrate eastward. Scatterometry data can also detect changes in land surface cover due to changes in the reflectivity of the surface, and also can observe sea-ice cover in the polar regions.

Although scatterometry provides wind speed data over the oceans, wind speed can also be derived from other instrumental observations. Most notable in this regard is the Special Sensor Microwave/Imager (SSM/I) on U.S. Department of Defense meteorological satellites[5]. Wind speed data are derived from microwave brightness temperatures measured by the SSM/I. The use of SSM/I data for wind speed estimation began in 1987 and is continuing at present.

Vocabulary

familiar: (adj) 熟悉的
imperceptible: (adj) 感觉不到的,极细微的
breeze: (n) 微风
violent: (adj) 暴力的,剧烈的
unimaginable: (adj) 难以想象的,难以置信的
eyewall: (n)(飓风)眼壁
hurricane: (n)(尤指西大西洋的)飓风
anemometer: (n) 风速计
deter: (vt) 阻止,制止
accomplish: (vt) 完成
scatterometry: (n) 散射测量
scatterometer: (n) 散射计
catastrophic: (adj) 灾难性的
sophisticated: (adj) 复杂的,先进的,精密的
flawless: (adj) 完美的,无瑕的
principle: (n) 原理
pulse: (n) 脉冲
capillary wave: 毛细波,表面张力波,界面波
roughness: (n) 粗糙度
unambiguous: (adj) 明确的,无歧义的
dish antenna: 碟形天线
wind vector: 风矢量
repetition: (n) 重复
circular: (adj) 环形的
tropical low: 热带低压
exchange: (n) 交换
falter: (vi) 衰退
migrate: (vi) 迁移

Notes

[1] anemometer 即风速计,又称风速仪,是一种专门用于测量空气流动速度的仪器。其种类繁多,在气象台站中,风杯风速计是最为常用的类型。风杯风速计的感应部分由 3 个抛物锥形的空杯构成,这 3 个空杯互成 120°并牢固地固定在支架上,且每个

空杯的凹面都朝向同一个方向。当风作用于风杯时,风杯便会以正比于风速的转速旋转,从而实现对风速的测量。

[2] QuikSCAT 卫星即“快速散射计”(Quick Scatterometer)卫星,是美国国家航空航天局(NASA)精心研发的一颗对地观测卫星,隶属于美国“地球观测系统”(EOS)计划。该卫星的核心使命是重启并推进 NASA 的“海洋风测量”计划,旨在满足提升天气预报精度和深化气候研究需求的迫切要求。卫星上搭载了先进的“海洋风场”微波散射计,具备全天候、连续作业的能力,能够高精度地测量并记录全球范围内海洋表面的风速和风向数据。

[3] roughness 即粗糙度,作为一个物理量,专门用于描述海面的空气动力学粗糙程度,它能够量化地反映出海面几何形态的起伏程度。

[4] 海气界面的热量、水汽以及各种气体的交换速率均与风速密切相关。通常情况下,风速越大,海气界面交换越快。

[5] 风场数据也可以由 Special Sensor Microwave/Imager(SSM/I)反演得到。

15. SAR (Synthetic-aperture Radar)

Text

Synthetic-aperture radar (SAR) is a form of radar that is used to create two-dimensional images or three-dimensional reconstructions of objects, such as landscapes. SAR uses the motion of the radar antenna over a target region to provide finer spatial resolution than conventional stationary beam-scanning radars[2]. SAR is typically mounted on a moving platform, such as an aircraft or spacecraft, and has its origins in an advanced form of side-looking airborne radar (SLAR)[3]. The distance the SAR device travels over a target during the period when the target scene is illuminated creates the large synthetic antenna aperture (the size of the antenna). Typically, the larger the aperture, the higher the image resolution will be, regardless of whether the aperture is physical (a large antenna) or synthetic (a moving antenna)—this allows SAR to create high-resolution images with comparatively small physical antennas. For a fixed antenna size and orientation, objects which are further away remain illuminated longer—therefore SAR has the property of creating larger synthetic apertures for more distant objects, which results in a consistent spatial resolution over a range of viewing distances.

To create a SAR image, successive pulses of radio waves are transmitted to "illuminate" a target scene, and the echo of each pulse is received and recorded. The pulses are transmitted and the echoes received using a single beam-forming antenna, with wavelengths of a meter down to several millimeters. As the SAR device on board the aircraft or spacecraft moves, the antenna location relative to the target changes with time. Signal processing of the successive recorded radar echoes allows the combining of the recordings from these multiple antenna positions. This process forms the synthetic antenna aperture and allows the creation of higher-resolution images than would otherwise be possible with a given physical antenna.

Motivation and Applications

SAR is capable of high-resolution remote sensing, independent of flight altitude and weather, as SAR can select frequencies to avoid weather-caused signal attenuation. SAR has day and night imaging capability as illumination is provided by itself[4].

SAR images have wide applications in remote sensing and mapping of surfaces of the Earth and other planets. Applications of SAR are numerous[5]. Examples include topography, oceanography, glaciology, geology (for example, terrain discrimination and subsurface imaging). SAR can also be used in forestry to determine forest height, biomass, and deforestation. Volcano and earthquake monitoring use differential interferometry. SAR can also be applied for monitoring civil infrastructure stability such as bridges. SAR is useful in environment monitoring such as oil spills, flooding, urban growth, military surveillance: including strategic policy and tactical assessment. SAR can be implemented as inverse SAR by observing a moving target over a substantial time with a stationary antenna.

Basic Principle

A synthetic-aperture radar is an imaging radar mounted on a moving platform. Electromagnetic waves are transmitted sequentially, the echoes are collected and the system electronics digitizes and stores the data for subsequent processing. As transmission and reception occur at different times, they map to different small positions. The well-ordered combination of the received signals builds a virtual aperture that is much longer than the physical antenna width. That is the source of the term "synthetic aperture," giving it the property of an imaging radar (Figure Ⅳ-15). The range direction is perpendicular to the flight track and perpendicular to the azimuth direction, which is also known as the along-track direction because it is in line with the position of the object within the antenna's field of view.

The 3-D processing is done in two stages. The azimuth and range direction are focused for the generation of 2-D (azimuth-range) high-resolution images, after which a digital elevation model (DEM)[6] is used to measure the phase differences between complex images, which is determined from different look angles to recover the height information. This height information, along with the azimuth-range coordinates provided by 2-D SAR focusing, gives the third dimension, which is the elevation. The first step requires only standard processing algorithms, for the second step, additional pre-processing such as image co-registration and phase calibration is used.

In addition, multiple baselines can be used to extend 3-D imaging to the time dimension. 4-D and multi-D SAR imaging allows imaging of complex scenarios, such as urban areas, and has improved performance with respect to classical interferometric techniques such as persistent scatterer interferometry (PSI).

Algorithm

SAR algorithms model the scene as a set of point targets that do not interact with each other (the Born approximation).

While the details of various SAR algorithms differ, SAR processing in each case is the application of a matched filter to the raw data, for each pixel in the output image, where the matched filter coefficients are the response from a single isolated point target. In the early days of

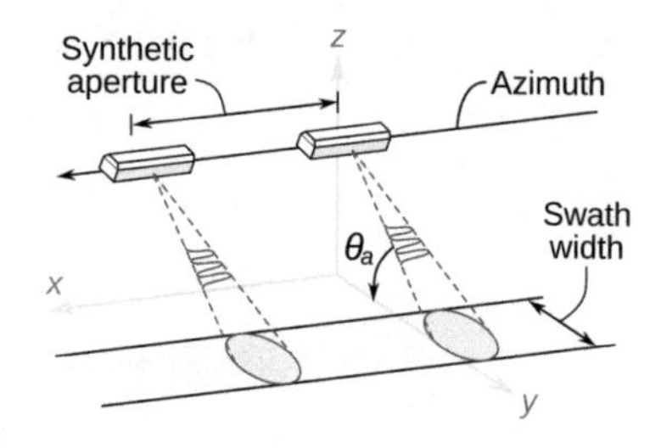

Figure Ⅳ-15: Basic principle of synthetic-aperture radar.

SAR processing, the raw data was recorded on film and the postprocessing by matched filter was implemented optically using lenses of conical, cylindrical and spherical shape. The Range-Doppler algorithm is an example of a more recent approach.

Synthetic-aperture radar determines the 3-D reflectivity from measured SAR data. It is basically a spectrum estimation, because for a specific cell of an image, the complex-value SAR measurements of the SAR image stack are a sampled version of the Fourier transform[7] of reflectivity in elevation direction, but the Fourier transform is irregular. Thus the spectral estimation techniques are used to improve the resolution and reduce speckle[8] compared to the results of conventional Fourier transform SAR imaging techniques.

Vocabulary

two-dimensional: (adj) 二维的
reconstruction: (n) 重构
landscape: (n) 景观,地形,地貌
motion: (n) 运动,移动
finer: (adj) 更精细的
resolution: (n) 分辨率
beam-scanning: (adj) 光束扫描的
mount: (vi) 被装载
side-looking: (adj) 侧视的
airborne: (adj) 机载的
radar: (n) 雷达
device: (n) 设备,装置
scene: (n) 场景,场面
illuminate: (vt) 照明,照亮
regardless: (adv) 不管,不顾
comparatively: (adv) 相对地
orientation: (n) 方向
echo: (n)(脉冲的)回波
relative: (adj) 相对的
signal processing: 信号处理
flight: (n) 飞行
capability: (n) 能力
glaciology: (n) 冰川学
discrimination: (n) 识别
subsurface: (adj) 次表层的
forestry: (n) 林业,林学
biomass: (n) 生物量
differential interferometry: 差分干涉术
oil spill:(海上)溢漏
digitize: (vt)(使数据)数字化

virtual：(adj) 虚拟的,模拟的
perpendicular：(adj) 垂直的
azimuth：(n) 方位角
digital elevation model：数字高程模型
phase difference：相位差
elevation：(n) 高程,高度
algorithm：(n) 算法
co-registration：(n) 配准
scenario：(n) 情景
interferometric：(adj) 干涉测量的
filter：(n) 滤波器
raw data：原始数据
postprocessing：(n) 后处理
  ant：pre-processing (n) 预处理
lens：(n) 透镜,镜片
cylindrical：(adj) 圆柱的
complex-value：(adj) 复数值的
stack：(n) 堆栈
Fourier transform：傅里叶变换
irregular：(adj) 不规则的
  ant：regular (a) 规则的
speckle：(n) 斑点

Notes

[1] Synthetic-Aperture Radar (SAR) 即合成孔径雷达,是一种利用微波进行成像的先进雷达系统,它既可以作为(航空)机载雷达,也可以作为(太空)星载雷达,具备生成高分辨率图像的能力。该系统通常被部署在移动的平台上,用于对静止目标进行成像,或者相反,利用静止平台对移动目标进行观测。自合成孔径雷达技术诞生以来,它已被广泛地应用于遥感探测和地图测绘领域。

[2] 对一个典型的机载合成孔径雷达系统而言,其天线通常被安装在飞机的侧面。该系统发射的电磁波波束宽度相对较大(可能达到几度),若依据衍射原理追求极窄波束,则需配置极为庞大的天线,这在现实中往往难以实现。在垂直方向上,波束宽度同样较大,导致天线波束覆盖的区域从飞机正下方一直延伸到遥远的天边。然而,当地表基本平坦或坡度变化保持在一定限度内时,位于载体正下方或卫星地面轨迹(星下点)不同距离的点就可以通过回声时延的差异进行有效区分。在沿运动方向进行区分时,使用较短小的天线面临较大挑战。但若能令飞行器在飞行过程中连续发射一系列脉冲,并精确记录回声的振幅与相位信息,则这些回声信号可通过特定算法进行组合处理。这一处理过程的效果,相当于这些信号是从一远超实际天线尺寸的巨大(长)天线中同时发射出来的。这种方法实质上“合成”了一个尺度远大于实际天线的虚拟天线。

[3] 侧视雷达简称 SLR,是一种工作于微波波段的成像雷达,其视场方向与飞行器的前进方向保持垂直,专门用于探测飞行器两侧的区域。该雷达系统主要由发射机、接收机、传感器、数据存储装置和数据处理单元等核心部件构成。在早期阶段,它通过直接加大天线孔径和发射窄脉冲的方式,来有效提高雷达图像的分辨率。自 20 世纪 60 年代起,合成孔径技术的引入使得雷达探测的分辨率实现了质的飞跃,提高了

几十倍乃至几百倍。时至今日,现代侧视雷达在 10 000 m 高空飞行时,其地面分辨率已达到 1 m 以内。

[4] SAR 具有以下特点:作为一种主动式的微波遥感设备,SAR 不受自然光照条件和气候状况的制约,能够实现不间断、全天候的地表/海表观测能力。尤为突出的是,它具备穿透地表/海表或植被层的能力,从而揭示出这些遮蔽物之下隐藏的信息。

[5] SAR 的应用场景:在农、林、水或地质、自然灾害等民用领域展现出了巨大的应用潜力,而在军事领域,其更具备无可比拟的独特优势。展望未来,战场空间不再局限于传统的陆、海、空,而是进一步拓展至太空,作为一种具有独特侦察能力的技术手段,合成孔径雷达卫星为夺取未来战场的制信息权,甚至对战争的最终走向具有决定性的影响。

[6] digital elevation model (DEM) 即数字高程模型,是一种通过有限的地形高程数据来精确模拟并数字化表达地面地形地貌的技术手段。具体而言,它是采用一组有序排列的数值阵列来构建地面高程的实体模型,是数字地形模型 (digital terrain model) 的一个重要分支。

[7] Fourier transform 即傅里叶变换,是一种线性积分变换方法,其核心功能在于实现信号在时域(或空间域)和频域之间的转换,在物理学和工程学等多个科学领域中具有广泛的应用价值。

[8] speckle 即斑点,指在 SAR 成像中观测到的一种现象。SAR 成像系统的工作原理是基于相干性原理。在雷达回波信号的处理过程中,相邻像素点的灰度值会由于相干效应而产生随机波动,这些波动围绕着某一平均值进行,从而在图像中形成所谓的斑点噪声。斑点噪声的产生源于 SAR 成像所依赖的相干原理本身固有的特性,因此它是不可避免的。

16. ATLAS Mooring (TAO array)

Text

Development of the Tropical Atmosphere Ocean (TAO) array[1] was motivated by the 1982–1983 El Nino event, the strongest of the century up to that time, which was neither predicted nor detected until nearly at its peak. The event highlighted the need for real-time data from the tropical Pacific for monitoring, prediction, and improved understanding of El Nino. As a result, with support from NOAA's Equatorial Pacific Ocean Climate Studies (EPOCS) program, PMEL began development of the ATLAS (Autonomous Temperature Line Acquisition System)[2] mooring (Figure Ⅳ–16). This low-cost deep ocean mooring was designed to measure surface meteorological and subsurface oceanic parameters, and to transmit all data to shore in real-time via satellite relay. The mooring was also designed to last one year in the water before needing to be recovered[3] for maintenance.

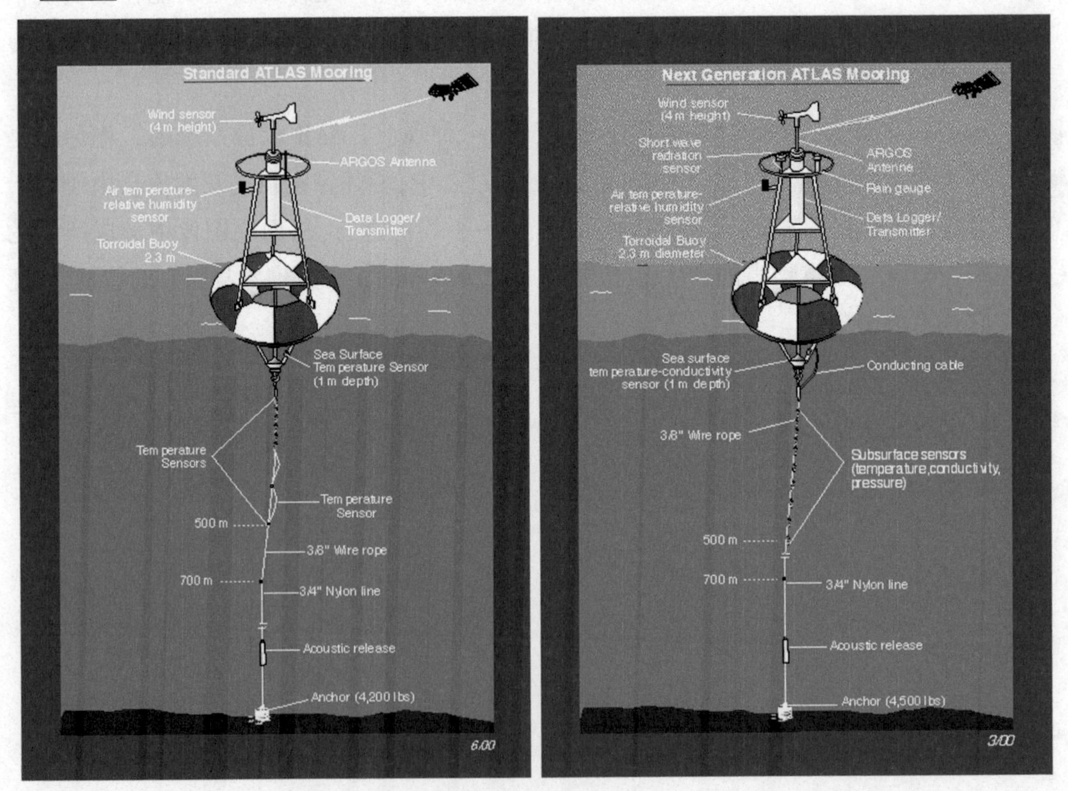

Figure Ⅳ–16: ATLAS moorings. [Source: NOAA]

Historical Standard ATLAS Moorings

After testing and deployment of prototype ATLAS moorings, the first elements of the large-scale monitoring TAO array were deployed in the eastern Pacific in November 1984. The full TAO array was eventually completed in December 1994. The standard ATLAS mooring had a design lifetime of one year, and the system proved to be robust and reliable. Over 500 Standard ATLAS moorings were deployed between 1984 and 2001. The final standard ATLAS was recovered in November 2001 and Next Generation ATLAS moorings[4] are now used exclusively in the TAO array.

Standard ATLAS moorings measured surface winds[5], air temperature[6], relative humidity[7], sea surface temperature[8], and ten subsurface temperatures from a 500 m long thermistor cable[9]. Daily-mean data were telemetered to shore in near real-time via NOAA's polar-orbiting satellites and Service Argos. A small subset of hourly values (2–3 per day) coinciding with satellite passes were also transmitted in real time. Hourly values of surface data were internally recorded and available after mooring recovery.

The TAO surface buoy is a 2.3 m diameter fiberglass-over-foam toroid, with an aluminum tower and a stainless steel bridle. When completely rigged, the system has an air weight of approximately 660 kg, a net buoyancy of nearly 2300 kg, and an overall height of 4.9 m. The buoy can be seen on radar from 4–8 miles depending on sea conditions.

Non-rotating 3/8" (0.92 cm) diameter wire rope jacketed to 1/2" (1.27 cm) is used in the upper 700 meters to guard against damage from fish bite. Standard ATLAS thermistor cables were fixed to the mooring wire with wire rope clamps. Plaited 8-strand 3/4" (1.9 cm) diameter nylon line is used for the remainder of the mooring. Anchors are fabricated from scrap railroad wheels, and typically weight 1900–2000 kg. All hardware is standard equipment as used in other PMEL taut-line moorings and deployments follow the traditional anchor last routine.

Moorings are deployed in water depths between 1500 and 6000m. To ensure that the upper section of the mooring is nearly vertical a nominal scope of 0.985 (ratio of mooring length to water depth)[10] is employed on the moorings in water depths of 1800m or more. At a few sites, slack moorings with scope 1.35 have been deployed due to either shallow bathymetry or severe current regimes. In these cases, the upper portion of the mooring keeps fairly vertical (but less so than taut-line moorings) by using a reverse catenary design.

Next Generation ATLAS Moorings

This effort used as many of the components and procedures of the existing system as possible, thereby minimizing the impact on the infrastructure that supported the array. The first Next

Generation systems were deployed in the array in May 1996. The transition to Next Generation systems throughout the array was completed in November 2001. By the mid-1990's, a reengineering effort was underway to modernize the ATLAS mooring with emphasis on:

- improving data quality
- adding new sensors
- increasing temporal resolution of internally recorded data
- improving reliability to extend system life
- simplifying fabrication procedures
- reducing costs

A significant Next Generation ATLAS improvement over the Standard ATLAS is the incorporation of inductively coupled sensors for subsurface data. The sensors clamp onto the wire rope strength member that serves as one of the inductive elements. This simplifies fabrication, eliminating the thermistor cable with its labor-intensive assembly and deployment procedures. Addressable modules on the cable allow the system to be expanded for new sensors by adding the appropriate hardware and software interfaces. Flexibility in the design also allows the interface of additional sensors including rainfall[11], short-wave and long-wave radiation[12], barometric pressure[13], ocean salinity[14] and currents[15]. Most measurements are made at a sample rate of 10 minutes, with the exception of barometric pressure (1 hour), short-wave and long-wave radiation (2 minutes), and rainfall (1 minute). These high temporal resolution data are recorded internally and available after mooring recovery.

Vocabulary

Tropical Atmosphere Ocean (TAO) array:(n) 热带大气海洋阵列

monitoring:(n) 监测

prediction:(n) 预测

mooring:(n) 锚系

meteorological:(adj) 气象的

oceanic:(adj) 海洋的

parameter:(n) 参数,变量

transmit:(vt) 传播,传送

satellite relay:(n) 卫星中继器

recover:(vt) 回收

deployment:(n) 布放

vt: deploy

prototype:(n) 原型

eventually:(adv) 最终

exclusively:(adv) 唯一地

relative humidity:(n) 相对湿度

thermistor cable:(n) 温度链

telemeter:(vt) 遥测,用遥测发射器传送

coincide:(vi) 同时发生,位置重合

diameter:(n) 直径,放大倍数

toroid:(n) 环体

aluminum:(n) 铝

rig:(vt) 安装

wire rope：(n) 钢丝绳
clamp：(n) 夹子，夹具
anchor：(n) 锚；(vt) 把……系住，使固定
fabricate：(vt) 组装
 n：fabrication 制作，制造
nominal：(adj) 名义上的
slack：(adj) 不紧的，懈怠的
infrastructure：(n) 基础设施
modernize：(vi) 使……现代化
reliability：(n) 可靠性
improvement：(n) 改善，改进
incorporation：(n) 结合，合并
inductively：(adv) 感应地
assembly：(n) 装配，集会
flexibility：(n) 灵活性
rainfall：(n) 降水
short-wave radiation：(n) 短波辐射
long-wave radiation：(n) 长波辐射

Notes

[1] tropical atmosphere ocean (TAO) array 即热带大气海洋阵列，涵盖了热带太平洋区域的大约 70 个 ATLAS 锚系浮标，这些浮标通过 Argos 卫星实时传送观测的大气与海洋数据。TAO 阵列的核心使命在于提升对厄尔尼诺现象的监测和预测能力，是全球气候观测系统的关键构成部分。

[2] ATLAS 锚系浮标作为 TAO 阵列的核心，负责观测海气界面的多项关键参数，包括海表风场、气温、湿度、气压、海表温度，以及向下短波和长波辐射，并且能探测次表层海温 (0~700 m 温度)。此外，部分 ATLAS 锚系浮标还具备测量海洋流场的能力。

[3] recover 在此处意指回收。ATLAS 锚系浮标的设计适用寿命为 1 a，期满后需执行回收作业，随后对其进行电池更换及必要配件的更新维护，以便进行再布放。

[4] next generation ATLAS mooring 相对于 standard ATLAS mooring 在性能上实现了显著提升，包括数据质量的优化、新传感器的集成、自容式传感器时间分辨率的提高、系统使用寿命的延长、装配步骤的简化，以及在成本控制上的显著降低。

[5] surface winds 即海表风场，包括了风速和风向两大要素。其中，风速通过螺旋桨式风速表 (propeller) 进行测量，而风向则依靠风向标 (vane) 和磁通门罗经 (fluxgate compass) 联合测定。

[6] air temperature 即海表气温，通过电阻温度探测器 (Pt-100 Resistance temperature detector) 进行测量。

[7] relative humidity 即相对湿度，通过电容湿度计进行测量。

[8] sea surface temperature 即海表温度，通过热敏电阻 (thermistor) 进行测量，其实际测量深度为 1.5 m。

[9] thermistor cable 为温度链，由多个热敏电阻测量次表层温度。

[10] 锚系长度和水深比值设为 0.985，可以保证锚系是接近垂直的。

[11] rainfall 即降水，通过雨量计进行测量。

[12] short-wave and long-wave radiation 即海表短波和长波辐射，通过辐射计（pyrgeometer）进行测量。

[13] barometric pressure 即大气压，通过压力传感器（pressure transducer）进行测量。

[14] ocean salinity 即海水盐度，通过电导率测定（internal field conductivity cell）得到。

[15] ocean currents 即海水流场，其单点流速和流向由多普勒海流计（Doppler Current Meter）测定，而流场的垂直剖面特征则通过声学多普勒海流剖面仪（Acoustic Doppler Current Profiler）进行测量。

17. Surface Drifter Program

Text

The Global Drifter Program (GDP)[1] (formerly known as the Surface Velocity Program (SVP) was conceived by Prof. Peter Niiler, with the objective of collecting measurements of surface ocean currents, sea surface temperature and sea-level atmospheric pressure using drifters. It is the principal component of the Global Surface Drifting Buoy Array, a branch of NOAA's Global Ocean Observations and a scientific project of the Data Buoy Cooperation Panel (DBCP). The project originated in February 1979 as part of the TOGA/Equatorial Pacific Ocean Circulation Experiment (EPOCS) and the first large-scale deployment of drifters was in 1988 with the goal of mapping the tropical Pacific Ocean's surface circulation. The current goal of the project is to use 1250 satellite-tracked surface drifting buoys to make accurate and globally dense in-situ observations of mixed layer currents[2], sea surface temperature, atmospheric pressure, winds and salinity, and to create a system to process the data. Horizontal transports in the oceanic mixed layer measured by the GDP are relevant to biological and chemical processes as well as physical ones[3].

SVP project drifter deployments began in 1979; the design continued to develop until reaching its current form in 1992. Each drifter consists of a spherical surface buoy tethered to a weighted nylon drogue that allows it to track the horizontal motion of water at a depth of 15 meters (Figure Ⅳ-17). If the drogue breaks off, the wind pushes the surface buoy through the water, creating erroneous current observations[4]. A tether strain gauge has been added to monitor tension of the buoy-drogue connection to resolve this issue. The original drifters are heavy, bulky (40 cm diameter), and expensive relative to the newer "mini" drifters that are smaller, (30.5 cm diameter) cheaper, and lighter because the hull contains fewer batteries. The surface float contains alkaline batteries, a satellite transmitter, a thermistor for sub-skin sea surface temperature[5], and sometimes other instruments that measure pressure, wind speed and direction, or salinity.

The drifters are deployed from research vessels, volunteer ships, and through air deployment. They typically transmit their data hourly and had an average lifetime of ~485 days in 2001. Presently, enough data is gathered to observe currents at a horizontal resolution of one degree (~100 km). Single drifters can be tracked with the name of the drifter.

The GDP consists of three components. The component at NOAA's Atlantic Oceanographic and Meteorological Laboratory (AOML) manages deployments, processes and archives the data,

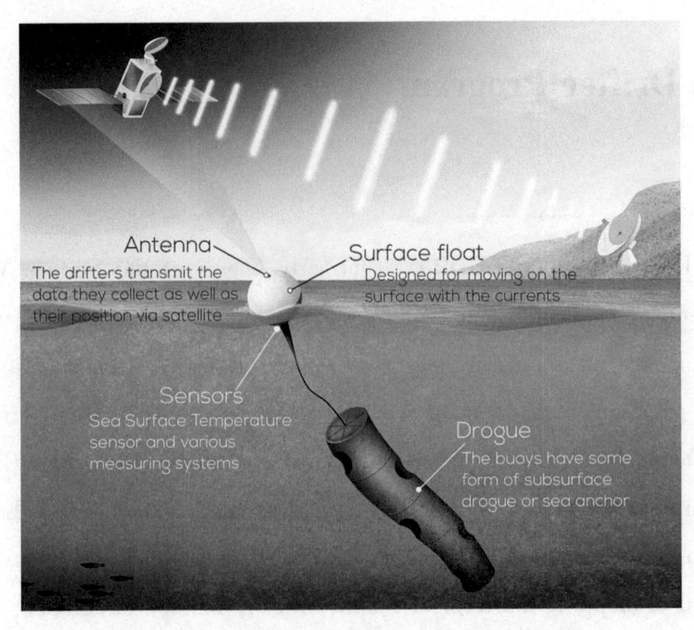

Figure Ⅳ-17: Drifter illustration. [Source: NOAA]

maintains META files describing each drifter deployed, develops and distributes data-based products, and updates the GDP website. The Lagrangian Drifter Laboratory at the Scripps Institution of Oceanography (SIO) leads the engineering aspects of the Lagrangian drifter technology, improves the existing designs, develops new drifters, manages the real-time data stream, including posting the drifter data to the Global Telecommunication System, supervises the industry, purchases and fabricates most drifters, and develops enhanced data sets. The third component is the manufacturers in private industry, who build drifters according to specifications. The GDP collaborates with partners from numerous countries including Argentina, Australia, Brazil, Canada, France, India, Italy, Republic of Korea, Mexico, New Zealand, South Africa, Spain, United Kingdom, and the United States.

Data from Global Drifter Program (GDP) buoys is available in various formats. (1) Real-Time Drifter Data: Real time drifter data and graphs are distributed on the Global Telecommunication System (GTS) for improved weather and climate forecasting and ocean state estimation. (2) Six-Hourly Interpolated Drifter Data: The Drifter Data Assembly Center (DAC) at AOML applies quality control procedures to drifter data (position and temperature) and interpolates them to 6-hour intervals using an optimum interpolation procedure called kriging, which is commonly used for two- and three-dimensional analyses. Interpolated data, and metadata, from more than 30,000 drifters are available for download at Six-Hourly Interpolated Drifter Data. These data go back to the earliest available GDP observations in 1979. (3) Hourly Interpolated Data[6]: The hourly drifter dataset is a valuable new tool for the study of relatively small-scale and high-frequency oceanic processes. In particular, velocity rotary spectra demonstrate that high frequency tidal and internal wave motions are detectable globally. These data are available primarily since the early 2000's when satellite coverage was sufficient to resolve hourly motion.

Vocabulary

global drifter program:(n) 全球漂流浮标计划
conceive:(vt) 构想,想出
objective:(n) 目的
principal:(adj) 主要的
component:(n) 组成
originate:(vi) 起源于
goal:(n) 目标
horizontal:(adj) 水平的
transport:(n) 输送
relevant:(adj) 相关的,有价值的
biological:(adj) 生物学的
chemical:(adj) 化学的
physical:(adj) 物理学的
spherical:(adj) 球状的
tether:(vt) 拴
drogue:(n) 帆
gauge:(n) 测量仪
tension:(n) 拉力,张力
bulky:(adj) 大块头的,笨重的
hull:(n) 壳体
battery:(n) 电池
alkaline:(adj) 碱性的
transmitter:(n) 传输器
thermistor:(n) 温度计
sub-skin sea surface temperature:次皮温
research vessel:科考船
volunteer ship:自愿船
archive:(vt) 把……存档
distribute:(vt) 分发,分配
aspect:(n) 方面,层面
technology:(n) 技术
data stream:数据流
supervise:(vt) 监督,指导
manufacturer:(n) 制造商,生产商
vt: manufacture 制造,生产
private industry:私营企业
specification:(n) 规范,说明书
collaborate:(vi) 合作,协作
forecasting:(n) 预报
estimation:(n) 估计
quality control:质量控制
procedure:(n) 程序,步骤
interpolate:(vt) 插值
interval:(n) 间隔
optimum interpolation:最优插值
small-scale:(adj) 小尺度的
high-frequency:(adj) 高频的
velocity rotary spectra:速度旋转谱
demonstrate:(vt) 证明,展示
tidal:(adj) 潮汐的
internal wave:(n) 内波
coverage:(n) 覆盖范围
sufficient:(adj) 足够的
resolve:(vt) 分辨

Notes

[1] global drifter program 即全球漂流浮标计划,致力于维持全球范围内大约 1300 个海表漂流浮标的运行,旨在通过这些浮标精确捕捉并提供海表流场、温度场以及其他关键海洋环境参数。

[2] mixed layer currents 即混合层流场。实际上,海表漂流浮标通常被用于观测海面下 15 m 处的流场状况。鉴于混合层的深度在多数情况下会超过 15 m,因此,通过海表漂流浮标所观测到的流场,也常被称作混合层流场。

[3] 混合层流场可改变物质、动量、能量等的再分布,故与生物、化学和物理过程都有关系。

[4] 漂流浮标观测流场的深度取决于其携带的水帆（drogue）所处的深度。一旦水帆掉落,漂流浮标的移动轨迹更易受海表风场的干扰,进而引起流场测量数据的误差。

[5] sub-skin sea surface temperature 即次皮温。实际上,海表漂流浮标测量的是深度约 20 cm 处的温度。

[6] hourly interpolated data 相对于 six-hourly interpolated drifter data 能展示更小尺度和更高频的海水运动特征。

18. Argo

Text

Argo[1] is an international program that measures water properties across the world's ocean using a fleet of robotic instruments that drift with ocean currents and move up and down between the surface and a mid-water level. Each instrument (float) spends almost all its life below the surface. The name Argo was chosen because the array of floats works in partnership with the Jason Earth observing satellites that measure the shape of the ocean surface. (In Greek mythology, Jason sailed on his ship, the Argo, in search of the golden fleece[2]).

The data that Argo collects describes the temperature and salinity of the water and some of the floats measure other properties that describe the biology/chemistry of the ocean[3]. The main reason for collecting these data is to help us understand the oceans' role in Earth's climate and so be able to make improved estimates of how it will change in the future.

For example, the changes in sea level (once the tides are averaged out) depend partly on the melting of icecaps and partly on the amount of heat stored in the oceans. Argo's temperature measurements allow us to calculate how much heat is stored and to monitor from year to year how the distribution of heat changes with depth and from area to area[4]. As ocean heat content increases, sea level rises, just like the mercury in a thermometer. Comparison of Argo's measurements with the Jason observations continue to give us new insights into how the oceans "work" that can be used to improve climate models.

At present (2020), Argo is collecting 12,000 data profiles each month (400 a day). This greatly exceeds the amount of data that can be collected from below the ocean surface by any other method. Argo plans to continue its data collection for as long as those data remain a vital tool for a wide range of ocean applications of which understanding and predicting climate change is but one.

Each Argo float (costing between $20,000 and $150,000 depending on the individual float's technical specification[5]) is launched from a ship. The float's weight is carefully adjusted so that, as it sinks, it eventually stabilizes at a pre-set level, usually 1 km. Ten days later, an internal battery-driven pump transfers oil between a reservoir inside the float and an external bladder. This makes the float first descend to 2 km and then return to the surface measuring ocean properties as it rises. The data and the float's position are relayed to satellites and then on to receiving stations on shore. The float then sinks again to repeat the 10-day cycle until its batteries are exhausted (Figure Ⅳ-18).

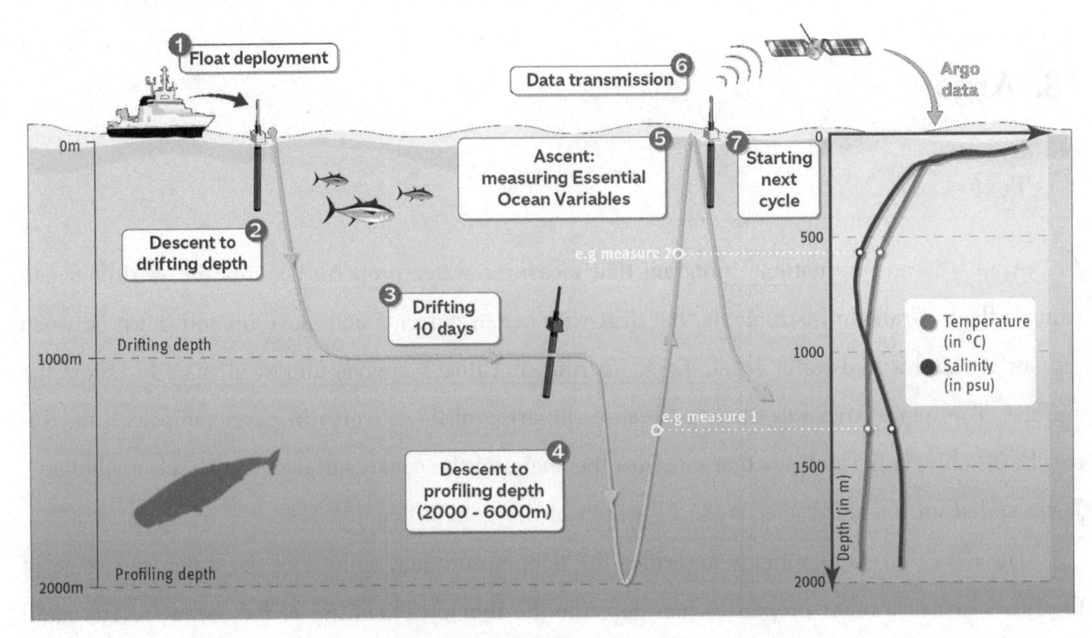

Figure Ⅳ-18: Argo float cycle[6]. [Source: UCSD]

The float measurements are sent to regional data centers where they are given rigorous quality checks and then passed to two global data centers from where they can be accessed by anyone wishing to use them. Some users have applications that require rapid access and so Argo aims for most "real-time" data to be available within 12 hrs. Other applications need a higher-quality version that replaces the real-time data after lengthy comparisons between nearby floats and between float data and information from research ships. A fundamental rule for Argo is that all data are freely and openly available.

There is no central funding for Argo. Each of the 30 countries that operate floats obtains their own national funding to buy floats, prepare and launch them and to process and distribute the data. The Argo Program is managed by teams of scientists and data experts[7] who ensure that the program is run as efficiently and effectively as possible and that standards are maintained at the highest possible level. Argo is part of the Global Ocean Observing System (GOOS) and Global Climate Observing System (GCOS). The total annual cost of Argo is estimated at $40 million each year.

Profiling floats similar to those used by Argo were first developed in the late 1990s. Float design has continuously incorporated new technologies to improve battery performance, satellite communication, and sensor stability. It has also had to keep track of changing demands made on Argo by scientists and by operational weather and climate forecasting centers. Because floats spend almost of their life (which may be as long as 10 years[8]) below the ocean surface and are unable to communicate, the technological challenge of Argo is in many ways greater than that of space science. Floats are manufactured by commercial companies and by research institutions but all must maintain common and high standards set by the international steering team. Since Argo started in 2000, innovative new float designs have allowed floats to profile deeper, to work in ice-covered regions[9], and to measure other ocean properties beyond the basic salinity, temperature, and pressure.

Argo has been described by New York Times science writer Justin Gillis as "one of the scientific triumphs of the age[10]". The triumphs have been technological and scientific. The free access to Argo data means that there is wide usage—over 4000 scientific publications have referred to or used Argo data as have 350 PhD theses.

Two science outcomes are worth highlighting[11]. Argo has greatly reduced the uncertainty of global heat storage estimates and hence projections of sea level rise. The changes in salinity monitored by Argo also allow changes in global rainfall patterns to be studied. Finally, the ready access to the data provides an educational resource to give the general public an ability to look inside the ocean.

Vocabulary

international: (adj) 国际性的
property: (n) 属性
robotic: (adj) 像机器人的
instrument: (n) 设备
mythology: (n) 神话
golden fleece: 金羊毛
melting: (n) 融合
icecap: (n) 冰盖
ocean heat content: 海洋热含量
mercury: (n) 水银
thermometer: (n) 温度计
insight: (n) 了解,领悟
exceed: (vt) 超过
vital: (adj) 至关重要的,必不可少的
launch: (vt) 发射,使……下水
sink: (vt) 下潜
stabilize: (vi) 稳定
pump: (n) 泵
transfer: (n/vt) 转移
reservoir: (n) 存储器
bladder: (n) 囊
descend: (vi) 下降
exhaust: (vt) 耗尽,用完
rigorous: (adj) 严格的
access: (vt) 访问,存取(计算机文件); (n) 通道
application: (n) 应用
replace: (vt) 用……替换,更新
funding: (n) 资金,基金
ensure: (vt) 确保,保证
efficiently: (adv) 有效率地,效率高地
effectively: (adv) 有效地
profiling: (n) 剖面
performance: (n) 表现,性能
sensor: (n) 传感器
stability: (n) 稳定性,稳定度

operational：(adj) 业务化的
commercial：(adj) 商业的
institution：(n) 研究所
steering：(n) 驾驶，控制方向，筹划，指导
  vt：steer 驾驶，转向，指导
innovative：(adj) 革新的
triumph：(n) 胜利，巨大成功
refer：(vi) 提及，涉及
thesis：(n) 学位论文
  pl：theses
outcome：(n) 成果
highlight：(vt) 强调，突出
reduce：(vt) 减小
uncertainty：(n) 不确定性
storage：(n) 存储
projection：(n) 预测

Notes

[1] Argo 计划 2000 年启动以来，经过 30 多个沿海国家的共同努力，已在全球海洋成功建立并持续维护了一个由近 4000 个活跃浮标组成的实时观测网络。该计划最初仅针对全球无冰覆盖区域的 0~2000 m 水深范围进行温度和盐度的观测（即核心 Argo），但现已拓展至极地、深海（2000~6000 m）以及生物地球化学（BGC）领域。

[2] golden fleece 即金羊毛，是古希腊神话中的稀世珍宝。

[3] 专门用于测量海水生物/化学属性的 Argo 浮标被称为 BGC-Argo。其主要测量参数包括叶绿素浓度、后向散射系数、溶解氧浓度、pH 值、有色溶解有机物（CDOM）、硝酸盐浓度以及下行辐照度等。

[4] Argo 浮标可测量 0~2000 m 深度范围内的海水温度剖面，这些数据可用于计算海水的热含量，进而分析其对海平面变化的影响。

[5] 单台 Argo 浮标的价格范围为 $ 20 000~ $ 150 000，具体取决于所搭载传感器的数量和类型。

[6] Argo float cycle 即 Argo 浮标循环，是指 Argo 浮标在大部分时间里会停留在 1000 m 水深位置；随后，在大约 10 d 之后，它会先下潜至 2000 m 深度；接着开始上浮，并在上浮过程中测量海水的各种属性；上浮至海面后，通过卫星通讯将所测量的数据传至陆地上的接收站；完成数据传输任务后，Argo 浮标会再次下潜回到 1000 m 处的水深位置，如此循环往复。

[7] Argo 计划由科学指导组和数据管理组共同管理。

[8] Argo 浮标的平均使用寿命为 4~5 a，最长的已超过 10 a。

[9] 冰下 Argo 浮标可用于观测冰下海水的各种属性，例如能够观测冰下的藻华现象。

[10] One of the scientific triumphs of the age：Argo 计划被誉为当代的一项重大科学成就。

[11] Argo 计划取得的重要科学成果包括：一是显著降低了全球海洋热含量估算的不确定性；二是为研究全球降水空间分布提供了重要数据支持。

19. Underwater Glider

Text

An underwater glider[1] is a type of autonomous mobile platform that employs variable-buoyancy propulsion instead of traditional propellers or thrusters. It employs variable buoyancy in a similar way to a profiling float, but unlike a float, which can move only up and down, an underwater glider is fitted with hydrofoils (underwater wings) that allow it to glide forward while ascending and descending through the water. At a certain depth, the glider switches to positive buoyancy to climb back up and forward, and the cycle is then repeated (Figure Ⅳ-19).

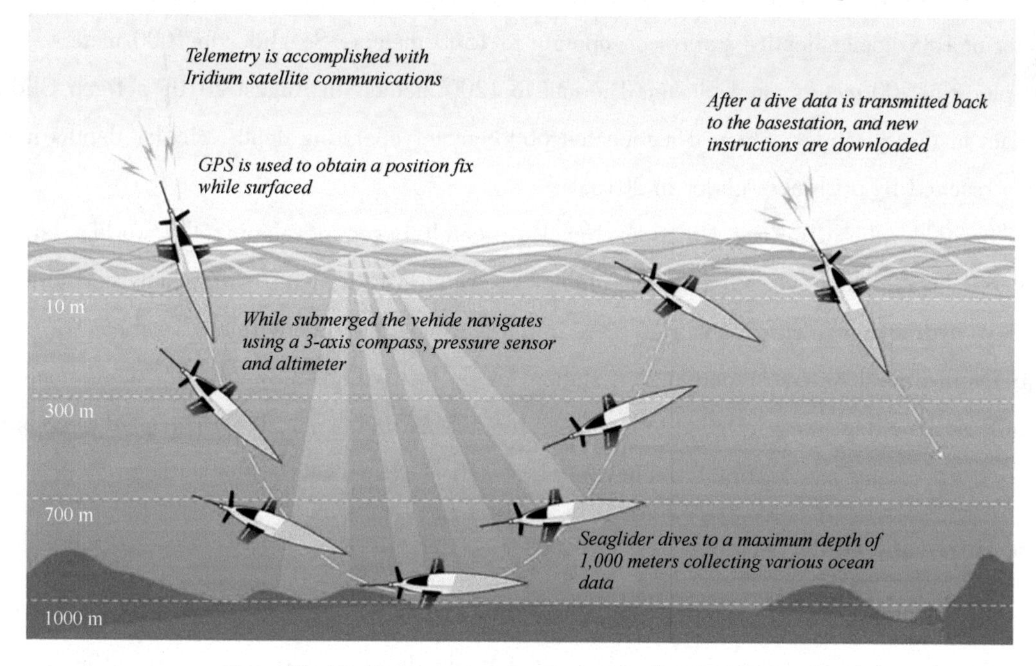

Figure Ⅳ-19: Underwater glider illustration. [Source: NOAA]

While not as fast as conventional AUVs, gliders offer significantly greater range and endurance compared to traditional AUVs, extending ocean sampling missions from hours to weeks or months, and to thousands of kilometers of range. The typical up-and-down, sawtooth-like profile followed by a glider can provide data on temporal and spatial scales unattainable by powered AUVs and much more costly to sample using traditional shipboard techniques[2]. A wide variety of glider designs are in use by navies and ocean research organizations, with gliders typically costing around US $ 100,000.

Gliders typically make measurements such as temperature, conductivity (to calculate salinity), currents, chlorophyll fluorescence, optical backscatter, bottom depth, and sometimes acoustic backscatter or ambient sound. They navigate with the help of periodic surface GPS fixes, pressure sensors, tilt sensors, and magnetic compasses[3]. Vehicle pitch is controllable by movable internal ballast (usually battery packs), and steering is accomplished either with a rudder (as in Slocum) or by moving internal ballast to control roll (as in SeaExplorer, Spray and Seaglider). Buoyancy is adjusted either by using a piston to flood/evacuate a compartment with seawater (Slocum) or by moving oil in/out of an external bladder (SeaExplorer, Seaglider, Spray, and Slocum Thermal). Because buoyancy adjustments are relatively small, a glider's ballast must typically be adjusted before the start of a mission to achieve an overall vehicle density close to that of the water it will be deployed in. Commands and data are relayed between gliders and shore by satellite[4].

Gliders vary in the pressure they are able to withstand. The Slocum model is rated for 200 meter or 1000 meter depths. Spray can operate to 1500 meters, Seaglider to 1000 meters, SeaExplorer to 700 meters, and Slocum Thermal to 1200 meters. In August 2010, a Deep Glider variant of the Seaglider achieved a repeated 6000-meter operating depth. Similar depths have been reached by a Chinese glider in 2016.

In 2004, the US Navy Office of Naval Research began developing the world's largest gliders, the Liberdade class flying wing gliders[5], which uses a blended wing body hull form to achieve hydrodynamic efficiency. They were initially designed to quietly track diesel electric submarines in littoral waters, remaining on station for up to 6 months. By 2012, a newer model, known as the ZRay, was designed to track and identify marine mammals for extended periods of time. It uses water jets for fine attitude control as well as propulsion on the surface.

The Hurricane Glider Project[6]

Since 2014, NOAA/AOML leads a multi-institutional effort that brings together the research and operational components within NOAA and the university community to implement and carry out sustained and targeted ocean observations from underwater gliders in the CaribbeanSea and tropical North Atlantic Ocean in support of hurricane studies and forecasts. Underwater gliders provide information about the properties of the seawater. All underwater gliders have the capability of measuring temperature, salinity, and pressure. Additional sensors may also be included, such as chlorophyll fluorometers, oxygen sensors, and ocean current profilers. After each dive, the vehicle surfaces to transmit the data collected and receive commands via satellite telemetry. The benefits of hurricane gliders are two-fold[7]. First, the gliders that are deployed in a given season will send back observations which are used to better represent the ocean conditions in NOAA's operational hurricane models. This ocean data can help improve the National Hurricane Center's forecasts for current storms. Secondly, the data collected from the gliders is used by our oceanographers and model experts to help validate how the ocean data improved the forecast guidance, and if changes to the model code or how data is assimilated is needed. We use our understanding of the ocean to help improve how the ocean is represented in operational models, with a goal that future model versions will more accurately forecast ocean conditions and help drive improvements to intensity forecasts.

Vocabulary

underwater glider: 水下滑翔机
autonomous: (adj) 自主的,自制的
employ: (vt) 运用,使用
variable-buoyancy: 可变的浮力
propulsion: (n) 推进,推动力
propeller: (n) 螺旋桨
thruster: (n) 推进器
fit: (vi) 安装;(adj) 合适的,强档的;
  (n) 合适,匹配
hydrofoil: (n) 水翼
switch: (vi) 转换;(n) 开关,转换器
positive: (adj) 正的
  ant: negative 负的
climb: (vt) 爬升
conventional: (adj) 传统的
significantly: (adv) 显著地
endurance: (n) 耐久性
sampling: (n) 采样
sawtooth: (n) 锯齿
unattainable: (adj) 无法得到的
wide: (adj) 广泛的,大量的
organization: (n) 组织
conductivity: (n) 传导率,导电率
navigate: (vi) 航行;(vt) 驾驶,导航
tilt: (n/vi) 倾斜
magnetic compass: 磁罗盘
pitch: (n) 上下颠簸,纵摇
ballast: (n) 压舱物
rudder: (n) 舵
piston: (n) 活塞
flood: (vi) 充满,大量涌入
  ant: evacuate 疏散,排空
compartment: (n) 隔间
adjust: (vt) 调整,校准
  n: adjustment
relatively: (adv) 相对地
achieve: (vt) 实现,达到
  n: achievement 成就,成绩
overall: (adj) 总体的,综合的
command: (n) 指令
relay: (vt) 转送,转播
withstand: (vt) 承受
reach: (vi) 达到,到达
hydrodynamic: (adj) 水动力的
initially: (adv) 最初,起初
submarine: (n) 潜艇
littoral: (adj) 浅海的;(n) 浅海地区
identify: (vt) 识别,找到
mammal: (n) 哺乳动物
attitude: (n) 高度
community: (n) 团体,社会
implement: (vt) 实施,执行
sustained: (adj) 持续的
targeted: (adj) 定向的

Caribbean Sea：（n）加勒比海
additional：（adj）额外的，附加的
dive：（n）潜水
receive：（vt）接收
telemetry：（n）遥测
benefit：（n）益处，优势
two-fold：（adj）双重的，两倍的
represent：（vt）代表
storm：（n）风暴
guidance：（n）指导
assimilate：（vt）同化
accurately：（adv）准确地

Notes

[1] underwater glider 即水下滑翔机，是一种新型的水下无人探测设备，它通过调节净浮力和改变姿态角来产生推进力，实现在下潜/上浮的同时向前移动，并在此过程中测量海水的各项属性。水下滑翔机能够搭载多种海洋传感器，包括但不限于温度、盐度、溶解氧、浊度、叶绿素和硝酸盐等传感器。

[2] 相较于传统的自主式水下航行器（AUV），滑翔机（glider）的航行速度较慢，但其能源消耗极低，因此，具有高效率、长续航力（可达上千千米）的特点。滑翔机可提供跨越多个时空尺度的海洋观测资料。

[3] glider 通过 GPS 装置、压力传感器、倾角传感器和磁罗盘导航。

[4] glider 利用卫星通信。

[5] Liberdade class underwater glider 是全球最大的水下滑翔机。"Liberdade"这个名字来源于 Joshua Slocum（首位独自驾船完成环球航行的人）所建造的一艘船的名字。

[6] The Hurricane Glider Project 旨在于加勒比海和热带北大西洋区域，利用 glider 进行持续且定向的观测，从而为飓风研究和预报提供支持。

[7] hurricane gliders 具有两方面优势：一方面，glider 观测的数据能够为 NOAA 的飓风模型提供更精确的海洋初始场，从而提升飓风预报的准确性；另一方面，这些数据也能为海洋学家及海洋模式研发人员提供有益的参考和帮助。

20. Saildrone

Text

A saildrone[1] is a type of unmanned surface vehicle (USV) used primarily in oceans for data collection. Saildrones are wind and solar powered and carry a suite of science sensors and navigational instruments. They can follow a set of remotely prescribed waypoints. The saildrone was invented by Richard Jenkins, a British engineer, founder and CEO of Saildrone, Inc[2]. Saildrones have been used by scientists and research organizations like the National Oceanic and Atmospheric Administration (NOAA) to survey the marine ecosystem, fisheries, and weather[3].

In January 2019, a small fleet of saildrones was launched to attempt the first autonomous circumnavigation of Antarctica[4]. One of the saildrones completed the mission, traveling 12,500 miles (20,100 km) over the seven-month journey while collecting a detailed data set using on board environmental monitoring instrumentation.

In August 2019, SD 1021 completed the fastest unmanned Atlantic crossing sailing from Bermuda to the UK, and in October, it completed the return trip to become the first autonomous vehicle to cross the Atlantic in both directions[5]. The University of Washington and the Saildrone company began a joint venture in 2019 called The Saildrone Pacific Sentinel Experiment, which positioned six saildrones along the west coast of the United States to gather atmospheric and ocean data.

Saildrone and NOAA deployed five modified hurricane-class vessels at key locations in the Atlantic Ocean prior to the June start of the 2021 hurricane season. In September, SD 1045 was in location to obtain video and data from inside Hurricane Sam. It was the first research vessel to ever venture into the middle of a major hurricane[6].

There are three Saildrone platforms: Explorer, Voyager, and Surveyor[7] (Figure Ⅳ-20). All three Saildrone uncrewed surface vehicles (USVs) combine wind-powered propulsion technology with solar-powered meteorological and oceanographic sensors.

Saildrone Explorer[8]

The Saildrone Explorer is a 23-foot-long (7.0 m) USV that can sail at an average speed of 3 knots (5.6 km/h; 3.5 mph) (depending on the wind) and stay at sea for up to 365 days. The Explorer is designed for fisheries missions, metocean data collection, ecosystem monitoring, and satellite calibration and validation missions.

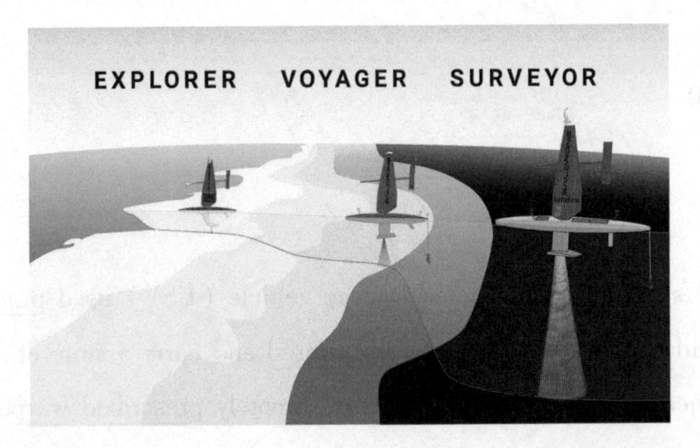

Figure Ⅳ-20: Three Saildrone platforms. [Source: saildrone.com]

Saildrone Voyager[9]

In August 2021, Seapower Magazine reported the company is adding a new mid-size USV to the fleet: The Voyager is a 33-foot-long (10 m) USV with primary wind power and auxiliary propulsion of a 4 kW electric motor for a wide variety of missions including bathymetry (ocean mapping) missions, border patrol and maritime domain awareness. The average speed is 5 knots.

Saildrone Surveyor[10]

At 72 feet (22 m) long and weighing 14 tons, the Surveyor is the largest vehicle in the Saildrone fleet. According to Wired, the Surveyor was first launched in January 2021 and is designed to carry multibeam echo sounders for IHO-compliant bathymetry surveys. The Surveyor's multibeam echo sounders can map the ocean seafloor to depths of 23,000 feet (7,000 m). It also carries an acoustic Doppler current profiler to measure the speed and direction of ocean currents.

In July 2021, the Surveyor completed its first trans-Pacific mapping mission sailing from San Francisco to Honolulu, Hawaii, and mapping 6,400 square nautical miles of seafloor along the way. Hawaii News Now reported that 20 Surveyors could map the entire ocean in less than 10 years[11].

In September 2022, it was announced that Austal USA signed an agreement with Saildrone, to build Saildrone Surveyor drones by year end for the US Navy, and other customers.

Vocabulary

unmanned: (adj) 无人操作的，自控的

powered: (adj) 由……驱动的

primarily: (adv) 主要

navigational: (adj) 导航的

prescribed：(adj) 规定的
waypoint：(n) 航路点
invent：(vt) 发明,创造
founder：(n) 创始人
survey：(vt) 调查
ecosystem：(n) 生态系统
fishery：(n) 渔业
　pl：fisheries
attempt：(vt) 尝试
circumnavigation：(n) 环球航行
Antarctica：(n) 南极洲
detailed：(adj) 详细的
environmental：(adj) 自然环境的
sailing：(n) 航行
direction：(n) 方向
joint：(adj) 联合的
venture：(n) 冒险旅行
position：(vt) 安置;(n) 位置
gather：(vt) 收集
modified：(adj) 改良的
prior to：在……之前
platform：(n) 平台
uncrewed：(adj) 无人驾驶的
combine：(vi) 结合
calibration：(n) 校准
auxiliary：(adj) 辅助的,备用的
motor：(n) 马达,发动机
border patrol：边境巡逻
multibeam：(n) 多波束
echo sounder：回声探测仪
map：(vt) 勘测;(n) 地图
acoustic Doppler current profiler：声学多普勒流速剖面仪
nautical mile：海里
announce：(vi) 宣布,宣告

Notes

[1] saildrone 是一种依靠风能和太阳能驱动的水上无人艇,用于收集海洋数据,它能够按照预先设定的路线进行自主航行。

[2] “Saildrone, Inc” 是一家总部位于美国的公司,专注于设计、制造并运营水上无人艇,以收集海洋数据。

[3] saildrone 能够搭载多种传感器,用于对海洋生态系统、渔业资源以及天气状况等进行综合调查和研究。

[4] saildrone 完成了环绕南极洲的航行任务:该无人艇于 2019 年 1 月 18 日从新西兰布拉夫南港出发,在历经 2 万多千米的环绕南极洲航行后,于 8 月 3 日安全返回出发港。在执行任务期间,该无人艇遭遇了极端环境条件的考验,包括冰冻低温、高达 15 m 的巨浪、130 km/h 的高速风,以及与巨型冰山的碰撞。

[5] saildrone 实现了大西洋的双向穿越:“SD 1021”号无人艇最初于 2019 年 1 月从纽波特出发,开启其大西洋航行之旅,主要执行测量墨西哥湾流热量和碳含量的科学任务。在航行过程中,“SD 1021”在百慕大进行了必要的维护停留,随后继续沿墨西哥湾流北上,直至抵达欧洲,并最终在英格兰南部海岸的索伦特完成靠岸。同年 8 月

15 日,“SD 1021”号无人艇从英国利明顿再次出发,沿北大西洋的直航航线航行,并于 10 月 22 日成功返回罗德岛州的纽波特,完成了横渡大西洋的回程。“SD 1021”号在顺利完成双向横渡后,不仅保持了无人横渡大西洋的最快纪录,还成了唯一实现大西洋双向横渡的无人水面航行技术解决方案。

[6] saildrone 深入飓风核心:2021 年 9 月,“Explorer SD 1045”号无人艇成功深入 4 级飓风“山姆”的内部,创下了无人艇首次进入飓风中心的历史记录。为了适应极端环境条件,该无人艇的机翼被特别缩短以增强其承受极端条件的能力。在风速高达 190 km/h 的极端环境下,这艘无人艇成功采集到了首批关于此类风暴内部的数据和图像资料。

[7] saildrone 平台有三类:Explorer、Voyager 和 Surveyor。

[8] Saildrone Explorer(探险者):该无人艇长度为 7 m (23 ft),具备在开阔海洋执行超过 12 个月长航时任务的能力,用于搜集海洋和气候数据。

[9] Saildrone Voyager(旅行者):这是一款新型的中型水上无人艇,长度为 10 m (33 ft),专为近岸海洋和湖床测绘任务而设计,同时能够有效应对非法、未报告和不管制捕捞活动。此外,Voyager 还能执行情报、监视和侦察任务,以及海上执法与安全、毒品拦截、边境和港口安全等领域的挑战。

[10] Saildrone Surveyor(测量者):这是最新且尺寸最大的水上无人艇,长度为 22 m (72 ft),具备在海上连续作业长达 12 个月的能力。它所搭载的传感器仪器设备中包含了声呐系统,该系统能够测绘水深至 7000 m(约 23 000 ft)的海域。

[11] Saildrone Surveyor 从旧金山到夏威夷的首次航行:Saildrone Surveyor 号无人艇总重量达 14 t,是全球规模最大、技术最先进的无人驾驶海洋测绘船。它能够观测水下生态系统,并以高分辨率绘制海底地图,测绘深度可达 23 000 ft。凭借风力驱动,该船在运行时极为安静,同时拥有进行深度测绘所必需的高精度声学测量能力。该船从旧金山起航,成功完成了跨越太平洋的首航。在此次航行中,该船成功测绘了 6400 n $mile^2$ 的海底地图。据预测,在未来 10 年内,仅需 20 艘该船即可完成全球所有海洋的测绘工作。

Text Sources

Acker et al., 2003. Remote Sensing from Satellites// Meyers (eds). Encyclopedia of Physical Science and Technology. Cambridge: Academic Press.

Argo. https://argo.ucsd.edu.

Dosits. https://dosits.org.

Leif Bjørnø, 2017. Absorption of Sound in Seawater// Neighbors, Bradley (eds). Applied Underwater Acoustics. Amsterdam Elsevier.

Mobley, 2020. Ocean Optics Web Book. https://www.oceanopticsbook.info.

National Oceanic and Atmospheric Administration. https://oceanexplorer.noaa.gov.

PART V

GENERAL KNOWLEDGE AND SKILLS

1.Introduction to Academic English

引言

海洋学专业英语属于学术英语（English for academic purposes，EAP）的范畴，而学术英语又是专门英语（English for special purposes）的一部分［另一部分是职业英语（English for professional purposes）］，专门英语与一般英语（English for general purposes）构成英语语言教学的整体。学术英语具有与职业英语、一般英语迥然不同的特点，广泛应用在学术论文、学位论文、国际学术会议上的报告交流等场合，是从事各专门学术领域的学者和学生需要熟练掌握的一种特殊英语。学术英语是信息型文本，关注学术思想内容的表达，具有客观、确切、明晰、严密、简练等特点。学术英语可细分为通用学术英语（English for general academic purposes）和专业学术英语（English for specific academic purposes），前者强调各学科学术语境所具有的共同特征，后者则强调各学科（如海洋学）的独特性，主要在专业术语的来源和构成方面彼此区分。学习海洋学专业英语，必须培养通用学术英语的语感和语言习惯，并与海洋学专业术语结合，才能臻于完善。本章我们结合文献资料和自己的理解向读者简要介绍通用学术英语（下文泛称"学术英语"）的词汇、语法、语义等方面的特点，辅助读者建立基本认识、明确学习方向。读者若对学术英语的详细信息感兴趣可参考专门书籍。

词汇

学术英语以少量专业词汇为主干,以较多的学术词汇为主体,以大量的非专业词汇为纽带。专业词汇即专业术语,不认识这些词汇则无法读懂专业论文的思想和观点。学术词汇就是通用科技词汇,虽不能传递主要的专业学术信息,但在提出研究问题、进行文献回顾、引用他人研究成果、描述研究方法、讨论研究结果、概况研究结论等方面经常使用。掌握学术词汇的多少是能否顺利阅读专业论文的重要指标。新西兰惠灵顿维多利亚大学的 Coxhead (2000) 编制的《学术词汇表》(academic word list, AWL) 含有 570 个词族、3100 个单词。

AWL 里的词汇并不特别针对某一学科,而是属于学术界的通用词汇。据 Coxhead 估算,AWL 里的单词占了学术文本词汇量的 10%,另有 80% 的词汇被一般目的词汇列表 (general service list, GSL) 覆盖,后者共含 2000 个单词,不限于学术情景,属于通用英语的核心词汇。因此,掌握了 GSL 和 AWL 就掌握了学术文本 90% 的词汇量,其余的词汇可能包含各专业的术语和不常用的日常词汇等。

AWL 里的 570 个词汇被编入 10 个子列表 (sublists),按照常用度从高到低排序。子列表 1 是最常用的,子列表 10 是最不常用的。前 9 个子列表各包含 60 个单词,子列表 10 只有 30 个。本书的线上资源中提供了 AWL 的全部内容,供读者参考、学习,获取方式请见 “本书的内容和用法”。

学术英语用词的惯例如下。

- 动词:倾向于用意义明确的单词动词,而少用与之相对应的动词短语(动词+副词或介词)。例如,用 implement 而不用 put … into practice,用 maintain 而不用 keep up,用 investigate 而不用 look into。

- 名词:常用名词组成很长的一连串,叫作名词化结构。学术英语存在一种用名词化结构代替动词的倾向。例如,

“The emergence of English as the international language of scientific communication has been widely documented.” 此句与 “English has emerged as the international language of scientific communication. This phenomenon has been widely documented.” 意思相同。

- 学术英语多用来源于法语、希腊语、拉丁语的以多音节为主的词汇,使文体显得庄重、高雅,而少用英国本土的盎格鲁-撒克逊来源的以单音节为主的词汇,这些词汇整体偏口语化。当然实际上两类词汇总是搭配使用。例如,

accomplish — do	considerable — much	interior — inner
additional — extra	demonstrate — show	regarding — about
approximately — about	identical — same	sufficient — enough

- 不用省略写法,如用 there is 而不是 there's,用 do not 而不是 don't。

- 常用由多个单词组成的词串。我们将 Hyland（2008）给出的部分常用词串整理如下：

in order to	in the following	with respect to
in terms of	the role of	in the context of
as well as	as a result of	as shown in figure
due to	on the other hand	it was found that
based on	at the same time	it should be noted that
in this study	in (the) case of	it can be seen that
a number of	in terms of	play an important role in
the fact that	on the basis of	from the point of view of
according to	in the form of	it is likely that
the present study	the nature of	it is possible that
the relationship between		

语法

语态方面，科技英语更多使用被动语态，以此来隐去主语，强调客观事实。被动语态是学术英语的重要特征。

与被动语态相配合，学术英语极少使用人称代词，特别是第二人称（you/your），很少用指人的第三人称（he/him/his/she/her/they/them/their），偶尔使用第一人称（I/me/my/we/us/our）。这是为了强调所发现的事实，而不是发现这些事实的人。然而现代有学者开始提倡使用第一人称，因为这可以增加阅读流畅度，提升亲和力。我们的观点是，第一人称并非不能使用，恰当的应用可产生正面效果，但其应用频率应该控制在较低的水平。

相应地，动词多用非限定形式，即不定式（to do）、现在分词（doing）、过去分词（done），少用限定形式，即原形（do）和过去式（did）。非限定形式往往没有主语，因此客观性强。其中不定式可表达目的，例如，

"Early attempts to predict cyclone development relied heavily upon the analysis of surface fronts."

现在分词和过去分词可引导从句，前者表达前述事实或对象主动、同时或将来进行的行为，后者表达前述事实或对象被动、事先或已经进行的行为。例如，

"The introduction of numerical weather prediction has greatly aided the forecasters, allowing predictions of increased accuracy many days ahead."

"The careful assessment of model output, combined with remote sensing and other observations, has opened new ways of understanding the weather."

学术英语中的长句较多，但句子结构相对简单，多为主谓宾结构或主谓补结构。其中

常常穿插一个或多个定语、宾语从句或多个插入语,使句子变长,但对句子结构没有影响,在理解句子成分时往往首先将之舍去。

时态方面,通常用一般过去时、一般现在时、现在完成时,但是学术英语的时态选择一般不反映简单的时间概念,仅反映作者的评论态度。通常,回顾某文献结果时用过去式。涉及一般性的规律时往往用一般现在时,即便这个规律是过去提出的,或通过分析过去发生的事件而得出的,这反映了学术英语对客观性和普适性的重视。现在完成时则通常用于强调过去研究对现在的影响。同一篇论文,甚至同一个句子中出现不同时态并不罕见。例如,

"Bunkers (2002) suggested that a measure of deep-layer shear is a better indication of supercell potential."

另外,记述作者自己在文中开展的研究工作时,可选择使用一般现在时或一般过去时,例如,

"To extract the anomalous interannual variability superimposed on the climatological annual cycle, a low-pass filter is (was) applied to all data prior to analysis."

"We therefore calculate(d) each term of the quasi-geostrophic vorticity budget equation to assess the driving factors quantitatively."

具体选择哪种时态是作者的自由,但应尽量保持一致。有人认为采用一般现在时稍有利于提升流畅性。但是在 conclusion 或 discussion 部分里面回顾本文做过的工作时,则往往用一般过去时。

语义

学术英语侧重于叙述和评论事实,不涉及情感,因此,大量使用陈述句,极少使用疑问句,感叹句和祈使句等其他句型则几乎从不使用。

学术英语中普遍存在不明朗或模糊的语言表达形式。这是基于对科学的敬畏,表明承认不确定性、临时性和其他可能性的存在,以及自己立论存在错误的可能,避免把话说得太满。例如,用 it seems/appears that…、probably、possibly、likely、presumably、indicate、suggest 等来降低肯定程度。下表总结了一些常用的模糊词汇。

肯定程度	数量	频率	程度	动词
complete	all/every/each no/none/not any	always never	definite certain undoubted clear	will/will not is/are (not) must have to
high	a majority (of) many/much	usual normal general as a rule on the whole	probable (most) likely presumable	should would ought to

续表

肯定程度	数量	频率	程度	动词
partial	some/several a number of	often frequent sometimes occasional	possible perhaps maybe	can/cannot could (not) may (not) might (not)
low	a minority a few/little	rare seldom hardly scarce ever	unlikely improbable	

参考资料:

章东华, 2016. 大气科学专业英语. 北京:气象出版社.

寿绍文,等, 2019. 气象科技英语教程. 北京:气象出版社.

Day, Sakaduski, 2011. Scientific English: A Guide for Scientists and other Professionals. New York: Greenwood Press.

Hamp-Lyons, Heasley, 2006. Study Writing A course in Written English for Academic Purposes. London: Cambridge University Press.

Hyland, 2008. As can be seen: Lexical Bundles and Disciplinary Variation. English for Specific Purposes, 27(1): 4-21.

2.Pronunciation Tips for Chinese Speakers

引言

科技英语多以书面的形式进行交流和传播,但口头的交流仍是不可欠缺的一环,例如,在国际学术会议、交流访问等场合,以及与国外专家个别讨论学术内容等场合,或者在朗读科技论文时。当然,口头上的学术交流必然大量引入日常英语口语的特点,但也同时混合科技英语的词汇和表达方式,成为与书面科技英语和日常英语口语都有区别的独特类型。本书不是日常英语口语教材,因此不会过多涉及此方面内容。但是为了帮助读者更好地将科技英语以口头的方式表达出来,我们编制了这份汉语母语者英语发音的特点和难点清单,供读者学习、参考。无论如何,语言的功能在于交流,而交流的基础是丰富的词汇量和熟练的语法。在此基础上,标准的发音和流利的表达是良好口语的特征,也是良好听力的基础。我们期望读者在提高词汇量和阅读理解能力的基础上,参考此资料和其他文字或视听资料,逐步纠正发音问题,练习标准朗读,力求流利表达,以期日后更流畅地进行国际学术交流。然而,我们愿强调流利的表达比标准的发音更重要,且应以正确的态度看待发音问题(见本节最后)。

元音

- 虽然汉语的元音个数(26 个)多于英语(20 个),但英语元音之间的差别较小,对于汉语普通话母语者其区分难度较大。

- 英语/iː/和/ɪ/之间的差别在汉语普通话中不存在,因此较难区分。汉语普通话存在类似/iː/的发音(即拼音 i, 音依),但不存在/ɪ/,后者较难发音。此二者不是发音时间长短的区别,/ɪ/介于/iː/和/ə/之间。例如,

— list /lɪst/ vs least /liːst/. This list is the least I could do.

— fit /fɪt/ vs feet /fiːt/. Feet ache when shoes do not fit.

— bit /bɪt/ vs beat /biːt/. Practice keeping the beat, it is a bit off.

— sit /sɪt/ vs seat /siːt/. The bike seat was so uncomfortable I couldn't sit for long.

— chip /tʃɪp/ vs cheap /tʃiːp/, grin /grɪn/ vs /griːn/. Do not grin, the green bin was cheap as chips.

— ship /ʃɪp/ vs sheep /ʃiːp/, sick /sɪk/ vs seek /siːk/. He tried to seek the sick sheep on the ship.

— slip /slɪp/ vs sleep /sliːp/. I am going to slip into bed and go to sleep.

— will /wɪl/ vs wheel /wiːl/. The wheel will turn.

- 汉语普通话不存在/æ/,常与/e/混淆。前者发音时口型比后者大,介于/e/和/a:/之间,后者类似于拼音 e(音埃)。我国英语教学中常把前者叫作"张口埃",后者叫作"扁口埃"。例如,

— bad /bæd/ vs bed /bed/. Whoever sleeps on this bed is bad.

— bat /bæt/ vs bet /bet/. I bet he will bat better this time around.

— pat /pæt/ vs pet /pet/. He's pissed just because I patted his pet on the head.

— bag /bæg/ vs beg /beg/. Please do not beg, I will lend you that bag.

— salary /sæləry/ vs celery /seləry/. You can try selling celery to earn a salary.

— dad /dæd/ vs dead /ded/.Their dad has been dead for 10 years.

— marry /mærɪ/ vs merry /merɪ/. I am going to marry a merry girl.

— latter /lætə(r)/ vs letter /letə(r)/. The latter half of the letter was quite offensive.

— sad /sæd/ vs said /sed/. John said he was sad.

- 汉语普通话不存在/ʌ/,常用/a:/代替。前者发音时口型小一些,后者接近汉语拼音的 a(音阿)。二者对应的字母组合不同,即发/ʌ/的通常是字母 o 或 u(如 bus、cup、luck),而发/a:/的通常是字母 a 或者 ar(如 after、ask、father、bar、car、far),因此相对较少混淆。

- /u:/和/ʊ/也较难区分。前者类似但不等于汉语普通话的拼音 u(音屋),后者类似但不等于汉语普通话的拼音 ou(音欧)。例如,

— fool /fu:/ vs full /fʊl/. That room is full of fools.

— Luke /lu:k/ vs look /lʊk/. Luke looked.

- /ɒ/和/ɔ:/在汉语普通话中都没有,常与汉语拼音的 ao(音凹)混淆。英式发音/ɒ/圆唇、短促,/ɔ:/嘟嘴、音长。美式发音/ɒ/直接变成/a:/,/ɔ:/发英式/ɒ/的延长版。例如,

— shot /ʃɒt/ vs short /ʃɔ:t/. Do not shot at such a short distance.

辅音

- 汉语普通话没有/v/音,因此十分普遍地用/w/代替。前者是咬唇爆破音(如 vortex、volume),后者类似于汉语拼音的 w(音屋,如 water、window)。注意 wave 的发音。

- /θ/和/s/经常混淆。前者多由字母组合 th 发出,类似于汉语拼音的 s(音思);后者多对应字母 s,在汉语普通话中没有,但有些方言中有。例如,

— think /θɪŋk/ vs sink /sɪŋk/. I think the ship is gonna sink.

— thank /θæŋk/ vs sank /sæŋk/. Thanks for the help when the ship sank.

— theories /θɪəri:z/ vs series /sɪəri:z/. He proposed a series of theories.

- /ð/和/z/经常混淆。前者多对应字母组合 th,类似于汉语拼音的 z(音兹);后者

则对应字母 z,汉语普通话中没有,但有些方言中有。例如,

— with /wɪð/ vs whizz /wɪz/. The missiles whizzed pass with deafening noise.

— that /ðæt/ vs "zat" /zæt/.

- 因为汉语没有以辅音结束音节(即"闭音节")的惯例,很多人倾向于在音节末尾的辅音后加一个元音,如 empty /emptɪ/变成 emp-e-ty /empətɪ/。
- 辅音/r/如果出现在音节末尾,英式英语不发音,而美式英语则发卷舌音,类似于汉语的儿化音。例如,important /ɪmˈpɔːtnt/(英式)/ɪmˈpɔrtnt/(美式)。但不少汉语母语者会滥用儿化音,即将儿化音应用在音节末尾没有/r/的情况下。

语调

英语常见的语调介绍如下。

- 陈述句——降调:I am going to the theatre.
- 一般疑问句——升调:Are there any apples on the table? 回答——降调:Yes, there are. / No, there aren't.
- 特殊疑问句——降调:Where have you been?
- 反义疑问句——可升可降。比较肯定或表达自己看法时用降调;不肯定、表示询问、征求意见时用升调。例如,

— He doesn't know her, does he? (可用升调,表示不太肯定他是否认识她)

— They are coming, aren't they? (可用降调,表示说话人预期他们会来,可理解成"他们还来吗?")

- 选择问句——前面升、最后降:Are we going by bike, bus, or taxi? (bike、bus 升,taxi 降)
- 感叹句——降调:What a nice day!
- 祈使句——可升可降。表示命令、不耐烦、语气强硬时用降调;表示勉励、态度和蔼亲切或客气的请求时用升调。例如,

— Cheer up.

科技英语不涉及情感,且其句子多为陈述句,少数情况下涉及一般疑问句或特殊疑问句,极少出现其他句子类型。而且,科技英语的句子往往较长,因此,上面提到的陈述句用降调的规则仅适用于句子末尾,而句子中间仍需要使用不同语调。一般而言,朗读较长的英语句子时常常在句子中间保持平调或插入升调来表示句子还没结束,使听者继续期待句子的剩余部分,直至读到句号(有时是逗号)时才降下来,有时甚至句子结尾也不降。例如,

- Many of the dress shoes→ I own↗ just sit in boxes↘.
- One of my favorite→ pair of boots↗ is from the Arctic Circle↘.

- Wind-driven water waves↗ can be thought of as comprised→ of the superposition of wavelets↗ with a continuous range→ of wavelengths and frequencies↗, and the amplitudes of waves↗ in a given band of wavelength or frequency→ can be characterized→ by an energy spectrum↘.

汉语的语调(注意不是阳平、阴平、上声、去声这四个声调)不用来区分句子意思,因此,在科技语言等不涉及情感的场合通常较平直,多用降调。通常在读很长的陈述句时,汉语母语者习惯于每个单词或词组都用降调直至句子结束;在读很长的疑问句时,往往也仅在句子最后才改成升调。而这在英语母语者听来容易造成理解困扰,因为他们听到降调就习惯性地感觉句子已经结束,后面要另开新句子了。

其他非英语母语者发音的常见特点

这里我们根据自己的经验并参考专业书籍,简单总结一些在国际学术交流场合常遇到的非英语母语人士的发音特点。这份列表显然不够全面,无法涵盖更多语言,甚至也不够精确,无法反映同一种语言内部不同变种和方言的区别(如高地德语和低地德语、欧洲大陆西班牙语和拉丁美洲西班牙语等)。而且,这里列举的仅是从不同语言带入英语的普遍的发音习惯,并非所有说同一种语言的人都会如此发音,不能一概而论。提供此列表的目的是帮助读者听懂非母语者的英语口语,当然这还依赖于在实际的语言环境中不断积累经验。

- 德语:

— 词尾的/b/、/d/、/g/常被发成/p/、/t/、/k/:pub→pup、aid→set、dog→dock。

— 有些人会把词尾的/g/发成类似于汉语拼音的 x(音西):Hamburg。

— /æ/与/e/常混淆:sat vs set。

— /ɔ:/与/əʊ/(类似汉语拼音 ou,音欧)常混淆:caught vs coat。

— 将/w/发成/v/:wine→vine。

— 词首和词尾的/v/发成/f/:very→ferry;leave→leaf。

— 将/ʒ/发成/ʃ/、/dʒ/发成/tʃ/:measure→“mesher”;Jane→chain。

— 词尾的/l/常被发得过强,听起来像/lə/刚摆好口型就强行停止:fill;full。

— /r/是小舌颤音,通常听起来像/h/:red→“hed”。

- 法语:

— /i:/和/ɪ/不分:leave vs live。

— /ʌ/发成/ə/:much→“mirch”。

— /u:/和/ʊ/不分:full vs fool。

— /ɔ:/与/əʊ/混淆:naught vs note。

— /æ/被发成/ʌ/、/a:/或/e/:bank→bunk;hand→“hahnd”;mad→med。

— /ei/发成/e/:paper→pepper;made→med。

— /tʃ/发成/ʃ/,/dʒ/发成/ʒ/:church→"shursh";joke→"roke"。

— /θ/和/ð/可能被发成/s/、/z/、/f/、/v/、/t/、/d/等:think→sink、fink、"tink"; that →"zat"、vat、"dat"。

— 词尾的/l/、/d/与/t/的区别、/g/与/k/的区别与德语相同。

- 西班牙语:

— /i:/和/ɪ/不分:seat vs sit。

— /æ/、/ʌ/、/a:/不分:cart vs cat vs cut。

— /ɔ:/与/ɒ/混淆:caught vs cot。

— /u:/和/ʊ/不分:full vs fool。

— 词首的/p/、/t/、/k/听起来像/b/、/d/、/g/。

— 词尾的/m/常被/n/和/ŋ/代替:dream→"drean""dreang"。词尾的/n/常作/ŋ/。

— 词尾的/ŋk/发成/ŋ/:sink→sing。

— /z/发成/s/:lazy→lacy;/s/接近/ʃ/:see→she。

— /b/与/v/混淆:bowel vs vowel。

— /ʒ/、/ʃ/、/dʒ/发成/tʃ/:sheep vs cheap vs jeep;pleasure→"pletcher"。

— /j/听起来像/dʒ/:year→jeer、cheer;you→Jew、chew。

— 音节开头的/w/可能发成/gw/或/g/:would→"gwould""gould"。元音之间的/w/听起来像/b/:Harry Walker→Arry Balker。

— /r/是大舌颤音,分单击和多击两种。

— 词尾的/b/、/d/、/g/与德语相同。

- 日语:

— 因为只有/a/、/i:/、/u:/、/e/、/o:/ 5 个元音且没有双元音,发生很多近似靠拢现象。

— /r/发成/l/:grammar→glamour、erection→election。

— /hu:/发成/fo:/:who→"foo"。

— /fɔ:/分成/ho:/:force→"horse"。

— /θ/和/ð/被发成/s/、/z/、/ʃ/、/dʒ/或汉语拼音的 x:thin→shin;then→zen。

— /v/发成/b/:very→berry。

— 两个元音之间的/g/发成/ŋg/:bigger→"binger"。

— /ti:/变成类似汉语拼音的 qi(音七),听起来接近/tʃ/:team→"cheam"。

— /si:/变成类似汉语拼音的 xi(音西),听起来接近/ʃ/:she→see。

— /di:/变成类似汉语拼音的 ji(音击),听起来接近/dʒ/:deep→jeep。

如何看待发音问题

- 无须过度紧张,语言的目的是交流,流畅的表达比准确的发音更重要。
- 母语人士的听力纠错容差较大,往往能够听懂不十分标准的发音。
- 了解其他非英语母语民族的常见发音特点有助于听懂对方。
- 要多听多练,尽量做到发音标准,以减少沟通障碍,特别是减少与非英语母语人士之间的理解困难。
- 口音是民族语言文化的反映,世界各民族的语言文化都有自己的特点和优点。中华文化博大精深,汉语语言文字具有很多其他语言无法比拟的优点,汉语的发音也十分有韵律性,这些特点或多或少地会被我们带入到英语的朗读和口语交流中。英语母语者享受着英语作为现行国际通用语言的红利,其他非母语民族在说英语时与我们一样存在自己的口音。我们应尽量练习流利而标准的英语发音,目的是促进科学思想和成果的沟通交流,但无须因发音不够标准而自卑,也不要因发音比某些其他非英语母语民族更标准些而自大。

参考资料:

Swan, Smith, 2001. Learner English: A Teacher's Guide to Interference and Other Problems. London: Cambridge University Press.

3.AMS Writing Skills

引言

虽然本书的重点是通过介绍海洋学专业词汇和常用表达方式来帮助读者提升阅读能力,但英文科技论文写作也是科技英语的重要组成部分,是提升阅读能力后的自然延伸。英文科技论文的写作技巧本身自成体系,且不同学科各有特点,非本书范围所能涵盖。作为本书的拓展内容,我们摘录并翻译了美国气象学会（American Meteorological Society,AMS）发表的《如何成为一名更优秀的作者》一文,供读者参考。美国气象学会是气象和海洋学界的顶级国际学术组织,旗下出版了《每月天气评论》(*Monthly Weather Review*)、《大气科学杂志》(*Journal of the Atmospheric Sciences*)、《物理海洋学杂志》(*Journal of Physical Oceanography*)、《气候学杂志》(*Journal of Climate*)、《水文气象学杂志》(*Journal of Hydrometeorology*)等多本高水平学术期刊,在国际学界享有盛誉。AMS 对科技英语的规范十分重视,编著了《流利的科学:成为更好的作者、演讲者和大气科学家的实用建议》一书,针对英文科技论文写作和口头交流的技巧提出了许多有用的建议。此处摘录的《如何成为一名更优秀的作者》一文于 2022 年 11 月发表在《每月天气评论》上,是该刊创刊 150 周年献礼的一部分,可视为《流利的科学:成为更好的作者、演讲者和大气科学家的实用建议》一书的浓缩版。虽然此文针对《每月天气评论》杂志而作,但其中提出的写作技巧和论文发表技巧具有很强的通用性(特别对同属地球科学的海洋学而言),可作为很好的参考。我们希望通过学习此文提升海洋学专业学生的专业英语应用能力,助力其将来以流畅、准确、规范的英文将科研成果表述出来。另外,此文的英文原文本身也具有很高的写作水准,是很好的学习材料。我们提供的中文译文经过精心打磨,采取意译的方式,在力求准确反映原文思想的同时兼顾中文的语言逻辑和思维习惯,保证译文的流畅性,自信基本达到了“信、达、雅”的翻译标准,可作为翻译方面的辅助学习材料,也可帮助读者理解英文原文。出于简洁考虑,译文删去或调整了一些参考文献的引用。

原文①

1. Paper preparation

Batchelor (1981) wrote, "Reading a paper is a voluntary and demanding task, and a reader needs to be enticed and helped and stimulated by the author." The authors' job is to make their paper as readable and comprehensible as possible for reviewers and future readers, thereby inviting a broad readership to understand their science. As such, the goal should be submission of a carefully written and proofread paper: one that is written for its readers. Specifically, scientists are naturally inquisitive and like a good research question to solve. Writing the paper in such a narrative can more readily engage readers. To achieve this objective, authors should remember that the readership of Monthly Weather Review may be very broad, given the global reach of publications on the internet and the increasing interdisciplinarity of our field: scientific results and research methods from one discipline are increasingly needed and being adapted in other disciplines. As such, taking several steps backward and considering a more general audience than was often considered in the past may yield benefits in terms of article visibility and enhancing the accessibility of our publications. Specific approaches that authors can take include the following:

1) Start with a storyboard of the paper. For example, compile the figures to be included in the paper, along with the key points for each of those figures. Then organize them into an order that will best tell the story of the research. As an author, the storyboard will help in writing the text of the paper more efficiently and will also make the paper more compelling to the readers.

2) Make the submitted paper similar in organization and style to recently published articles. Follow the format requirements for AMS submissions. Reviewers will more easily recognize the paper as suitable for publication if they can visualize it in the pages of Monthly Weather Review. The best way to do that? Read recently published articles and emulate them.

3) Make the paper as easy to read and interpret as possible through effective layout and structure, using an appropriate number of numbered sections and lettered subsections while avoiding fragmentation of the text into too many short subsections and subsubsections. Indent paragraphs or use blank space between paragraphs.

4) Make the paper flow smoothly from one topic to another through coherent-writing approaches such as the placement of old information at the beginning of each sentence and new information at the end (Gopen and Swan 1990). The lack of coherent writing is one of the most common

① Schultz et al., 2022. How to Be a More Effective Author. Monthly Weather Review (150): 2819-2828. (节选)

weaknesses of submitted papers. One way to ensure coherence is to make a separate list of the topic sentences and see if the progression is logical. Keep each paragraph to a single topic. Start each paragraph with the topic sentence and end with the point to emphasize (Gopen and Swan 1990).

5) Make the paper as easy to read and interpret as possible by using precise, accurate, and objective language. Use engaging verbs and avoid unnecessary adjectives and adverbs. Choose precise words wherever possible and avoid jargon (i.e., scientific terminology designed for insiders). Terminology that may be unfamiliar to the readers should be defined. It doesn't hurt to define even familiar terminology. A comparison (e.g., between two plots, current results vs past results, or observations vs model output) should be carefully documented rather than just passing off the comparison as "excellent agreement."

6) Be specific in descriptions of physical processes. Consider the statement "the low-level jet played a primary role in the heavy-precipitation event". Many problems exist with this sentence. First, what does it mean for the jet to play "a primary role" in a process (convection) that requires three ingredients to occur (lift, instability, and moisture; Doswell et al. 1996)? Is the author saying that one is more important than the other? Second, is it even possible to quantify this statement? What is the relative importance of the low-level jet to the convection? If the jet were 5 m s−1 weaker, would the convection not occur? Third, the exact role that the low-level jet played is unstated. Was it the supply of moisture that was important? Was it the low-level wind shear that was important? Such phrases hinder precise scientific communication.

2. Introduction: Motivation for the research and purpose of the paper

The introduction section is the location where the author engages potential readers, motivates the research question, defines the purpose of the paper, and informs readers of its contents. A reader should quickly be able to assess "why is this important?" and "why should I care?". Importantly, the introduction revolves around the problem statement, the source of tension that motivates the research and draws the reader into the paper. As scientists, we like a good mystery to solve and the problem statement creates that mystery. Therefore, the best problem statements are those that engage the reader through a paradox, error, missing data or information, or inconsistency in the literature. A source of conflict between research groups/publications or differing interpretations of research results can also serve as a compelling problem statement.

The problem statement leads to a specific research question or questions that the paper will address. The more specific these are, the better. Studying a case because it was a heavy-precipitation event is certainly worthy, but it alone is an insufficient reason to publish in Monthly

Weather Review. Instead, what is the scientific question that motivates this research? If a testable hypothesis can be articulated (not all research projects can), then this approach may motivate the paper, as well. Stating that this study will "explore" or "examine" is not precise enough. What new knowledge will be gained by study of this case? How does this paper differ from others, presenting new and original research results? Furthermore, the problem statement and the research questions should directly follow from an orderly evidenced argument from the literature, if possible.

Stating that there is a gap in the literature and that "very few studies have examined this topic" is not a particularly compelling problem statement. Either one of two things is typically the case: the author has not examined the literature with sufficient depth to discover that the literature has indeed touched upon this topic (even if outside the particular geographic region of interest of the author) or the research question may not be worth investigating (e.g., "very few studies have examined the relationship between pickle prices in the United States and the frequency of Indian Ocean tropical cyclogenesis" would be an exaggerated version of such).

The problem statement also allows a more focused presentation of the previous literature through only citing literature that builds up to the problem statement and avoiding the common tendency for authors to wander around various topics and cite broad swaths of literature. The synthesis of the previous literature should tell a story, interweaving themes from one or multiple studies into a cohesive narrative rather than linearly describing the results one paper at a time. In this sense, literature syntheses should typically be science-focused rather than paper- or person-focused. Literature syntheses should help to introduce the research problem and the motivation for the paper.

Following the problem statement is the response to the problem: a brief exposition of what the author is going to do to address the problem statement. Such a response can sometimes be achieved by the last paragraph of the introduction, which often serves as a roadmap for the rest of the paper. Although not necessary, such paragraphs can be effective to explain how the paper addresses the problem statement as well as providing the structure of the paper to the reader beyond the basic introduction-methods-results-discussion-conclusions structure of a standard journal article. In addition, the paper should have an explicit statement akin to "The purpose of this paper is to…". Such a statement allows the reader to compare the results at the end of the paper with the authors' intent. That way, the author can be judged by their own stated goal. Other problems with introductions and how to fix them have been discussed by Schultz (2022b).

3. Take-home message

Another weakness of some papers is that the take-home message is unclear. What new results should readers remember? How are these new contributions to the scientific literature? The take-home message should feature in the abstract and the conclusion section, as well as the results section. The readers should be clear on why these results are important to understanding weather or improving numerical weather prediction, for example. The take-home message should not overstate the importance of the research, however.

4. Other parts of the paper

This section gives suggestions for handling the rest of the sections found in a typical AMS paper. Familiarize yourself with the AMS Author Resources online. Follow the guidelines for the elements of the paper, including abstracts, section types, figures, and references (refer often to the examples of reference types). Note that AMS has a limited set of figure widths that are used for print and PDF versions, a simple table style, and a specific way to set matrices, vectors, and scalar variables—authors can familiarize themselves with these styles by viewing past published AMS papers.

1) Titles should provide specific and accurate information about the paper but in a concise package. The title should be grammatically correct, easy to read, and easy to understand. Avoid uncommon acronyms.

2) Abstracts in AMS journals should be one paragraph and a maximum of 250 words. Because titles and abstracts may be the only parts of the paper read by potential readers, the resourceful author gets the most out of those 250 words through minimizing introductory material and maximizing results. In any case, more specific abstracts are more likely to be found via search engines (Weinberger et al. 2015), thus ensuring that the article receives a larger potential audience. AMS does not currently allow citations to figures or references in abstracts.

3) Given the easy access to the scientific literature and the broadening of disciplines, the audiences for our articles are greater than ever before. Significance statements help to reach those nontraditional audiences and are aimed at the educated layperson without formal training or education in the atmospheric and related sciences (Huntington and Lackmann 2020; Schultz et al. 2020a). Although optional, we encourage all authors to write a significance statement, limited to a maximum of 120 words. Examples are provided by the AMS online author information.

4) Keywords for AMS journals are taken from those selected by the author when uploading the paper to the online Editorial Manager paper-management system. Thus, all keywords will be taken from the standard list that AMS provides. AMS requires at least three, but no more than six, keywords. Select keywords that reflect the main topic(s) of the paper, are frequently used in the text, and would help others doing web searches of the literature find the article. Avoid words that are used only one or two times.

5) The principal test of a data and methods section is whether independent readers can replicate the study from the information in this section. If they cannot, then more detail is needed. Indeed, insufficient information about the data and methods is a typical reviewer comment. Try to remember and describe all of the steps involved in the research, including any preprocessing or filtering of the data.

6) The results section may be one section or several. Results section(s) provide appropriate numerical or experimental evidence to demonstrate the validity of the claims in the paper. The results section should give a factual and objective description of the experiments performed or data collected to evaluate the proposed solution to the problem stated in the introduction. Where possible, the authors should state how their results relate to other literature, other cases, other models, and other parts of the world but should not overstate the generality of their results. The text style here should aim at guiding the reader through the succession of tests and evidence that support the ideas behind the paper. Use and compare the results with existing benchmarks and quality-assurance measures available in the literature whenever possible; new tests should only be proposed when they give insight that existing tests cannot provide. Do not include unnecessary results; do prioritize evidence that is essential for understanding the scientific significance of the paper. Longer portions of text and argumentative discussion of results are best left for a later discussion section.

7) Authors are often unclear about the different purposes of the discussion and conclusions (or summary) sections. The discussion and conclusions sections in Monthly Weather Review should be separate sections. The discussion section occurs after the results section but before the conclusions section, which summarizes the paper. The discussion section is best for longer pieces of text and for content that does not belong in the results section, such as generalizing the results to other cases, elaborations on interpretations, synthesis of results, applications, implications, unresolved issues, or alternative interpretations. In contrast, the conclusions section should consist of a summary of the paper that is longer than the abstract (roughly 500–800 words so that it is not simply a restatement of the abstract) and perhaps some brief concluding remarks about the significance or application of the paper's results. The conclusions section should revisit the research questions raised in the introduction and bring closure to the paper. In short, elaboration and extrapolation belong in a discussion section; summarizing belongs in the conclusions section.

8) Please recognize in the acknowledgments section all who have helped in the research and preparation of the paper, including funding agencies and data/software providers. Acknowledge the contribution of the anonymous reviewers, whose comments helped to improve the paper. Such collegiality is not necessary but is generally appreciated.

9) Provide formal citations to datasets used in the paper in the data availability statement

(Schultz et al. 2020b), if not also in the body of the paper. Many dataset archives now make this easy with DOIs. Clear dataset citations credit the dataset creators and make the science more readily reproducible. In the case in which authors create their own datasets, data and computational workflows underpinning the results should be shared in an appropriate repository, even if it is not required by the funding agency. The more data and code that are provided, the more that reviewers and readers will better understand the results, assess their veracity, and replicate the approach to facilitate further advances in understanding. Not all of a project's data, particularly in the case of simulation-based studies, needs to be saved. Although formal guidance as to how much of a dataset and how many workflows to save has yet to be developed, we recommend that studies geared primarily toward knowledge production (as is often the case in Monthly Weather Review) prioritize retaining and sharing computational workflows, notebooks, and the data necessary to replicate the study's findings (e.g., Mullendore et al. 2021; Erdmann et al. 2022), as previously discussed for the data and methods section above. Authors should confirm that any URLs to data sources are still active and correct. More on citing datasets and the data availability statement can be found at AMS (2022c,d).

10) References should appear in a complete and consistent format upon submission for the assurance and convenience of the reviewers. Upon acceptance of the paper, the reference list will go through more rigorous checking and formatting during the construction of the page proofs. Try to avoid references that cannot be found anywhere online or in print, like a conference presentation that does not exist as a preprint or in published proceedings. References to sources not in English should include the English translation of the title in parentheses and type of document being referenced. Citation guidelines for many common reference types can be found at AMS (2022e).

11) Figures should present the results clearly, should avoid too many overlapping fields that are difficult to interpret, and should omit unnecessary figure panels. Figures and their embedded text (e.g., labels on axes, contours, and color bars) should be large enough to be readable, and the quality should be sufficient to avoid pixelation. This requirement often necessitates choosing a font size that appears too large when viewing the figure in isolation, as the rendered version in the typeset paper will be smaller. Choose sans serif fonts, if possible. Avoid the rainbow color scheme, because it exaggerates gradients and is not well-suited for color-blind individuals (e.g., Stauffer et al. 2015). Captions should be complete and explain all features of the figure. If figure panels have letter labels, then these should be used in the caption for the reader's benefit. Figures (and tables) must be cited sequentially, unless there is a parenthetical note such as "(see Fig. 7 later in the paper)." The text accompanying the figures in the main body of the paper should be more than a mere description of the figures without any interpretation. Ensure that

the text in the paper says why the figure is needed, sufficiently explains the figure, and says what the figure means.

5. Decision processes and handling a rejected paper

Peer reviewers at AMS journals are not referees as in a football game. Although reviewers make recommendations to the editors on the suitability of the paper for publication in Monthly Weather Review, only editors make decisions. These decisions might not be based on majority rule, and these decisions may be different from the recommendations provided by the reviewers. In addition, editors often weigh in on the suitability of the paper for Monthly Weather Review, add their own comments as subject-matter experts, interpret reviewer comments, and synthesize reviewer comments.

Rejection at Monthly Weather Review is not a decision that generally prevents authors from revising the paper and resubmitting it to Monthly Weather Review or that prevents submission to another journal. Some rejections are because the anticipated revisions would take longer than the 2 months allotted by AMS for revisions. Although authors may be disappointed with the rejection, such rejections allow the authors unlimited time to consider the reviewers' and editor's comments and determine the best way to modify the paper that addresses the reviews before resubmission to Monthly Weather Review or submission at a different journal. Such flexibility can ultimately benefit the published article.

6. Responding to reviewers and editors

Getting a paper accepted at a journal can typically take one to three rounds of back-and-forth with reviewers and the editor, with each round taking 1–3 months. The number of rounds that revisions take can be minimized when authors and reviewers engage positively and constructively during each round, resulting in an efficient and effective peer-review process. From the authors' perspective, this means making optimal use of the comments and recommendations made by the reviewers through careful, comprehensive revisions to the paper and complete, well-reasoned responses to the reviewers and the editor. Because authors have invested a lot of time, effort, and emotion into their paper, it can be natural to be on edge when taking in the reviewers' and editor's comments. Rather than letting it spawn a "fight or flight" response, let this edge sharpen the authors' focus during revision. Consider how an outside observer who lacks the depth of knowledge about the research would perceive what has been written. Remember that the reviewers have also invested a considerable amount of time and energy in making recommendations on how to improve the presentation of the research. The goal of everyone involved in the peer-review process is to end up with the best possible outcome for the work. In this framework, assume positive motives behind each comment to the extent possible, bringing any concerns about unconstructive comments to the editor's attention during revision. Approaching revision and reviewer responses in this way generally enables authors to address issues in the paper that may not have been immediately apparent and facilitates constructive discussions with reviewers throughout the peer-review process.

Perhaps the easiest way to see how this might work is for the authors to put themselves in the reviewers' shoes. Reviewers are volunteers whose goal is to help improve the paper. Being anonymous, they receive no credit for their efforts, so their reward is seeing authors benefit from their guidance and input. If you (the author) were the reviewer of somebody else's paper and spent 10 hours (the typical time that reviewers spend with your submitted paper; Golden and Schultz 2012) reading the paper and writing up your review, you would probably want to see that the authors acknowledged that they read and understood each comment; acknowledged that they took your comment seriously, even if they disagreed with it; and responded to the comment and revised the paper. The exchange between authors and reviewers can go off the rails when the author fails in one or more of these three things. Thus, here is a proposed framework for responding to reviewers' and editor's comments.

1) Acknowledge whether you, as author, agree or disagree with the basis of the comment.

2) If you agree, say what you have done to the paper to revise it in accordance with the reviewer's wishes and where in the paper you have done so. Please be as precise as possible to ensure that you understood the comment and have responded appropriately. It often helps to copy the exact revised text into your responses and provide line numbers of where these changes were implemented to aid the reviewer and editor in evaluating the modifications. If you address the comment in your responses but no change is needed to the paper, please indicate that in your response as well. If the reviewer's concern has been alleviated or eliminated through major rewriting of the text, then a simple acknowledgment of this fact is sufficient.

3) If you disagree, provide an argument, using evidence from new calculations or citations from previous literature that show the validity of your argument, and explain why you disagree. You are free to disagree with the reviewers—differences of opinion, interpretation, and conclusions are inevitable. Regardless, respond point by point to all of the comments—not just specifically to the individual points by the reviewer, but also to the spirit that all of the comments taken together are trying to convey. Addressing specific comments in your responses but not fixing the underlying issues with the paper is a suboptimal outcome of the peer-review process. Remember that future readers do not have access to your exchanges with the reviewers, so addressing sources of possible concern within the paper is preferable to an extended response. If additional calculations or figures are required to address a reviewer's comment and those figures

are included in the response to the reviewers, then such figures should be strong candidates for inclusion in the revised paper, even if included in a published online supplemental file.

4) Finish with a clear statement that you acted on their concern, either something like "No revision to the paper" or, preferably, some revision to the paper so that future readers will not ask the same questions as the reviewer. Do not write vague responses such as "The paper has been revised" or "The introduction has been rewritten." Do not make reviewers track down where in the paper those changes are or make them wonder if the substance of their concern has been addressed. For example, the decision letter that AMS journals send out specifically states: "If you have made a change to the paper, please indicate where in the paper the change has been made. (Indicating the line number where the change has been made would be one way, but is not the only way.)"

5) Copy and paste the reviews and editor comments into a file. Interspersing responses in between the original comments in a different font or color makes it easier for the reviewers and editor to assess how well their concerns were addressed and contributes to a smoother, more effective peer-review process.

6) Authors should do their best to ensure that they understand the reviewers' comments. If the comment is unclear, it may be useful to ask the editor for clarification. Authors could also respond, "If I understood what you are saying, then you mean…" If it is a genuine misunderstanding from the lack of clarity in the writing, revise the paper so that other readers do not misinterpret the paper in the same way.

7) Often when reviewers feel that their comments have not been adequately addressed, they will bring that issue up again in the next round of review. Inadequately addressed comments slow down the peer-review process as well as frustrate the reviewer and often also the editor. Avoid this by following the four-step process enumerated above. If editors find that authors are not addressing the reviewers' comments, then they can send the paper back to the authors for further revision or they can reject the paper because it is not progressing toward a publishable outcome.

8) Do not pit reviewers against each other in the response (e.g., "Reviewer 2 said that this section was well-written, so I do not have to make the changes requested by Reviewer 1."). Just because another reviewer agrees with the authors on a point (or does not mention a concern) does not mean that it is necessarily the correct one. Respond to each reviewer's concerns individually.

9) It is not uncommon to see divergent recommendations sent to the editor by the reviewers of a submitted paper. When arriving at the decision delivered to the authors, the editor has taken the content of the reviews and expertise of the reviewers into account. The best strategy for revisions and responses under these conditions remains the same as the framework laid out above,

with each comment from each reviewer being diligently addressed. If authors are uncertain about the editor's expectations or require additional guidance under these conditions, contact the handling editor for advice. Submitting an additional document with tracked changes is not mandatory but does make it easier for some reviewers to check how their comments have been addressed. Making reviewers' and editor's work easier can result in a faster reviewing process.

7. Frequently asked questions

1) I've heard that AMS considered banning case studies. Is there truth to that rumor?

AMS journals welcome insightful case studies that contribute to our scientific knowledge. Like all submissions, case studies must meet minimum standards for publication. To avoid overgeneralizing results from a single case, submissions should discuss why this particular case was chosen and discuss the generality of results beyond a single case. More guidance about writing case studies can be found in Schultz (2010b).

2) I reused text in my present submission from an earlier article that I published. The editor sent it back to me for revisions because of self-plagiarism. What does that mean?

Although it is unethical to publish the same content twice, effective communication in science requires clear and precise descriptions. As such, authors may duplicate text, especially from data and methods sections, from earlier publications. To avoid self-plagiarism or copyright violations, duplicated text should cite the original source and indicate that the text largely follows from that source (e.g., "The description of the methods is similar to that of Jones et al. (2021), and the following two paragraphs are derived from there with minor modifications."). More on plagiarism and self-plagiarism can be found in Schultz et al. (2015).

3) Reviewers will sometimes write, "I've noticed grammatical problems, but the copy editors will fix those in the construction of the page proofs." Is this right?

Although it is part of copy editors' jobs to catch errors authors and reviewers miss, accepted papers that require extensive copyediting are expensive to prepare for publication and eventually increase publishing fees for all authors. Furthermore, reviewers will have to review a paper that is not in its final form, frustrating them and slowing down the peer-review process. Authors, and not AMS staff, should therefore be taking on the bulk of the proofreading effort. As W. J. Steenburgh said, authors should only submit a paper that they would be comfortable publishing as is (Schultz 2009, p. 167).

4) Can I use parentheses to shorten sentences, as in the example: "warm (cold) advection at 700 hPa was east (west) of the cyclone center"?

Although such constructions have been common in published articles in the past, avoid these parenthetical constructions in submissions to Monthly Weather Review. Such sentence structure makes it difficult for a reader to follow the meaning of the text and may be confusing when parentheses are used correctly in other sentences (Robock 2010).

5) May I introduce new acronyms in my paper?

Papers with unfamiliar acronyms make it difficult for readers, as readers encountering an unfamiliar or forgotten acronym must flip back through the paper to track it down. Also, acronyms challenge readers who do not read the paper linearly from introduction to conclusion, but instead skip around through the paper, to get the relevant information they require. Introducing acronyms may save the author time during paper generation, but it slows down the readers. Please eliminate many, if not all, nonstandard acronyms to improve the readability of the paper. The list of AMS-approved acronyms that do not need to be defined in AMS journals can be found at AMS (2022f). Any others should probably not be introduced (e.g., Schultz 2022c). If a new acronym must be introduced and is not on the AMS list, authors must provide the full name and set the acronym after it in parentheses.

8. Guidance for nonnative English speakers

We recognize the challenges that nonnative English speakers face—some of us editors are nonnative English speakers ourselves. Not only do they have to get the science right and communicate it in the right style, but they have to do it in a language that is not their mother tongue. Poor writing style in a submitted paper generally does not lead to rejection, but it can inhibit the ability of a reader to understand the paper. Thus, improving the submission at the start will ensure a smoother path through peer review. AMS reviewers, editors, and staff do not have the time available to edit papers that require extensive grammatical changes. Although AMS wishes to encourage the international exchange of scientific results through its journals, it requests that such authors make their own arrangements to ensure that submitted papers are already in correct English. If not, their submissions may be returned unreviewed or rejected.

There are a number of different ways to get assistance:

1) Ensure that spell checkers and grammar checkers are turned on in the word-processing program.

2) Specialty grammatical tools and translation apps can be installed on browsers or laptops.

3) Google Docs and other word-processing programs such as Microsoft Word offer predictive tools that can assist in improving sentence-writing.

4) If unsure about a word or phrase, put it into a search engine. There is no guarantee that search results will mean it is correct; but only a few hits being returned may mean that it is not well used or is incorrect. Alternatively, search AMS published articles for common usage. Another useful resource is the English Collocations Dictionary to assist with choosing the right word

pairs or context (http://ozdic.com).

5) Writing down useful words, phrases, or sentences to maintain a digital and personalized document to search can also be helpful.

6) English-language editing services can be purchased to improve papers before submission. The AMS offers a web page listing some of them (AMS 2022g).

9. Summary

Peer review does not ensure that all the science published is correct, but it does try to ensure a basic level of quality assurance set by the standards of the journal's editors with the assistance of the reviewers. Given that only 56.7% of submitted papers are accepted for publication at Monthly Weather Review, acceptance is not inevitable. We hope that the guidance in this editorial helps authors improve their papers prior to publication, thereby ensuring greater success for publication. The rigorous peer-review process at Monthly Weather Review can be challenging for authors, but we hope that the feedback from reviewers and the dialogue with editors improves each paper and provides us all with an opportunity for continued learning.

10. Additional Resources for Authors

Naturally, one resource to which we are partial has been published by the American Meteorological Society: Eloquent Science: A Practical Guide to Becoming a Better Writer, Speaker, and Atmospheric Scientist (Schultz 2009). The book has been translated into Chinese and published by the China Meteorological Press (Schultz 2021). In addition, we recommend the following books:

1) Day and Gastel (2006): How to Write and Publish a Scientific Paper. 6th ed. Cambridge University Press, 302 pp.

2) Glasman-Deal (2021): Science Research Writing for Native and Non-Native Speakers of English, 2nd ed. World Scientific, 356 pp.

3) Schimel (2012): Writing Science: How to Write Papers that Get Cited and Proposals that Get Funded. Oxford University Press, 221 pp.

4) Strunk and White (2000): The Elements of Style. 4th ed. Allyn and Bacon, 105 pp.

5) Sword (2012): Stylish Academic Writing. Harvard University Press, 220 pp.

6) Zinsser (2012): On Writing Well: The Classic Guide to Writing Nonfiction, 30th Anniversary Edition. Harper, 335 pp.

An essential article to read about the expectations of readers of scientific writing: Gopen and Swan (1990): The science of scientific writing. (https://www.usenix.org/sites/default/files/gopen_and_swan_science_of_scientific_writing.pdf)

In addition, we recommend a number of online sources for improving writing:

1) Chicago Manual of Style online (https://www.chicagomanualofstyle.org)

2) Purdue Online Writing Laboratory (https://owl.purdue.edu/owl)

3) Kathleen Jones White Writing Center of the Indiana University of Pennsylvania (https://www.iup.edu/writingcenter/writing-resources/index.html)

4) Grammarly (http://www.grammarly.com)

5) English Style Book (https://www.litencyc.com/stylebook/stylebook.php)

译文

怎样成为一名更优秀的作者①
（节选）

1.论文的准备

Batchelor（1981）写道："阅读一篇论文是一件自愿而且消耗精力的任务，读者需要被作者引导、帮助、激励"。作者的任务是尽量使论文对审稿人和未来的读者而言变得更可读、更易于理解，从而吸引更广阔的读者群来理解他们的科学发现。因此，目标应该是提交一篇经过认真写作和审阅的论文，一篇为其读者而作的论文。特别是考虑到科学家们自然而然地具有旺盛的好奇心，且喜欢解决好的科学问题，以这样一种叙述方式来写作可以轻而易举地调动读者的兴趣。为达到此目标，作者需要记住，《每月天气评论》的读者群非常广泛。这得益于全球读者都可以在网上获取本期刊的论文，也因为我们的研究领域越来越趋向于多学科交叉。一个学科的科学结论和研究方法正越来越多地被另一个学科需要，也被引入到其他学科中。因此，应该退一步考虑如何适应比过去更广泛的读者群，这往往对提升论文的显示度、降低阅读门槛大有裨益。作者可以采取的具体措施包括以下方面。

1）为论文创建一份情节梗概。例如，编制一份文章中要插入的图的列表，并标注每张图要说明的核心论点。然后将这份列表的顺序编排得最适宜用来讲述此项研究的"故事"。这样的情节梗概可以帮助作者更有效地写作论文的正文，也可以使得论文对读者而言更有吸引力。

2）将论文的格式和组织方式调整得与近期发表的论文相似。遵从 AMS 对论文的格式要求。如果审稿人能想象出论文发表在《每月天气评论》上的样子，他们更容易认为此论文适合发表。怎样才能做到这样呢？阅读最近发表的论文并模仿它们。

3）采用有效的排版和布局，以便尽可能地使论文易读、易解释：使用一定数量的有数字编号的小节和有字母编号的子节，但避免将论文碎片化，即分隔成太多很短的小节和子节；使用缩进或空行来分隔各段落。

4）用有条理的方法写作，例如，将旧的信息放在每个句子的开头，而将新的信息放在后面，从而使不同主题之间的过渡自然流畅。缺乏条理性是所有提交的论文最常见的缺点。一种保证条理性的方法是，将所有主题句放到一个单独的列表中，然后看其推进顺序是否符合逻辑。确保每个段落仅涉及一个主题。每个段落都从主题句开始，并以需要强调的论点结束。

5）使用准确、精确、客观的语言，以便尽可能地使论文易读、易解释。使用有吸引力的

① 译自 Schultz et al., 2022. How to Be a More Effective Author. Monthly Weather Review (150): 2819-2828.

动词，避免没有必要的形容词和副词。尽量使用准确的词汇，避免使用专业术语（即内行人士才懂的科学词汇）。对读者可能不熟悉的术语需要进行定义，甚至把熟悉的术语也进行定义。若将两张图表、现在的结果和过去的结果，或观测事实与模式结果等进行对比，需要认真描述其异同点，而不是简单地抛出“符合得很好”等粗略结论。

6）描述物理过程时要具体。以下面这句话为例：“低空急流对强对流降雨事件起主要作用”。这句话存在许多问题。首先，对流降雨过程需要三个组分（上升、不稳定、水汽）才能发生，那么所谓急流起主要作用是什么意思？作者是说这三个组分之一比其他两者更重要吗？其次，这个结论有办法进行量化吗？低空急流对对流降雨的相对重要性有多大？如果急流减弱 5 m/s，对流降雨是否就不会发生？再次，没有论证低空急流的具体作用，是通过提供水汽起作用还是通过风速剪切起作用？这样的表述阻碍了科学的精确交流。

2.引言：研究的动机和论文的目的

在引言部分，作者引起潜在读者的注意，介绍研究问题的动机，定义论文的目的，向读者介绍论文的内容。读者应该可以很快了解“此问题为什么重要”，以及“我为什么需要关心此问题”。引言部分最重要的内容是问题陈述，即研究动机所在、吸引读者进入论文的张力来源。作为科学家，我们喜欢解决疑问，而问题陈述就创造了这样一个疑问。所以，最好的问题陈述能够把读者引入文献中存在的悖论、错误、缺失的数据或信息，或者不协调之处。科研派别或不同论文之间的争议、研究结果的不同解读等也可作为有吸引力的问题陈述。

问题陈述引入本文要解决的一个或几个具体的科学问题。这些问题越具体越好。对某一次强降雨事件的个例研究虽然也是有意义的，但并不足以单独成为在《每月天气评论》上发表的理由。除了对个例事件的研究，本项研究工作的动机是为了解决哪个科学问题？如果能够总结出一个可检验的假设（不是所有研究项目都能做到），那么也可以成为一个很好的研究动机。“本研究将探讨/调查……”这种表述不够准确。通过研究此特例能够得到什么新知识？这篇论文在展示新的原创研究结果方面与其他论文有什么区别？而且如果可能，问题陈述和本文的科学问题应该从文献中经过有条理、有证据的论证得出来。

诸如“文献存在缺口，而此问题很少被研究”之类的表述并不是很吸引人的问题陈述。通常，真实情况是以下二者之一：作者没有足够深入地钻研文献，因此没有发现文献确实涉及过这个问题，尽管可能这些文献超出了作者的地理范围或兴趣范围；或者此问题并不值得研究（例如，“很少有研究关注美国的腌黄瓜价格与印度洋热带气旋生成过程的关系”就是一个夸张的例子）。

通过问题陈述，我们也可以对现存文献进行更有针对性的展示。为此应该仅引用那些可逐步归结到问题陈述的文献，而不要在不同主题之间游走并引用很多范围很宽泛的文献，这是作者们常犯的错误。文献总结应该讲述一个故事，将一篇或多篇文献的主题编织成一个自成一体的叙事逻辑,而不是平铺直叙地描述一篇又一篇文献的结果。从这个角度来说,文献总结通常应该聚焦于科学,而不是聚焦于论文或人。文献总结应该助力于本文研究问题和动机的引入。

问题陈述后面应该是对问题的回应,即简要阐述作者为解决问题而将要做的事。这种对问题的回应有时可以在引言部分的最后一段中实现,因此,这一段常常起到路线图的作用。虽然不是必需的,此种段落可以有效地解释本论文是如何解决问题陈述的,并向读者介绍文章的结构,特别是与标准期刊文章的基本结构“引言—方法—结果—讨论—结论”的不同之处。另外,论文应该有类似“本文的目的是……”的明确陈述。这种陈述使读者可以将文章末尾的结果与作者的意图进行比较。这样就可以用作者自己表述的目标来对其进行评判。Schultz(2022b)探讨了引言部分可能存在的其他问题,以及解决这些问题的方法。

3.核心信息

有些论文还可能有另一个弱点:核心信息不明确。读者应该记住哪些新结果?这些结果怎样成为学术界的新贡献?核心信息应该出现在摘要、结论、结果等部分。例如,应该使读者明确:为什么这些结果对于理解天气或改善数值天气预报而言是重要的?当然,核心信息也不能将此项研究的重要性拔得太高。

4.论文的其他部分

本节对 AMS 期刊论文里的其他小节的处理方法提出建议。请熟知 AMS 网站上的作者资源。遵守 AMS 对文章各部分写法的指导意见,如摘要、小节类型、图、参考文献(通常可参考不同引文类型的示例)。注意,AMS 明确指定了印刷版和 PDF 版里的图可采取的几种宽度、表格的简单样式,以及矩阵、矢量、标量等的格式。作者可以通过浏览过去发表的 AMS 文章来熟悉这些样式。

1)标题应该以一种简明的方式提供关于论文的具体、准确的信息。标题应该语法正确、易读易懂。不要使用不常见的缩写。

2)AMS 期刊的摘要应该是一个段落,不超过 250 字。因为潜在读者很可能只会读完标题和摘要而不读其他部分,聪明的作者会在这 250 个字内减少介绍性的文字而将结果部分最大化,以便得到最佳效果。不管怎样,越具体化的摘要越有可能被搜索引擎找到,因此能拥有更广阔的读者范围。AMS 目前不支持在摘要里引用图或文献。

3)得益于更易获取的科学文献以及日益拓宽的学科领域,我们论文的读者范围也比以前更广。意义陈述有助于文章被那些非传统的读者群接触到,其具体目标是那些受过良好教育但没有大气科学或其他相关学科背景的外行人。我们鼓励所有作者都写一个不超过 120 字的意义陈述,当然这不是必需的。AMS 网站的作者信息栏目有示例。

4)AMS 期刊论文的关键词是由作者在线上论文管理系统“Editorial Manager”里提交

论文时选定的。因此,所有关键词都来自 AMS 提供的标准列表。AMS 要求列出 3~6 个关键词。关键词的选择标准是反映论文的主题、在文中经常使用、有助于被网络搜索找到。不要选择那些只用了一两次的单词作关键词。

5)检验“数据与方法”部分的标准是独立的读者是否能基于本节的内容复现本文的研究。如果他们不能,那么需要更多细节信息。事实上,审稿人经常指出数据与方法部分的信息缺失。尽量记住将研究过程中涉及的所有步骤都描述清楚,包括数据的预处理或滤波等操作。

6)结果部分可以是一节也可以是多个小节。结果部分通过提供恰当的数值证据或实验证据,来展示论文主要结论的合理性。结果部分应当在事实基础上对开展的实验或收集的数据进行客观描述,从而解决引言部分提出的问题,并对解决方法进行检验。如果可能,作者应该阐述他们的结果与其他文献资料、其他情况、其他模式、世界其他地方等的区别,但不应该过于强调结果的普适性。此部分的行文风格应该是引导性的,帮助读者理解论文所做的一系列检验和得出的证据,以支持论文的主要思想。尽可能用结果与文献中现存的性能评估或质量检测方法进行比较。新的测试方法的提出应该以提供现存方法不能提供的新认知为条件。不要包含非必要的结果。应该强调那些对理解论文的科学意义而言至关重要的证据。长篇的讨论性文字最好留到后面的讨论部分。

7)作者们常常不清楚讨论与结论(或总结)两部分的目的有何不同。《每月天气评论》的讨论和结论应该放在不同的小节里。讨论部分应该出现在结果部分之后、结论部分之前,结论部分是对论文的总结。讨论部分最好由较长的文字组成,内容通常不适合放在结果部分,例如,将本文结果推广到其他情况,对结果的解释进行展开论述,对结果的综合讨论、应用、隐含意义、未解决的问题或其他可能的解释等。相反,结论部分应该包含对论文的总结,且应比摘要长(约 500~800 字),但不是摘要的简单重复。也可以对论文的意义或应用价值进行简要的最终评述。结论部分应该回顾引言中提出的科学问题,并终结整篇论文。总而言之,详尽的阐述和扩展属于讨论部分,而总结属于结论部分。

8)在致谢部分对所有曾助力本文科研工作和撰写工作的人进行感谢,包括资助机构和数据或软件的提供者。对匿名审稿人进行感谢,因为他们曾帮助提升论文的质量。此种合作精神并不是必需的,但通常会被认可。

9)如果在正文里未对所用数据集进行正式引用,请在数据可用性声明中进行引用。现今许多数据存档机构都提供 DOI,使对数据的引用变得简单。清晰的数据引用是对数据提供者的认可,也提高了论文科学内容的可复现性。如果作者创建了自己的数据集,则应该在一个合适的数据库上将本文结果所依赖的所有数据和计算机程序进行公开,尽管资助机构可能没有要求这样做。数据和代码公开得越多,就越能帮助审稿人和读者理解论文的结果、评估其真实性、复现所用方法等,从而促进对科学问题的理解。并非所有数据都要存档,尤其是对于基于数值模拟的研究而言。虽然现在仍没有关于应该保存多少数据和代码的正式指导意见,我们建议那些以创造新知识为特征的研究工作(《每月天气评论》上的论文常属于此类）更加重视程序、记录、数据等的存档和公开，以保证论文结果的可复现性，正如上文讲到数据和方法部分时所论述的。作者应该确认所有指向数据源的链接仍然正确、有效。关于引用数据和数据可用性声明的更多信息请见 AMS 网站上的作者信息栏目。

10）论文提交时应保证参考文献以完整、统一的方式编排，以方便作者查阅和确认。论文被接受后，参考文献列表会在校对时进行更严格的检查和格式调整。尽量避免引用既不能在网络上找到也没有印刷版的文献，例如，没有预印本也没有在会议论文集上发表过的会议报告。非英文的参考文献应该在括号里提供题目的英文翻译和文献的类型。AMS 网站上的作者信息栏目提供了很多常见文献类型的引用格式规范。

11）图的作用是清晰地展示研究结果。应该避免太多重叠的、难以辨认的区域，也不要包含不必要的子图。图和图上的标注文字（如坐标轴、等值线、色标等的标记）的大小应该保证其可读性，图的质量应该足以避免像素化。这一要求常常意味着作图时应该选择较大的字体，使得单独看图的时候字体显得偏大，因为完成排版后图会变小。尽量在图上使用无衬线字体。不要使用彩虹色标，因为它过于强化数据的梯度，且对色盲的读者不友好。图的标题应该完整地描述图上的所有特征。如果子图采用了字母进行编号，那么图标题中用该使用这些编号进行描述，使读者更容易知道所描述之物在哪里。图和表格应该按顺序引用，除非用括号进行类似"（见下文图 7）"这样的注释。正文里提到图的时候不应该仅仅对其进行描述而不进行解读。确保在正文里说明为什么需要这张图，并对图所展示的信息进行充分解释，且说明这张图意味着什么。

5.录用流程和拒稿处理

AMS 期刊的同行评议专家与足球比赛的裁判不同。审稿人向编辑做出关于文章是否适合发表在《每月天气评论》上的建议，但只有编辑有最终决定权。编辑做出的决定不一定基于少数服从多数的原则，也不一定与审稿人的建议一致。编辑还会对稿件的合适度和主题做出自己的评断，并对审稿人给出的意见进行解读和综合。

《每月天气评论》的拒稿决定并不意味着作者不可以把论文修改后再次投稿到本期刊或其他期刊。有时候拒稿决定是因为预期的修改工作量超过了 AMS 设置的两个月修改时长上限。虽然作者可能感到失望，但其实这样的拒稿决定是有好处的，因为这给了作者无限长的时间来思考审稿人和编辑给出的意见，并采取最佳方式来修改论文，最终作者仍可将论文投稿到《每月天气评论》或其他期刊。这种灵活性可提高最终发表的论文质量。

6.对审稿人和编辑的回复

论文被期刊接收之前，通常要与审稿人和编辑进行 1~3 轮反复的修改和审稿，每一轮需要 1~3 个月。如果作者与审稿人每一轮都可以积极地、建设性地配合，那么审稿的轮数可能可以缩减，使得同行评议的过程效率高、效果好。对于作者来说，这种良好的审稿过程意味着最大程度上利用审稿人的意见和建议认真、完备地优化论文,并向审稿人和编辑做出详尽而合理的答复。因为作者为论文投入了大量的时间、努力和情感,在看到审稿人和编辑的意见时当然会高度紧张。不要让这种紧张情绪发展为“要么战斗,要么逃走”的反应,要利用它来更加专注地修改论文。设想一下一个对此项研究知之甚少的外来观察者会如何看待这篇稿件。切记,审稿人为了提供好的建议来帮助改善此项研究的呈现方式同样付出了大量时间和精力。审稿过程中涉及的每个人都想使此项工作获得最好的产出。基于这种思想,请对每一条审稿意见的初衷做出最好的预设。如果确实出现了没有建设性的审稿意见,应该让编辑知道。以这种态度对待审稿意见和文章的修改需求,可以使作者发现不那么明显的问题,并得以与审稿人进行建设性的交流。

也许最能使作者明白如何做到上述要求的办法就是换位思考。审稿人是自愿承担起帮助改善论文的目标的。既然是匿名的,他们的努力得不到任何赞许,唯一的回报就是看到作者在他们的帮助下受益。如果你(作者)是别人的审稿人,你花了10个小时(审稿人花费在论文上的典型时间)的时间来阅读论文并写评论,你最想看到的应该是作者告诉你他们读了、理解了你的每一条意见并且认真对待了(即便他们可能不一定同意),而且回应了你的评论并修改了论文。如果作者不按照这种态度来对待,那么作者和审稿人之间的交流很可能偏离方向。为此,我们对如何回应审稿人和编辑的意见提出了如下建议。

1)说明你作为作者是否同意这条意见的立论基础。

2)如果你同意,告诉审稿人你根据他的建议对论文做了何种修改,修改在论文的哪个地方。要用尽可能精确的表达方式,以确保你理解了这条意见并恰当地进行了回应。通常,为了让审稿人和编辑更容易评估你的修改,可以把修改后的文本复制到你的答复中,并提供修改之处的行号。如果你在答复中解决了审稿人的问题但不需要修改论文,也要让审稿人明确知道。如果经过大幅重写后审稿人顾虑的点已经不存在或被避免了,那么对这条意见进行简单致谢就足够了。

3)如果你不同意,则要在新的计算结果或现有文献的支持下阐明你不同意的原因。你并不是必须要同意审稿人的意见,因为人与人之间的观点、理解和结论的分歧是不可避免的。无论如何,要逐条答复所有审稿意见,不只是答复每个单独的论点,也要对这些论点总体表达的思想进行答复。只解决某些具体的修改意见而不对背后存在的问题进行处理不是最佳的同行评议。要记住,将来的读者不知道作者与审稿人之间是如何交流的,所以,把文章中可能的问题来源解决远比写很长的答复信更重要。如果需要新的计算结果或图来解决审稿人的疑问,而且这些图已经放在了答复信中,那么应该认真考虑将它们加入修改后的论文中,即使不然,也应放在在线发表的附加材料中。

4)在每一条答复的最后明确声明你已经处理了这条审稿意见,如“没有改动论文”之类,或者最好实际修改一下论文,免得读者将来发出与审稿人一样的疑问。不要含糊其词地回复,如“已修改”或“重写了引言”等。不要让审稿人自己去在论文里找修改之处,也不要让他们怀疑他们的顾虑是否被解决。AMS期刊发出的录用决定中明确写了“如果你对论文做了修改,请指明修改之处在哪里。(方法之一是指出修改之处所在的行号,但不是唯一的方法。)”

5)将审稿人和编辑的意见复制粘贴到一个文件中。将你的答复穿插放在每一条审稿意见之后并用不同的字体或颜色标明,这样更方便审稿人和编辑判断他们的意见是怎样被处理的,使同行评议的过程更流畅更有效。

6)作者应该尽最大努力确保理解审稿人的意见。如果审稿意见本身不明确,可以询问编辑以求澄清。作者也可以这样回复:“如果我理解得没错,您的意思是……”。如果确实是由于论文写得不明确而导致审稿人误解了作者的意思,那么应该修改论文,免得其他读者也同样被误导。

7)如果审稿人觉得他们的意见没有被彻底解决,他们通常会在下一轮审稿中再次提出。未充分解决的意见能拖慢同行评议的流程,也会让审稿人甚至编辑感到懊恼。请按照上面给出的四个步骤来避免此种情况的发生。如果编辑发现作者没有认真解决审稿人的意见,他们可以将论文发回给作者要求继续修改,或者直接拒绝这篇论文,因为它没有朝着可以发表的方向前进。

8)不要在答复信中将各位审稿人置于对抗之中(例如,“审稿人 2 认为这一小节写得很好,所以我不必做审稿人 1 要求的修改”)。即便某位审稿人认可作者的某一观点(或没有提出质疑)也不代表那就是对的。要单独答复每位审稿人的意见。

9)有时审稿人给出的推荐意见彼此相左,这其实并不罕见。编辑向作者发出录用、大修、小修、拒稿等决定时,他们已经充分考虑了审稿人的意见和审稿人本身对此领域的熟悉度。这种情况下论文的修改和意见的答复最好还是按照上面说过的方式进行,即每名审稿人的每条意见都充分地进行处理。如果这种情况下作者不知道审稿人的期望是什么,或者需要更多的指导意见,可联系经手的编辑。提交一份含有修订痕迹的论文不是必需的,但确实可以帮助一部分审稿人检查他们的意见是如何被处理的。让审稿人和编辑的工作简单些,这样可以加速审稿流程。

7.常见问题

1)我听说 AMS 正考虑禁止个例研究,这是真的还是谣言?

AMS 欢迎那些有独到见解的个例研究,它们可以对我们的科学知识做出贡献。个例研究与其他类型一样也需要满足发表的最低要求。为了避免从一个个例得出过于一般化的结论,提交的论文中的讨论部分需要涉及选择这个个例的原因,并讨论研究结果在这个个例以外的适用性。关于个例研究论文的写作方法请见 Schultz(2010b)。

2)我提交的论文里用了我自己之前发表的文章中的一部分文字,编辑发回给我让我修改,说存在自我剽窃。这是什么意思?

虽然将同一内容发表两次不符合学术规范,清晰而准确的描述的确是有效的学术交流所必需的。因此,作者可以复用过去文章里的一些文字,特别是对于数据和方法部分。为避免自我剽窃或侵犯版权,复用的文字必须引用原文,并说明本文基本沿袭了原文[如“方法的描述与 Jones 等(2021)相似,接下来的两段从彼处经小幅修改而来”]。关于剽窃和自我剽窃的更多信息可在 Schultz 等(2015)找到。

3)审稿人有时会这样写:“我注意到一些语法问题,但文字编辑人员在制作校样的时候会改正。”是这样的吗?

文字编辑人员的工作职责确实包括纠正作者和审稿人没有发现的错误,但出版之前繁重的文字编辑任务将最终导致作者需支付更高的版面费。而且这还意味着审稿人需要审阅一篇“草稿”,这会使他们烦恼,继而拖慢同行评议的速度。因此,承担起主要校对任务的人应该是作者而不是 AMS 员工。正如 W. J. Steenburgh 说过的,作者提交的论文应该是让自己感到十分满意的,满意到如果直接原样发表也能接受。

4)我可以用括号来使句子缩短吗?如“700 hPa 上的暖(冷)平流在气旋中心东(西)侧”?

这种句子结构确实在过去发表的论文中很常见,但在提交给《每月天气评论》的论文里应该尽量避免。这种用法增加了读者的阅读困难,使之难以理解句子的含义,而且如果其他句子里面正确地使用了括号,反倒会造成混淆。

5)我可以在论文里引入新的缩写吗?

使用不常见的缩写会给读者造成困难,因为读者遇到这些缩写词的时候不得不到论文前面去翻找它们的定义。对那些不是从头到尾读完而是跳跃式地查找他们想要的信息的读者而言,缩写词更加不友好。引入缩写词能节省作者写作的时间,但延长读者的阅读时间。即便不能删除所有的缩写,也请尽量删除大部分非公认的缩写,这样可以提升论文的可读性。AMS 网站的读者信息栏目里给出了 AMS 期刊认可的、不需要进行定义的缩写列表。若必须定义一个新的缩写,而且 AMS 列表里没有,作者必须给出全称并在后面用括号定义其缩写。

8.给非英语母语者的建议

我们理解非英语母语者面临的困难,我们的编辑中也有一些非母语者。非英语母语的作者不但需要确保科学正确性并用规范的方式进行交流,而且这一切还需要用英语这样一种非母语来进行。低劣的写作质量一般不会导致拒稿,但确实可能阻碍读者对论文内容的理解。因此,从一开始就提升稿件的质量可以保证同行评议的顺利进行。AMS 的审稿人、编辑、员工没有那么多时间来彻底修改一篇充斥着语法错误的论文。AMS 殷切期望我们的期刊成为国际科学交流的平台,但必须要求作者自行采取措施确保提交的论文以正确的英语写就。否则,论文可能不经过同行评议就退回或拒稿。

下面是一些获得帮助的方法:

1)开启文字处理软件提供的拼写检查和语法检查功能。

2)在浏览器或笔记本电脑上安装专业的语法工具或翻译软件。

3）Google Doc 和 Microsoft Word 等文字处理软件提供了输入预测功能，有助于提升句子写作质量。

4）如果对某个单词或词组把握不准，把它放到搜索引擎里去搜索。虽然不能保证搜索到的结果一定符合单词的准确意思，但如果搜索得到的结果很少，往往意味着此单词不常用或不正确。或者也可以在 AMS 期刊上搜索，这样可以得知单词的通常用法。另一个有用的工具是英语搭配词典（http://ozdic.com），可以帮助作者选择正确的单词组合方式和语境。

5）把有用的单词、词组、句子等记录下来，形成一个自己的电子词汇库，需要时在这个词汇库里查找想要的单词。

6）购买英语润色服务，以提升稿件质量。AMS 网站上提供了一些英语润色服务的列表。

9.总结

虽然同行评议不能保证所有发表的科学成果都是正确的，但确实有助于维持基本的质量水平。这是通过期刊编辑设立的标准和审稿人的帮助而实现的。《每月天气评论》论文的接收率仅为 56.7%，因此投稿的论文不总是被接收。我们希望这篇文章提供的指导意见能够帮助作者在投稿之前提升稿件质量，以便提高发表成功的可能性。《每月天气评论》采取的严格的同行评议对作者来说确实比较有挑战性，但我们希望审稿人的反馈和与编辑的对话可以提升每一篇论文的质量，并借此为人人都提供持续学习的机会。

10.为作者提供的附加资源

我们参与编写了由 AMS 出版的《流利的科学：成为更好的作者、演讲者和大气科学家的实用建议》一书，可供作者们参考。此书已被翻译成中文，并由气象出版社出版。我们另外推荐下列书籍：

1）Day，Gastel，2006. How to Write and Publish a Scientific Paper. 6th ed. London：Cambridge University Press，302.

2）Glasman-Deal，2021. Science Research Writing for Native and Non-Native Speakers of English，2nd ed. Singapore：World Scientific，356.

3）Schimel，2012. Writing Science：How to Write Papers that Get Cited and Proposals that Get Funded. Oxford：Oxford University Press，221.

4）Strunk，White，2000. The Elements of Style. 4th ed. Allyn and Bacon，105.

5）Sword，2012. Stylish Academic Writing. Boston：Harvard University Press，220.

6）Zinsser，2012. On Writing Well：The Classic Guide to Writing Nonfiction，30th Anniversary Edition. Harper，335.

关于读者对科技论文期待什么的核心文章：Gopen，Swan，1990. The science of

scientific writing. https://www.usenix.org/sites/default/files/gopen_and_sw an_science_of_scientific_writing.pdf.

另外,我们推荐一系列可帮助改善写作质量的线上资源:

1) Chicago Manual of Style online. https://www.chicagomanualofstyle.org.

2) Purdue Online Writing Laboratory. https://owl.purdue.edu/owl.

3) Kathleen Jones White Writing Center of the Indiana University of Pennsylvania. https://www.iup. edu/writingcenter/writing-resources/index.html.

4) Grammarly. https://www.grammarly.com.

5) English Style Book. https://www.litencyc.com/stylebook/stylebook.php.